SMALL ENGINES
Second Edition

Authorized by
Briggs & Stratton Corporation

AMERICAN TECHNICAL PUBLISHERS, INC.
HOMEWOOD, ILLINOIS 60430-4600

R. Bruce Radcliff
Dann L. Roark

Small Engines, 2nd Edition and CD-ROM contain procedures commonly practiced in industry. Specific procedures used in small engine operation and service vary and must be performed by a qualified person. For maximum safety, always refer to specific manufacturer recommendations; insurance regulations; specific company procedures; applicable federal, state, and local regulations; and any authority having jurisdiction. American Technical Publishers, Inc. assumes no responsibility or liability in connection with this material or its use by any individual or organization.

©2004 by American Technical Publishers, Inc.
All rights reserved

3 4 5 6 7 8 9 – 04 – 9 8 7 6 5 4 3 2

Printed in the United States of America

T 42609

ISBN 0-8269-0012-7

ACKNOWLEDGMENTS

The author and publisher are grateful to the following companies and organizations for providing technical information and assistance.

- American Petroleum Institute
- American Society of Agricultural Engineers (ASAE)
- Ariens Company
- Bacharach, Inc.
- Billy Goat Industries, Inc.
- Coleman Powermate, Inc.
- Corbis-Bettmann
- Eagle Manufacturing Company
- EasyRake®/EverGreen International, Inc.
- Fluke Corporation
- GardenWay, Inc./Bolens®
- GardenWay, Inc./Troy-Bilt®
- Generac Corporation
- W.W. Grainger, Inc.
- Hobart Brothers Company
- Homelite, Inc.
- Jacobsen Division of Textron Inc.
- JLG Industries, Inc.
- John Deere Worldwide Commercial & Consumer Equipment Division
- Justrite Manufacturing Company
- Lab Safety Supply, Inc.
- Mi-T-M Corporation
- MTD Products Inc.
- Murray Inc.
- National Fire Protection Association
- Outdoor Power Equipment Institute
- Professional Chemicals Corporation
- Ransomes America Corporation
- Safety-Kleen Corp.
- Snapper, Inc.
- The Lincoln Electric Company
- The L.S. Starrett Company
- Thermo King
- The Toro Company
- Wacker Corporation
- Whiteman Industries Inc.

CONTENTS

INTERNAL COMBUSTION ENGINES	**1**	Engine Classification • Small Engine Development History • Energy Conversion Principles • Small Engine Industry	... 1
SAFETY AND TOOLS	**2**	Small Engine Operation Safety • Industry and Standards Organizations • Fire Safety • Carbon Monoxide • Personal Protective Equipment • Hazardous Materials • Emergency Plans • Accident Reports • Tools	... 21
ENGINE OPERATION	**3**	Engine Components • Four-Stroke Cycle Engines • Two-Stroke Cycle Engines • Valving Systems • Diesel Engines • Rotary Engines • Engine Output	... 47
COMPRESSION SYSTEM	**4**	Compression • Valves • Valve Guides • Valve Seats • Pistons • Cylinder Bore • Crankcase Breather System • Compression Release System • Valve Resurfacing Service Procedures	... 75
FUEL SYSTEM	**5**	Fuel • Air Pressure Dynamics • Carburetor Operation Principles • Carburetor Design • Carburetor Service Procedures	... 97
GOVERNOR SYSTEM	**6**	Governor System Operation Principles • Governor Droop • Governor Sensitivity • Service Procedures	...125
ELECTRICAL SYSTEM	**7**	Electrical Principles • Charging System • Ignition System • Starting System	...145
COOLING AND LUBRICATION SYSTEMS	**8**	Engine Heat • Engine Materials and Characteristics • Air-Cooled Engine Cooling Systems • Liquid-Cooled Engine Cooling Systems • Lubrication • Cooling and Lubrication System Service Procedures	...173
MULTIPLE-CYLINDER ENGINES	**9**	Multiple-Cylinder Engine Design • Multiple-Cylinder Engine Systems • Multiple-Cylinder Diesel Engines • Multiple-Cylinder Engine Service Procedures	...193

TROUBLESHOOTING	**10**	Troubleshooting Methods • Troubleshooting Steps • Troubleshooting Compression Systems • Troubleshooting Fuel and Governor Systems • Troubleshooting Electrical Systems • Troubleshooting Cooling Systems • Troubleshooting Lubrication Systems	...207
FAILURE ANALYSIS	**11**	Engine Failure • Premature Wear	...233
ENGINE APPLICATION AND SELECTION	**12**	Repowering • Engine Selection • Fuel Systems • Electrical Systems • Vibration • Cooling System • Maintenance and Service • Safety Considerations	...249
APPENDIX			...273
GLOSSARY			...301
INDEX			...315

CD-ROM CONTENTS

- Using This CD-ROM
- Quick Quizzes™
- Illustrated Glossary
- Media Clips
- Reference Material

INTRODUCTION

Small Engines, 2nd Edition, is a comprehensive full-color textbook that provides the basis for a complete small engine technician training program. Fundamental small engine operation principles are presented using concise text, detailed illustrations, and informative factoids. Information is supplemented with photographs from recognized outdoor power equipment manufacturers. The evolutionary development and the scientific principles of small engine operation are presented. Building on these concepts, all small engine systems are covered including the compression, fuel, governor, electrical, cooling, and lubrication systems. Troubleshooting methods and failure analysis techniques are presented and reinforced with common industry applications. Information on engine selection offers an overview of pertinent engine selection factors. All key terms are italicized in the text and defined in the Glossary, and the Appendix contains many useful charts and tables.

Small Engines, 2nd Edition, is designed for technicians in the small engines industry, for those preparing for the Master Service Technician Exam, and for students in power technology, automotive, and engineering programs. This new edition includes updated and expanded coverage of the following:

- two-stroke cycle engines
- turbocharged diesel engines
- liquefied petroleum gas (LPG) fueled engines
- outdoor power equipment industry data
- small engine troubleshooting and service tools

Small Engines, 2nd Edition, is authorized by the Briggs & Stratton Corporation, the leading manufacturer of small air-cooled engines in the world. The text was developed and written by R. Bruce Radcliff and Dann L. Roark in conjunction with the technical staff at American Technical Publishers, Inc. Both authors have extensive industry and instructional experience which is reflected in the content of the book and the approach of the material presented. R. Bruce Radcliff has served as Director of Customer Education and currently serves as Director - Global Business-to-Business Communications for Briggs & Stratton Corporation. Dann L. Roark currently serves as Manager, Service Training for Briggs & Stratton Corporation.

The *Small Engines,* 2nd Edition, CD-ROM is located in the back of the book. This CD-ROM is designed as a self-study aid to enhance information presented in the book and includes Quick Quizzes™, Illustrated Glossary, Media Clips, and Reference Material. The Quick Quizzes™ provide an interactive review of topics in each chapter. The Illustrated Glossary provides a helpful reference to commonly used terms in the industry. The Media Clips button accesses a collection of video clips and animated graphics. The Reference Material button provides links to related industry and standards organizations. Clicking on the American Tech web site button (www.go2atp.com) or the American Tech logo accesses information on related small engine training products.

The Publisher

INTERNAL COMBUSTION ENGINES

CHAPTER

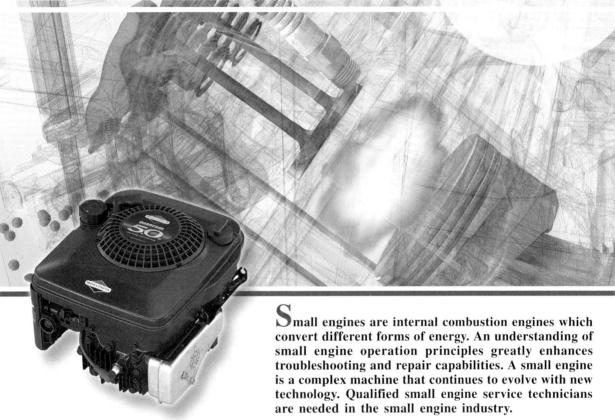

Small engines are internal combustion engines which convert different forms of energy. An understanding of small engine operation principles greatly enhances troubleshooting and repair capabilities. A small engine is a complex machine that continues to evolve with new technology. Qualified small engine service technicians are needed in the small engine industry.

ENGINE CLASSIFICATION

An *engine* is a machine that converts a form of energy into mechanical force. Engines can be broadly classified as external combustion or internal combustion. An *external combustion engine* is an engine that generates heat energy from the combustion of a fuel outside the engine, such as a steam engine. An *internal combustion engine* is an engine that generates heat energy from the combustion of a fuel inside the engine, such as a gasoline engine.

A *small engine* is an internal combustion engine that converts heat energy from the combustion of a fuel into mechanical energy generally rated up to 25 horsepower (HP). Small engines are sometimes referred to as air-cooled engines and are further classified by ignition, number of strokes, cylinder design, shaft orientation, and cooling system. See Figure 1-1.

Ignition

A small engine is either a spark ignition or compression ignition engine based on how the fuel is ignited. A *spark ignition engine* is an engine that ignites an air-fuel mixture with an electrical spark. A *compression ignition engine* is an engine that ignites fuel by compression. Spark ignition engines commonly use gasoline. Compression ignition engines commonly use diesel fuel. Both spark ignition and compression ignition engines are available as either four-stroke cycle or two-stroke cycle engines.

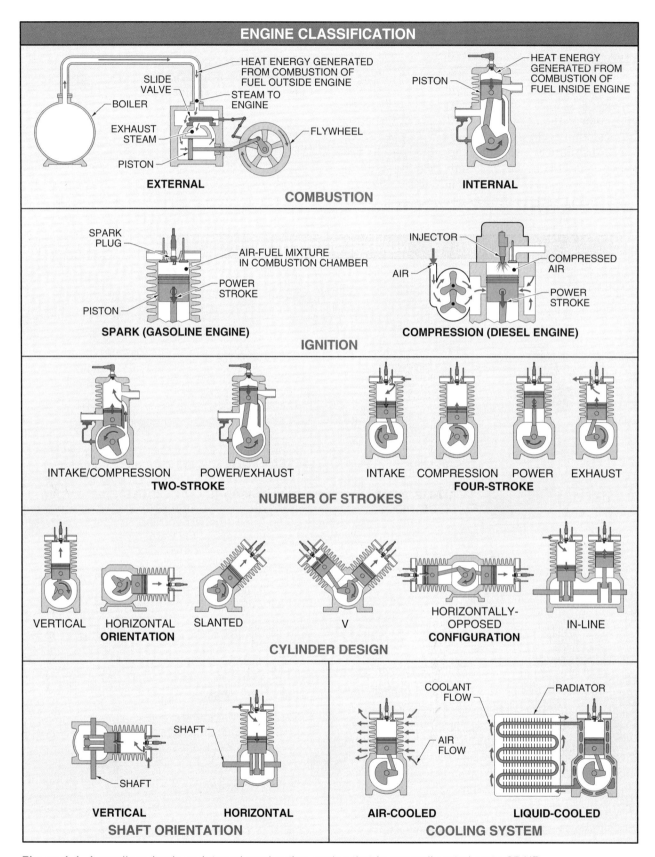

Figure 1-1. A small engine is an internal combustion engine that is generally rated up to 25 HP.

Number of Strokes

Small engines are commonly classified as either four-stroke cycle engines or two-stroke cycle engines. A *four-stroke cycle engine* is an internal combustion engine that utilizes four distinct piston strokes (intake, compression, power, and exhaust) to complete one operating cycle. A *two-stroke cycle engine* is an engine that utilizes two strokes to complete one operating cycle of the engine. Both four-stroke cycle and two-stroke cycle engines complete five distinct events during each cycle. These events include the intake, compression, ignition, power, and exhaust events. Engine components function separately and/or together to complete these events. Engine components vary in two-stroke cycle engines and four-stroke cycle engines. See Figure 1-2.

The term charge, which is used to describe the trapped air-fuel mixture inside a cylinder, is named after the compacted gunpowder in the Huygens gunpowder engine, developed in 1680.

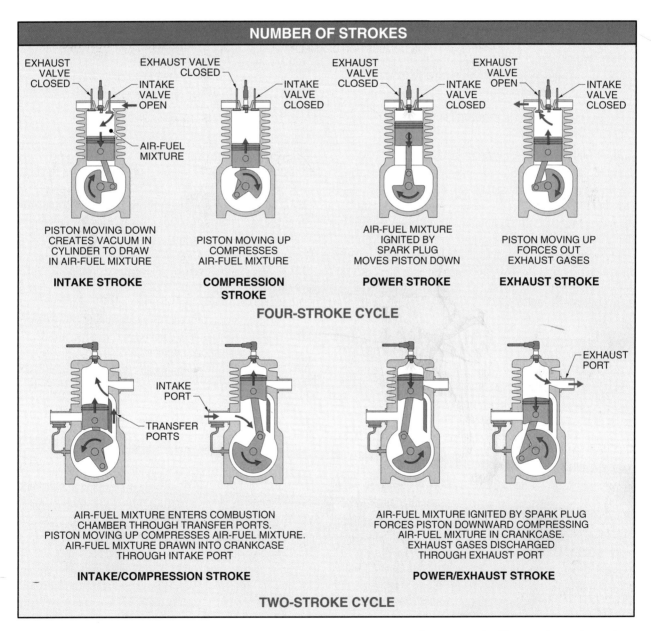

Figure 1-2. Four-stroke cycle engines operate by completing four strokes for each cycle. Two-stroke cycle engines operate by completing two strokes for each cycle.

Cylinder Design

Small engines typically contain one or two cylinders, but can have as many as four cylinders. Cylinder orientation is vertical, horizontal, or slanted, depending on the axis of the cylinder. Cylinder configuration in multiple cylinder engines is V, horizontally-opposed, or in-line. Cylinder orientation and configuration are selected for power and for requirements of the application. For example, a two-cylinder V engine provides the power of two cylinders in a compact space.

Horizontally-opposed engines have a low profile and produce a low degree of vibration. An in-line vertical or horizontal engine provides ease of manufacture but requires more space when multiple cylinders are required. A slanted engine is used when the engine application requires an angled cylinder.

Shaft Orientation

Shaft orientation is the axis of the crankshaft as vertical or horizontal. Shaft orientation is selected for efficiency in driving components of the application. For example, most walk-behind rotary lawn mowers use a vertical shaft, horizontal cylinder engine. The blade is driven directly from the engine and rotates parallel to the ground. Generators and snow throwers commonly use a horizontal shaft, vertical cylinder engine. This allows for efficient power transfer to driven components.

Cooling System

When fuel is oxidized (ignited providing combustion) in a typical small engine, approximately 30% of the energy released is converted into useful work. The remaining energy is lost in the form of heat to cooling air, exhaust system, radiation, and friction. See Figure 1-3. Heat generated by the engine must be removed to maintain operating efficiency and to prevent damage to engine components.

The cooling system on a small engine removes heat using air or liquid. An *air-cooled engine* is an engine that circulates air around the cylinder block and cylinder head to maintain desired engine temperature. A *liquid-cooled engine* is an engine that circulates coolant through cavities in the cylinder block and cylinder head to maintain desired engine temperature. In addition to the cooling system, heat is also removed from the engine by the exhaust system and by radiant heat emitted from engine components.

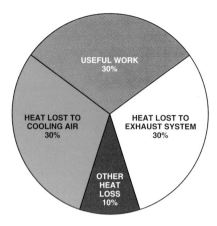

Figure 1-3. Energy conversion of fuel in a small engine results in useful work, and energy lost to heat and friction.

SMALL ENGINE DEVELOPMENT HISTORY

The small engine has evolved with changes in demand and technology over the years. See Figure 1-4. Although the steam engine is seldom used today, the design and development of the steam engine played an important role in the evolution of internal combustion engines. The utilization of a piston inside of the cylinder dates back to the late 1600s. In 1680, Christian Huygens developed a gunpowder engine that consisted of a tube sealed on one end with a close-fitting slug located inside. His design was derived from the operating principles of firearms. The tube was filled with a volume of gunpowder. The gunpowder was then compacted and ignited, driving the slug to the opposite end of the tube. The compacted gunpowder in the tube was known as the charge.

The design of the gunpowder engine was not practical due to the amount of time required to recharge the engine. Consequently, the idea of utilizing gunpowder for energy inside of an engine was passed over and remained dormant for many years. The principles used in the Huygens gunpowder engine made it the first recorded internal combustion engine.

During the next 100 years, inventors focused on utilizing steam along with atmospheric pressure to perform work. In 1698, Thomas Savery developed an engine designed to pump water from underground mines. The Savery pump utilized low pressure generated by condensed steam to fill a vessel with water. The Savery pump then pressurized the vessel with steam to force water to the surface.

SMALL ENGINE DEVELOPMENT		
Year	Engine	Designer/Developer
1680	Gunpowder	Christian Huygens
1698	Savery Pump	Thomas Savery
1712	Newcomen Steam	Thomas Newcomen
1763	Watt Double-Acting Steam	James Watt
1801	Coal Gas with Electric Ignition	Eugene Lebon
1859	Pre-Mixed Coal Gas and Air	Etienne Lenoir
1862	Gasoline	Nikolaus Otto
1876	Four-Stroke Cycle Gasoline	Nikolaus Otto
1892	Diesel	Rudolf Diesel
1920	Flyer Vehicle, Model P	Briggs & Stratton
1931	Model Y L-Head	Briggs & Stratton
1953	Die-Cast Aluminum for Consumer Use	Briggs & Stratton

Figure 1-4. The design and development of the steam engine played an important role in the development of small engines used today.

In 1712, Thomas Newcomen developed another steam engine used to pump water from mines. Newcomen utilized many of the same components found in engines today. See Figure 1-5. The Newcomen steam engine consisted of a piston located inside a cylinder, valves, a boiler, and a pivot arm. The engine used condensed steam and atmospheric pressure to operate. The piston was lifted to the top of the cylinder by steam pressure generated inside a boiler located below the cylinder. With the piston at the top position, the cylinder was sealed and steam was condensed by water injection. A low pressure was created inside the cylinder and the piston was pushed down by atmospheric pressure. The up and down movement of the piston inside the cylinder actuated a lever arm that operated a water pump.

Beginning in 1763, James Watt made several significant improvements that greatly increased efficiency of the Newcomen steam engine. The first improvement was the incorporation of an outside condenser. This increased efficiency by eliminating the constant thermal cycles and using steam directly. Another improvement was a valve system that allowed the piston to be pushed by steam in both directions. This design was known as a double-acting steam engine. See Figure 1-6. Watt also expanded the usefulness of the steam engine by converting the reciprocating (linear) motion of the piston into rotary motion using sun-and-planet gearing.

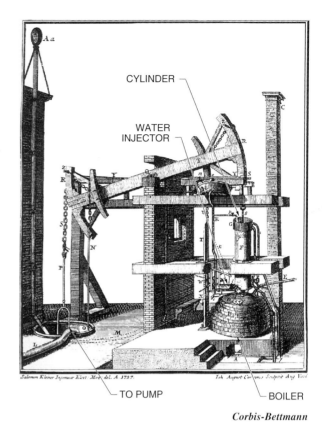

Corbis-Bettmann

Figure 1-5. The Newcomen steam engine had many of the same components found in internal combustion engines today.

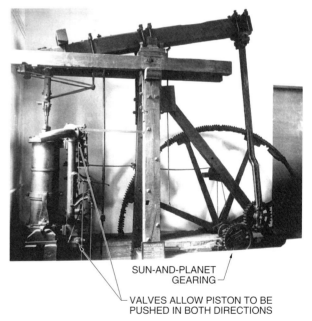

Corbis-Bettmann

Figure 1-6. The Watt double-acting steam engine allowed the piston to be pushed by steam in both directions for greater efficiency.

Lawn tractors are powered by engines capable of operating accessories such as mower, rear bagger, and snow blower attachments.

Watt continued refining the steam engine by developing an engine speed control device. The engine speed control device, or governor, used a set of flyweights driven by centrifugal force to maintain a constant speed. Watt also discovered that a full cylinder of steam was not required to produce a full stroke of the piston in a steam engine. Expansion of the steam in the cylinder provided enough energy to complete the stroke.

These advancements increased engine efficiency and established the steam engine as a dependable power source required for the industrial revolution. The steam engine was used to power land and water vehicles and factories. It was also used to generate electricity. Despite its many uses, the steam engine was still too inefficient and bulky to meet growing demands. This led other inventors to pursue power source alternatives.

In 1801, Eugene Lebon developed an internal combustion engine that used coal gas with electric ignition. In 1859, Etienne Lenoir introduced an internal combustion engine that mixed coal gas and air. However, a resurgence of steam power occurred during this time with the application of high-pressure steam by Richard Trevithick in 1802, and with further development of the steam turbine by Charles Parsons.

These developments rejuvenated steam as a power source until 1862, when Nikolaus Otto developed the first successful gasoline engine. See Figure 1-7. Otto and his colleague Eugen Langen discovered that engine efficiency could be greatly improved if the air-fuel mixture (charge) was compressed before ignition. In their engine, the combustion in the engine forced the piston up the cylinder. When the expanding gases cooled, atmospheric pressure pushed the piston back down. The operation of the engine was similar to that of the Newcomen steam engine. Otto continued to modify the engine until 1876, when he introduced the four-stroke cycle engine. The four-stroke cycle engine was his greatest contribution.

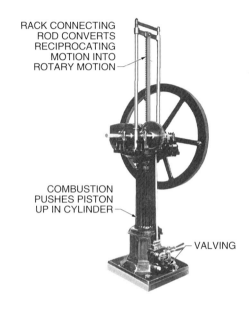

Corbis-Bettmann

Figure 1-7. The first successful gasoline engine was developed in 1862 by Nikolaus Otto.

In 1892, Rudolf Diesel patented a new type of internal combustion reciprocating engine that ignited fuel by high compression. This high compression engine became known as the Diesel engine. High compression produced superheated air in the combustion chamber. Fuel introduced was ignited on contact. Some of the first experimental Diesel engines utilized coal dust as a fuel source.

The turn of the century was a productive and eventful time in the history of internal combustion engines. Most efforts were directed at the new automobile industry, which would soon provide transportation for the masses. Although a great deal of effort was put into multiple cylinder liquid-cooled engines for automobiles, Briggs & Stratton introduced the Briggs & Stratton Flyer. The Flyer was a flexible-frame vehicle with a small, air-cooled Model D gasoline engine. See Figure 1-8. Although the Flyer was not as successful as some other full-featured means of personal transportation, the engine was instrumental in efforts to provide economical portable power to rural America.

Figure 1-8. Founders Harry Stratton and Steve Briggs introduced the Briggs & Stratton Flyer in 1920 as an inexpensive vehicle designed for road use.

In 1920, the Briggs & Stratton Corporation introduced the portable Model P engine. The Model P engine operated at 2200 revolutions per minute (rpm) and developed 1 HP. The engine was very successful and was widely used in stationary and portable applications such as washing machines, garden tractors, electric generators, cream separators, lawn mowers, and air compressors. See Figure 1-9.

As new materials and technologies were developed, new engine designs and applications were introduced. Washing machines were put on stilts to fit the engine under the tub. In 1931, a low-profile Briggs & Stratton L-head, $\frac{1}{2}$ HP Model Y engine was introduced to fit under washing machine tubs. In 1953, Briggs & Stratton introduced the first die-cast aluminum engine designed for consumer use. The small engine industry continued to grow in order to accommodate demand. The first 50,000,000 Briggs & Stratton engines were produced between 1924 and 1967. The second 50,000,000 engines were produced within the next eight years.

Figure 1-9. The Briggs & Stratton Model P engine, which was introduced in 1920, was widely used in agricultural applications.

ENERGY CONVERSION PRINCIPLES

All internal combustion engines exhibit and convert different forms of energy. *Energy* is the resource that provides the capacity to do work. Two forms of energy are potential energy and kinetic energy. *Potential energy* is stored energy a body has due to its position, chemical state, or condition. For example, water behind a dam has potential energy due to its position. Gasoline or any other fossil fuel has potential energy based on its chemical state. A compressed spring has potential energy due to its mechanical condition. See Figure 1-10.

Kinetic energy is energy of motion. Examples of kinetic energy include water falling over a dam, a speeding automobile, and a released spring. Kinetic energy is, in effect, released potential energy. Small gasoline engines convert the potential energy in gasoline to the kinetic energy of a rotating shaft. The rotating shaft of the engine is used to do work required by the application.

The operation of an internal combustion engine is based on basic energy conversion principles. An understanding of these principles is required for successful troubleshooting and service by the small engine service technician. All internal combustion engines operate by utilizing basic principles of heat, force, pressure, torque, work, power, and chemistry.

8 SMALL ENGINES

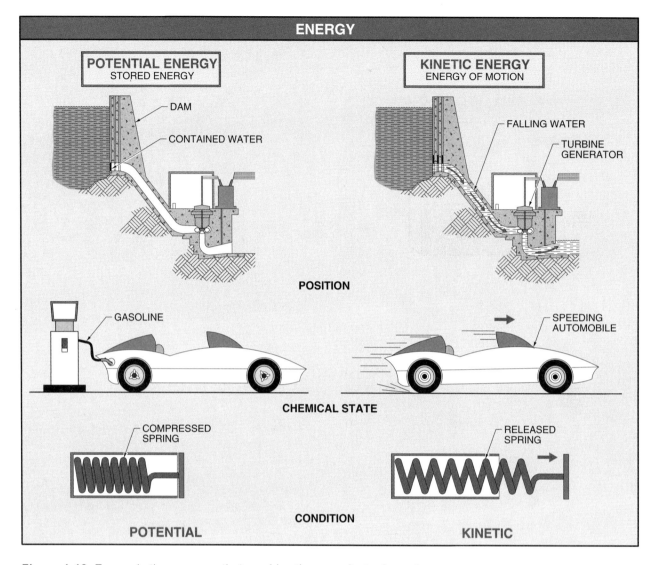

Figure 1-10. Energy is the resource that provides the capacity to do work.

Heat

All matter is composed of atoms and molecules that are in a constant state of motion. *Heat* is kinetic energy caused by atoms and molecules in motion within a substance. Heat added to a substance causes molecule velocity to increase. This results in an increase of internal energy. Heat removed from a substance causes molecule velocity to decrease, which results in a decrease of internal energy. These events occur during the compression and power strokes of an internal combustion engine. As the air-fuel charge is compressed, internal energy increases, producing heat. When the charge is ignited and the burning gases expand, internal energy decreases and heat is given up. See Figure 1-11.

A substance can be in solid, liquid, or gas (vapor) state. For example, water in its solid state is ice, in its liquid state is water, and in its gaseous state is steam. The state of a substance as solid, liquid, or gas depends on the intensity of the vibration (movement) or energy of the molecules.

The change to or from one state to another requires the addition or removal of heat energy. When heat is added to ice, it changes to water. When heat is added to water, it changes to steam. A change in state occurs with gasoline in an internal combustion engine during the compression stroke. As compression begins, heat increases and causes liquid droplets in the gasoline vapor to change to a gaseous state. This change in state prepares the fuel for efficient combustion.

Figure 1-11. Compression of the air-fuel mixture in the combustion chamber increases heat, and causes liquid droplets in gasoline vapor to change to a gaseous state for efficient combustion.

Heat Transfer. Heat flows or is transferred from one substance to another when a temperature difference exists. Heat is always transferred from a substance with a higher temperature to a substance with a lower temperature. Heat transfer rates are proportional and they increase with the temperature difference between two substances. Three methods of heat transfer are conduction, convection, and radiation. See Figure 1-12.

Conduction is heat transfer that occurs from atom to atom when molecules come in direct contact with each other, and through vibration, when kinetic energy is passed from atom to atom. For example, if one end of a metal rod is heated, heat is transferred by conduction to the other end. By heating the end of the metal rod, molecules are heated and move faster. The faster moving molecules transfer energy from molecule to molecule across the metal rod.

Heat conduction occurs in small engines through the medium of lubricating oil. The oil comes in direct contact with engine parts that have a much higher temperature than the oil. When this occurs, the oil conducts the heat away from the part and into the crankcase. Once the oil reaches the crankcase oil reservoir, heat conduction occurs as the cooler crankcase assembly conducts heat from the oil to the air contacting the outside of the engine.

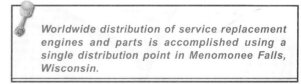

Worldwide distribution of service replacement engines and parts is accomplished using a single distribution point in Menomonee Falls, Wisconsin.

10 SMALL ENGINES

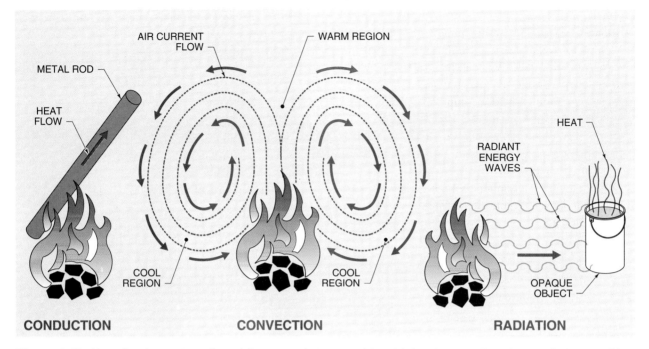

Figure 1-12. Heat is always transferred from a substance with a higher temperature to a substance with a lower temperature.

Convection is heat transfer that occurs when heat is transferred by currents in a fluid. For example, as air is warmed by a fire, the warm air rises and is replaced by cooler air. The movement of air creates a current that continues as long as heat is applied. Heat transfer by convection occurs in a liquid-cooled engine radiator. A *radiator* is a multi-channeled container that allows air to pass around the channels to remove heat from the liquid within. Fins on the radiator channels increase the surface area in contact with passing air to improve heat transfer efficiency from the liquid in the radiator.

Liquid gives up heat as it flows through the surface area. The warm liquid from the engine is pumped into the top of the radiator. The liquid gives up its heat to the air as it passes through the radiator. Cooler liquid is then drawn from the bottom and returned to the engine to repeat the cooling cycle.

Radiation is heat transfer that occurs as radiant energy without a material carrier. The amount of heat transferred depends on the intensity of heat (temperature) of the substance. Radiant energy waves move through space without producing heat. Heat is produced when the radiant energy waves contact an opaque object. Heat produced on earth by light waves from the sun is a form of radiation.

Heat radiation occurs in small engines as the engine block, cylinder head, and other metal engine components have heat passed through them into the atmosphere. For example, the bottom of an engine transfers a significant amount of heat by radiation because there is little air passing across the surface to transfer the heat by conduction.

Temperature. *Temperature* is the intensity of heat. The temperature of matter is a comparison of the degree of hotness or coldness on a certain scale. Temperature can be measured with a glass thermometer. A *glass thermometer* is a graduated glass tube that is filled with a material such as alcohol or mercury which expands when heated and contracts when cooled. A scale or gradient is applied to the glass tube and the degree of heat is measured based on the known expansion rate of the material in the glass thermometer.

The quantity of heat is the amount of energy needed to produce an accepted standard of physical change in matter. A common unit for quantity of heat measurement is the British thermal unit (Btu). A *British thermal unit (Btu)* is the amount of heat energy required to raise the temperature of 1 pound (lb) of water 1°F (Fahrenheit). Another common unit of heat measurement is the calorie. A *calorie* is the amount of heat energy required to change the temperature of one gram of water 1°C.

Temperature Scales. Temperature is commonly expressed in the small engine industry by using the Fahrenheit or Celsius scale. The Fahrenheit scale has no real base, but it provides a larger number of increments between the freezing point (32°F) and boiling point of water (212°F) compared to the Celsius scale. The Celsius scale uses 0°C as the freezing point of water and 100°C as the boiling point of water. Sometimes it is necessary to convert temperature readings. See Figure 1-13.

When degrees Fahrenheit is known, degrees Celsius is found by applying the formula:

$$°C = \frac{°F - 32}{1.8}$$

where

°C = degrees Celsius

°F = degrees Fahrenheit

32 = freezing point on Fahrenheit scale

1.8 = conversion ratio

For example, what is 180°F on the Celsius scale?

$$°C = \frac{°F - 32}{1.8}$$

$$°C = \frac{180 - 32}{1.8}$$

$$°C = \frac{148}{1.8}$$

°C = **82.22°C**

When degrees Celsius is known, degrees Fahrenheit is found by applying the formula:

$$°F = (1.8 \times °C) + 32$$

where

°F = degrees Fahrenheit

°C = degrees Celsius

32 = freezing point on Fahrenheit scale

1.8 = conversion ratio

For example, what is 60°C on the Fahrenheit scale?

°F = (1.8 × °C) + 32

°F = (1.8 × 60) + 32

°F = 108 + 32

°F = **140°F**

Force

Force is anything that changes or tends to change the state of rest or motion of a body. A *body*, when referring to force, is anything with mass. See Figure 1-14. For example, if an individual pushes against a vehicle trying to move it, a force has been exerted on that vehicle (object or mass). Force is measured in pounds (lb) in the English system and newtons (N) in the SI metric system. See Appendix.

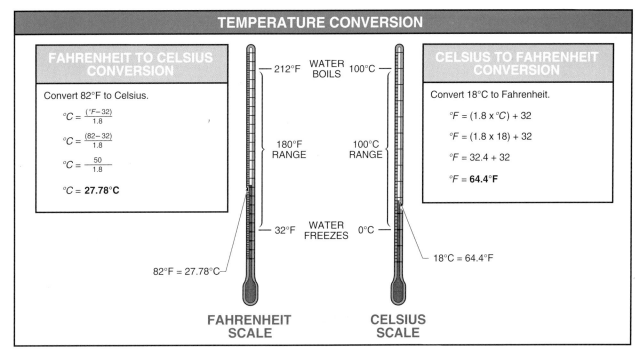

Figure 1-13. Temperature is commonly expressed by using Fahrenheit or Celsius scales.

12 SMALL ENGINES

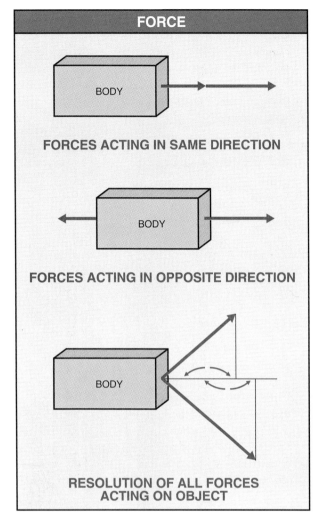

Figure 1-14. Force or forces acting on a body may not always result in motion.

One or more forces can act on a body in several ways. Force acting on a body does not always result in motion. Force applied in different ways produces pressure, torque, or work.

Pressure

Pressure is a force acting on a unit of area. *Area* is the number of unit squares equal to the surface of an object. Area is determined by applying the proper formula. In the cylinder of an internal combustion engine, force is exerted by combustion pressure applied to the area of the piston head. Piston motion is transferred through the connecting rod to the crankshaft. In internal combustion engines, the amount of force exerted on the top of a piston is determined by the cylinder pressure generated during the combustion process. See Figure 1-15.

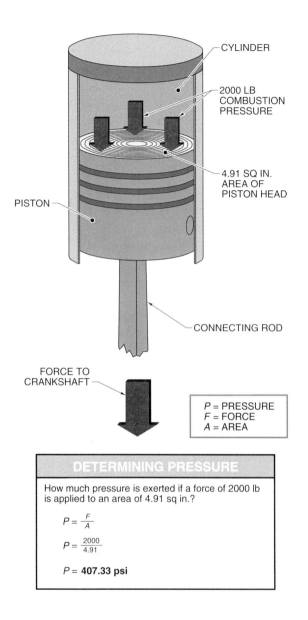

Figure 1-15. During the combustion process, combustion pressure applied to the area of the piston head results in force transferred through the connecting rod to the crankshaft.

When force and area are known, pressure is found by applying the formula:

$$P = \frac{F}{A}$$

where
P = pressure (psi)
F = force (in lb)
A = area (in sq in.)

For example, what is the pressure exerted if a 60 lb force is applied to an area of 4 sq in.?

$P = \dfrac{F}{A}$

$P = \dfrac{60}{4}$

$P =$ **15 psi**

For example, what is the pressure if a 1000 lb force is applied to an area of 5 sq in.?

$P = \dfrac{F}{A}$

$P = \dfrac{1000}{5}$

$P =$ **200 psi**

Torque

Torque is a force acting on a perpendicular radial distance from a point of rotation. It is equal to force times radius. See Figure 1-16. The result is a twisting or turning force, which is expressed as pound-feet (lb-ft) or in newton-meters (Nm) that may or may not result in motion.

When force and radius (distance) are known, torque is found by applying the formula:

$T = F \times r$

where

$T =$ torque (in lb-ft or Nm)
$F =$ force (in lb or N)
$r =$ radius (distance in ft or M)

For example, what is the torque developed if a 60 lb force is applied at the end of a 2′ lever arm?

$T = F \times r$

$T = 60 \times 2$

$T =$ **120 lb-ft**

The same amount of torque results if a 120 lb force is placed at the end of a 1′ lever.

Radius length can be adjusted for specific task requirements. For example, a 1′ wrench is used to tighten a bolt to 49 lb-ft. To tighten the bolt further, more torque is required. Additional torque is obtained by applying a 50 lb force at a distance of 1′ for a 50 lb-ft torque.

The same amount of torque can be obtained by increasing the length of the wrench to 2′ and applying a 25 lb force. However, with a length of 2′, the wrench must travel a greater distance to move the bolt head the same amount. The time required to move the wrench is also increased with the greater distance from the bolt head. An equivalent amount of torque is obtained with a 1″ wrench with a force of 600 lb. The force applied moves the bolt in less time. Change in mechanical advantage achieved by increasing wrench handle length is based on the principles of a lever.

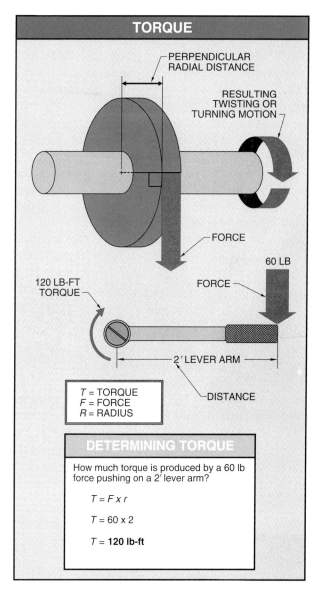

Figure 1-16. Torque is a force acting on a perpendicular radial distance from a point of rotation and is equal to force times distance.

 Briggs & Stratton tested chrome-plated pistons successfully in cast iron engines with aluminum alloy cylinder sleeves before introducing the die cast aluminum alloy engine.

Levers. A *lever* is a simple machine that consists of a rigid bar which pivots on a fulcrum (pivot point) with both resistance and effort applied. The purpose of a lever is to obtain mechanical advantage to overcome large resistance with reduced effort. The lever principle is used on several rigid and semi-rigid components in a small engine. See Figure 1-17.

A major lever used in small engine design is located on the crankshaft. The offset of the crankshaft provides a lever distance from the centerline of the crankshaft to the centerline of the connecting rod. This converts force applied by the piston, which results in rotation of the crankshaft. The *stroke* is the linear distance a piston travels inside the cylinder.

Stroke distance is determined by the throw of the crankshaft. The *throw* is the measurement from the center of the crankshaft to the center of the crankpin journal, which is used to determine the stroke of an engine. The larger the stroke, the more force applied at the crankshaft. Levers are also used in small engine governor systems. Levers and springs provide the mechanical advantage required to sense and control engine performance and speed.

Work

Work is the movement of an object by a constant force to a specific distance. Work occurs only when the force results in motion. For example, work is done by lifting a weight from the ground against gravity. Work is not done if the weight is not lifted from the ground. Work is measured in lb-ft in the English system or Nm in the SI metric system. See Figure 1-18.

Ariens Company
Dual-handle steering controls allow a zero turning radius.

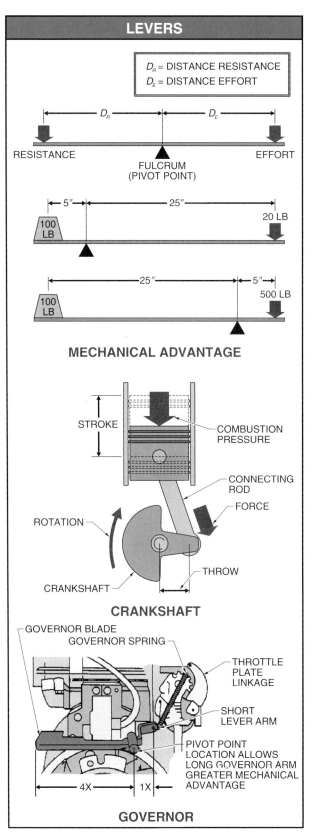

Figure 1-17. Levers are used in small engine components to obtain mechanical advantage.

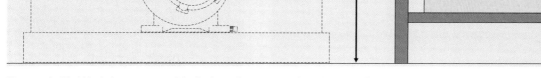

Figure 1-18. Work is measured in lb-ft and occurs only when the force results in motion.

Work requires only enough force to accomplish the desired task. For example, if a person lifts a 50 lb weight, only enough force is exerted to lift the weight. Additional force applied results in acceleration. Work and torque are very similar. The only true difference is that torque does not always result in perceptible motion.

When force and distance are known, work is found by applying the formula:

$W = F \times D$
where
W = work (in lb-ft)
F = force (in lb)
D = distance (in ft)

For example, what is the amount of work performed if a horse pulled a container that weighed 330 lb 100′?

$W = F \times D$
$W = 330 \times 100$
$W = $ **33,000 lb-ft**

Power

Power is the rate at which work is done. Power adds a time factor. Therefore, power is work divided by time. Power can be expressed by force, distance, and speed. Typical examples of power ratings include horsepower (HP) and the watt (W) or kilowatt (kW). Although they are different numerical scales, both the watt and horsepower measure how fast work is done.

16 SMALL ENGINES

When work and time are known, power is found by applying the formula:

$$P = \frac{W}{T}$$

where
P = power (in lb-ft/min)
W = work (force × distance) (in lb-ft)
T = time (in min)

For example, what is the power output of an engine that performs 100,000 lb-ft of work in 6 min?

$$P = \frac{W}{T}$$

$$P = \frac{100,000}{6}$$

$$P = \mathbf{16,666.67\ lb\text{-}ft/min}$$

Horsepower. *Horsepower (HP)* is a unit of power equal to 746 watts (W) or 33,000 lb-ft per min (550 lb-ft per sec). See Figure 1-19. Horsepower is commonly used to rate and rank the power produced by an engine based on a finite engine speed. However, variation in some engine testing specifications has resulted in vague and inconsistent horsepower measurement.

The evolution of HP as a measurement used today is deeply rooted in the history of the combustion engine. James Watt, after developing a steam engine for the mining industry in the 1800s, soon realized that his steam engine produced more power than any one human. He needed a reference point to compare the power of his new steam engine. After researching alternatives, he selected the horse.

Through observation, Watt determined that an average horse could move 33,000 lb on a linear plane 1' in 1 min. This is the basis for the standard 550 lb-ft per sec that is still used today.

Horsepower is found by applying the formula:

$$HP = \frac{W}{T \times 33,000}$$

where
HP = horsepower
W = work (force × distance) (in lb-ft)
T = time (in min)
33,000 = HP constant (in lb-ft/min)

The first governor system was developed by James Watt. It utilized the variable centrifugal force produced by rotating flyweights at different rpm.

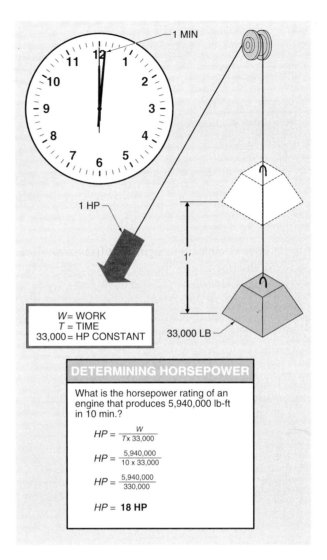

Figure 1-19. Horsepower is used to rate the power produced by a small engine.

Horsepower as a standard

When selecting a reference point to compare the power of his new steam engine, James Watt selected a horse rather than a buffalo or an elephant. The horse was selected because most of the population at the time was familiar with horses and their capacity to do work. The intent in establishing horsepower as a reference point was to provide a consistent standard when comparing the rate of work produced.

For example, what is the horsepower rating of an engine that produces 412,500 lb-ft in 2½ min?

$$HP = \frac{W}{T \times 33{,}000}$$

$$HP = \frac{412{,}500}{2.5 \times 33{,}000}$$

$$HP = \frac{412{,}500}{82{,}500}$$

$$HP = \mathbf{5\ HP}$$

Chemistry

All internal combustion engines utilize some form of fossil (hydrocarbon) fuel as a source of energy. A basic understanding of combustion chemistry of the ignition of a hydrocarbon fuel can offer service technicians greater insight to the cause of a symptom or failure in a small engine. Combustion chemistry involves the combining of hydrocarbon fuel with oxygen from the atmosphere. A hydrocarbon fuel contains a number of hydrogen and carbon atoms. The composition of the air and fuel and the air-fuel ratio changes the chemical equation.

When ignition of the air-fuel mixture occurs in an engine, a chemical reaction between the hydrocarbon molecule and atmospheric oxygen causes an exchange of elements which releases heat energy. See Figure 1-20. For example, if gasoline (C_8H_{18}) is combined with atmospheric oxygen (O_2) in a perfect ratio, the result is water (H_2O) and carbon dioxide (CO_2). The energy released and rapid gas expansion from this chemical reaction provides the force to move the piston in a cylinder. An understanding of the effect of different gasoline blends and/or additives and the effect of altitude in combustion chemistry can expedite troubleshooting efforts.

The Lincoln Electric Company
Portable arc welders generate continuous power and are rated for a specific duty cycle.

SMALL ENGINE INDUSTRY

Small engines are a major part of the outdoor power equipment industry, and offer a reliable, portable power source for many applications. Small engine applications are as varied as the people who use them. According to the Outdoor Power Equipment Institute (OPEI), outdoor power equipment is an $8.3 billion industry that has a major impact on the American economy. The industry also provides 28,000 jobs in original equipment manufacturing (OEM) plants. Outdoor power equipment OEMs produce products that serve some or all markets classified as consumer lawn and garden, commercial turf care, golf course maintenance, and gasoline powered hand-held equipment. Sales of outdoor power equipment vary in seasonal patterns. See Figure 1-21.

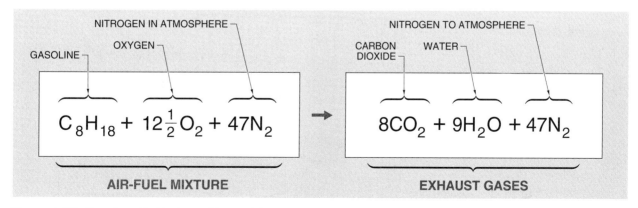

Figure 1-20. Gasoline (C_8H_{18}) combined with atmospheric oxygen (O_2) in a perfect ratio results in water (H_2O) and carbon dioxide (CO_2).

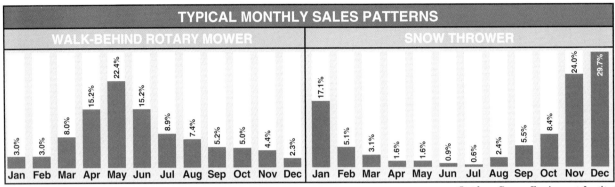

Figure 1-21. Sales of outdoor power equipment vary by the month depending on the product and climate.

A conservative estimate of the average useful life of outdoor power equipment includes six years for walk-behind lawn mowers, rear engine riding mowers, and lawn tractors; and nine years for garden tractors, snow throwers, and tillers. Rear engine riding mowers are larger in size than walk-behind mowers but smaller than lawn tractors. They are commonly used to cut medium-sized lawns.

Lawn tractors are designed for mowing but can be used for other tasks with additional attachments. The size of a lawn tractor engine commonly ranges from 11 HP to 18 HP. Garden tractors are larger than lawn tractors, usually have more horsepower, and incorporate a more sophisticated frame to accommodate ground-engaging attachments such as tillers, plows, dozer blades, backhoes, and other accessories.

On a given day, millions of small engines are in use throughout the world. In addition, each year, more than 8,000,000 new engines are manufactured by Briggs & Stratton Corporation to meet industry needs. Support for manufacturing and servicing these engines requires 35,000 parts maintained in active service inventory. Production of these parts and engines during a day of operation requires 270 t ductile and gray cast iron, 175 t aluminum, 125 t steel, and 35 t zinc.

Future Trends

The small engine industry is changing as new technology is developed for greater efficiency and new application requirements. As new products are designed and manufactured, qualified service technicians are needed to meet the increasingly more complex service requirements. A recent survey indicated that the average small engine service technician is 45 years old.

The Mola Agitator, a wringer washer made by the Modern Laundry Machine Company in the 1920s, was put on stilts to accommodate the Briggs & Stratton Model FH engine.

With a growing list of new products and technology coupled with a maturing workforce, the need for qualified service technicians is expected to grow. This has made the supply of trained service technicians the number one issue facing the small engine industry. Competent service technicians at local dealers are crucial for meeting the needs of a service network.

Advancement in Field

Successful small engine service technicians practice lifelong learning. Once basic skills are mastered, higher levels of skill mastery are required. Small engine service technicians make the most of each task by evaluating and refining strategies and skills. In addition, new products, tools, and equipment are being developed regularly.

The ability to adapt as the industry changes is crucial for success. Attending classes, technical update seminars, and factory schools; reading trade publications; and participating in professional organizations are activities which provide valuable information on current topics and trends. See Figure 1-22. New skills are required to remain current with advancing technology and to grow in the small engine service field.

In an effort to establish a benchmark of small engine service technician competency, the Briggs & Stratton Corporation has instituted the Master Service Technician (MST) program. The MST program is based on the successful completion of the Briggs & Stratton MST exam. The MST exam is a comprehensive test of Briggs & Stratton Corporation product and service knowledge. See Figure 1-23. The exam is administered through the regional Central Sales and Service Distributor (CSSD) and consists of 311 questions completed during a $4\frac{1}{2}$ hour period.

Master Service Technicians offer proven knowledge and skills to potential customers. The MST exam is widely recognized in the United States and Canada. Other countries participating in the MST program include Australia and the United Kingdom. This type of worldwide recognition is invaluable to the small engine service technician. Information about MST exam dates and times is available from the local Central Sales and Service Distributor. See Appendix.

Members of OPEI manufacture over 95% of gasoline and electrical-powered lawn and garden maintenance products for the consumer retail market.

Figure 1-22. Small engine service technicians must constantly acquire new knowledge and skills to remain current with advancing technology in the field.

20 SMALL ENGINES

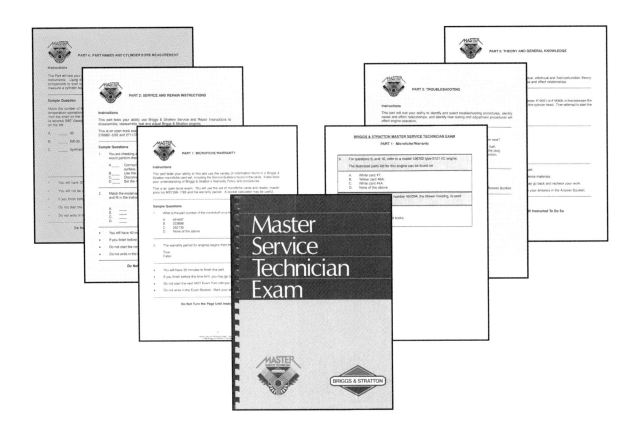

Figure 1-23. The Master Service Technician exam is a comprehensive test of Briggs & Stratton Corporation product and service knowledge.

EasyRake®/EverGreen International, Inc.
A lawn vacuum accessory can be used to collect yard debris.

Thermo King
The engine used to power this refrigeration unit has a compression ratio of 8.25:1 and delivers 11.9 HP (continuous duty) at 2450 rpm.

SAFETY AND TOOLS

CHAPTER 2

Safety procedures reduce the possibility of accidents and injuries when working on small engines. Industry and standards organizations promote efforts in the design, manufacture, and service of small engines for maximum safety and efficiency. Tools used to service small engines may be standard hand and power tools or specialty tools from the engine manufacturer.

SMALL ENGINE OPERATION SAFETY

Small engine operation involves risks that may be reduced by safety precautions and a safe working environment. Specific tasks on small engines and equipment may require special safety procedures. General safety rules which apply to engine operation procedures include:

- Never operate an engine in an enclosed area. The work area must be properly ventilated to remove hazardous carbon monoxide gases from the work area or other areas of the building.
- Work in a space with ample room for maneuvering around the engine and application.
- Engines should be operated only by qualified service personnel.
- Use an approved spark tester to check for spark.
- Do not have an open flame or smoke near flammable liquids.
- Do not refuel an engine while it is operating. Allow the engine to adequately cool before refueling.
- Do not operate an engine if gas has spilled. Move the engine away from the spill and avoid creating any ignition until the gas has evaporated.
- Be familiar with engine shutoff procedures for quick response in an emergency.
- Do not operate an engine with the air cleaner removed (excluding snow thrower engines).
- Refuel an engine outdoors or in an approved area with proper ventilation and fire safety equipment.
- Do not operate an engine without a muffler.

- Disengage an engine from driven equipment as required before starting.
- Do not operate an engine with an accumulation of grass, leaves, or other combustible materials near the muffler area.
- Never leave an operating engine unattended.
- Do not operate an engine at excessive speeds or adjust the governor to exceed manufacturer's specifications.
- Avoid contact with hot engine parts such as the muffler, cylinder head, or cooling fins.
- Do not operate an engine in dry grass or other combustible materials that could be ignited.
- Keep feet, hands, and clothing away from moving engine and equipment components. Perform service and maintenance procedures with the engine not operating if possible.
- Remove grass and other debris from cylinder fins and governor parts.
- Do not attempt to crank an engine with the spark plug removed.
- Always have an approved fire extinguisher near the work area.

Eagle Manufacturing Company
Safety storage cabinets are constructed to meet OSHA requirements for flammable liquid storage.

INDUSTRY AND STANDARDS ORGANIZATIONS

The small engine industry has evolved over the years through the efforts of many organizations involved in the manufacture, safe operation, and service of small engines and related equipment. These organizations have sought to establish safety standards, provide quality and consistency from manufacturer to manufacturer, and provide a vehicle for product improvement. The small engine service technician can utilize the resources of these organizations to ensure product safety, quality, and efficiency. These organizations can be broadly classified as government agencies, standards organizations, technical societies, private organizations, and trade associations. See Figure 2-1.

Government Agencies

Government agencies are federal, state, and local government organizations and departments which establish rules and regulations related to safety, health, and equipment installation and operation. Company procedures may exceed, but must comply with, minimum federal, state, and local agency rules and regulations. Federal agencies which commonly affect small engine service technicians include the Occupational Safety and Health Administration (OSHA), the National Institute for Occupational Safety and Health (NIOSH), the Environmental Protection Agency (EPA), the Department of Transportation (DOT), the Department of Defense (DOD), and the Consumer Products Safety Commission (CPSC).

The *Occupational Safety and Health Administration (OSHA)* is a federal agency that requires all employers to provide a safe environment for their employees. OSHA was established under the Occupational Safety and Health Act of 1970. Under OSHA guidance, states may develop and operate state job safety and health plans. State plans may exceed, but must be revised to comply with, minimum federal OSHA standards.

The Office of the Federal Register publishes all adopted OSHA standards and required amendments, corrections, insertions, or deletions. All current OSHA standards are reproduced annually in the Code of Federal Regulations (CFR). OSHA standards are included in Title 29 of the CFR, Part 1900-1999. CFRs are available at many libraries and at Government Printing Offices in major cities. See Figure 2-2.

INDUSTRY AND STANDARDS ORGANIZATIONS

 CPSC
Consumer Products Safety Commission
4330 East West Hwy
Bethesda, MD 20814

 EPA
Environmental Protection Agency
401 M Street SW
Washington, DC 20460

 DOD
Department of Defense
Defense Printing Service
700 Robbins Avenue
Philadelphia, PA 19111-5094

 DOT
Department of Transportation
400 7th Street SW
Washington, DC 20590

 NIOSH
National Institute for Occupational
Safety and Health
4676 Columbia Parkway
Cincinnati, OH 45226

 OSHA
Occupation Safety and
Health Administration
230 South Dearborn Street
Chicago, IL 60604

GOVERNMENT AGENCIES

  ANSI
American National Standards Institute
11 West 42nd Street
New York, NY 10036

 CSA
Canadian Standards Association
178 Rexdale Blvd
Rexdale, ON M9W 1R

 ISO
International Organization for Standardization
Case Postale 56 CH - 1211
Geneve 20 Switzerland

STANDARDS ORGANIZATIONS

 SAE
Society of Automotive Engineers
400 Commonwealth Dr.
Warrendale, PA 15096

 ASAE
American Society of Agricultural Engineers
2950 Niles Rd.
St. Joseph, MI 49085

 ASTM International
American Society for Testing and Materials
1916 Race St.
Philadelphia, PA 19103

TECHNICAL SOCIETIES

 NFPA
National Fire Protection Association
Batterymarch Park
Quincy, MA 02269

 UL
Underwriters Laboratories Inc.
333 Pfingsten Rd.
Northbrook, IL 60062

TRADE ORGANIZATIONS

 API
American Petroleum Institute
1220 L St. NW
Washington, DC 20005

 OPEI
Outdoor Power Equipment Institute
341 S. Patrick St.
Old Town Alexandria, VA 22314

PRIVATE ORGANIZATIONS

 EETC
Equipment & Engine Training Council
P.O. Box 648
W307 N5480 Anderson Road
Hartland, WI, 53029

 OPEESA
Outdoor Power Equipment and
Engine Service Association
210 Allen Drive
Exton, PA 19341

 AED
AED Foundation
Associated Equipment Distributors, Inc.
615 W. 22nd Street
Oak Brook, IL 60523

TRAINING-RELATED ORGANIZATIONS

 SkillsUSA
SkillsUSA–VICA
P.O. Box 3000
Leesburg, VA 20177

 FFA
National FFA Center
P.O. Box 68960, 6060 FFA Drive
Indianapolis, IN 46268

STUDENT-RELATED ORGANIZATIONS

Figure 2-1. Industry and standards organizations have been instrumental in the development of small engines.

24 SMALL ENGINES

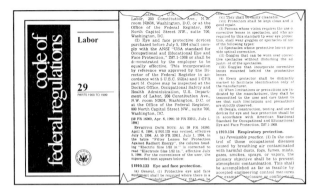

Figure 2-2. All current OSHA standards are reproduced in the Code of Federal Regulations (CFR).

For example, OSHA has established a color code to designate certain cautions and dangers and to increase efficiency in work areas. Red is used for fire protection equipment and apparatus, portable containers, and emergency stop buttons and switches. Orange is used to identify potentially dangerous parts of machines including exposed pulleys, gears, rollers, cutting devices, and power jaws. Yellow is used to indicate caution and potential physical hazards. This includes the starting point or power source of machinery, caution signs for equipment under repair, and safety lines. Purple is used to indicate radiation hazards. Green is used to indicate safety devices and locations of first aid equipment. Black and white are used to designate traffic and housekeeping areas. See Figure 2-3.

The *National Institute for Occupational Safety and Health (NIOSH)* is a national organization that acts in conjunction with OSHA to develop and periodically revise recommended exposure limits for hazardous substances or conditions in the workplace. NIOSH also recommends preventive measures to reduce or eliminate the adverse health and safety effects of these hazards. NIOSH is primarily concerned with research activities, while OSHA is responsible for enforcement.

The *Environmental Protection Agency (EPA)* is a federal agency established in 1970 to control and abate pollution in the areas of air, water, solid waste, pesticides, radiation, and toxic substances. The primary responsibilities of the EPA include creating and establishing

Color		Use	Applications
SAFETY COLOR CODING			
Red		Identify: 1. Fire protection equipment and apparatus 2. Danger 3. Stop	1. Fire exit signs, fire alarm boxes, fire extinguishers, fire hose locations, fire hydrants 2. Safety cans with a flash point of 100°F or less, danger signs 3. Emergency stop box on hazardous machines and stop buttons used for emergency stopping of machinery
Orange		Designate dangerous parts of machines or energized equipment which may cut, crush, shock, or otherwise cause injury and to emphasize such hazards when doors are open or guards are removed	Inside mowing guards; safety starting buttons; inside transmission guards for gears, pulleys, chains, etc.; exposed parts of pulleys, gears, rollers, etc.
Yellow		Designate caution and mark physical hazards such as striking against, stumbling, falling, and tripping. Solid yellow, yellow and black stripes, or yellow and black checks may be used in any combination to attract the most attention	Construction equipment such as bulldozers, tractors, carryalls, etc.; coverings or guards for guy wires; exposed and unguarded edges of platforms, pits, and walls; handrails, guardrails, on top and bottom treads of stairways where caution is needed; markings for projections, doorways, etc.; pillars, posts, or columns
Purple		Designate radiation hazards. Yellow is used in combination with purple for markers such as tags, labels, signs, etc.	Rooms and areas where radioactive materials are stored or handled, burial grounds, disposable cans for contaminated materials, containers of radioactive materials, etc.
Green		Designate safety and location of first aid equipment	Safety bulletin boards, gas masks, first aid kits, stretchers, etc.
Black, White, or B/W		Designate: 1. Traffic 2. Housekeeping areas	1. Dead ends of aisles or passageways; location and width of aisleways, stairways, and direction signs 2. Location of refuse cans, food dispensing equipment, etc.

Figure 2-3. Colors are used to designate safety hazards and to increase efficiency in a small engine service facility.

standards affecting public health and the environment. This includes procedures for handling hazardous material used in small engine work such as gasoline, solvents, oil, battery electrolyte, and engine coolant.

The *Department of Transportation (DOT)* is a federal agency responsible for traffic control, enforcement of safety regulations, and aids to navigation. The small engine service technician is primarily concerned with DOT regulations related to vehicle requirements.

The *Department of Defense (DOD)* is a federal agency responsible for developing United States Military Standards (MIL Standards). MIL standards cover the specifications used by the armed forces, but are not restricted to them.

The *Consumer Product Safety Commission (CPSC)* is a federal commission empowered to implement consumer safety standards throughout the United States. This commission oversees the design and safe operation of many consumer goods manufactured domestically and internationally.

In June 1982, the CPSC implemented a standard for all rotary lawn mowing equipment based on the mowing blade tip speed and stopping time. This standard mandated a maximum blade tip speed of 19,000 fpm (feet per minute). Blade tip speed is based on blade length and engine speed. The standard also mandated that rotary lawn equipment blades must stop within 3 sec of leaving the operator presence zone. See Figure 2-4. All consumer rotary lawn equipment must conform to this standard. Standards for commercial lawn equipment vary. For specific equipment requirements, consult the manufacturer's specifications.

John Deere Worldwide Commercial & Consumer Equipment Division

Figure 2-4. All consumer rotary lawn equipment must conform to CPSC standards for blade stopping time.

Standards Organizations

Standards organizations are organizations affiliated with their (and often other) governmental organizations. These organizations coordinate the development of codes and standards among member organizations. Codes and standards are used to protect people and property from potential hazards. A *code* is a regulation or minimum requirement. A *standard* is an accepted reference or practice. Codes and standards ensure that equipment is manufactured, installed, and operated for maximum safety and efficiency. Codes and standards also detail requirements for worker safety in different work settings. Organizations specializing in codes and standards include the American National Standards Institute (ANSI), the Canadian Standards Association (CSA), and the International Organization for Standardization (ISO).

The *American National Standards Institute (ANSI)* is a national organization that helps identify industrial and public needs for national standards. ANSI standards are produced by professional and technical societies, trade associations, and consumer and labor groups. ANSI is the national coordinator of voluntary standards activities and serves as an approval organization and clearinghouse for consensus standards.

The *Canadian Standards Association (CSA)* is a Canadian national organization that develops standards and provides facilities for certification testing to national and international standards. The CSA uses volunteers from industry to develop consensus standards in areas such as occupational health and safety, manufacturing materials and processes, and electrical and electronic products.

The *International Organization for Standardization (ISO)* is a nongovernmental international organization comprised of national standards institutions of over 90 countries (one per country). The ISO provides a worldwide forum for the standards developing process. ANSI is the United States representative to the ISO.

Technical Societies

Technical societies are organizations composed of groups of engineers and technical personnel united by professional interest. The *American Society of Agricultural Engineers (ASAE)* is a professional and technical organization of members interested in engineering knowledge and technology in food and agriculture, associated industries, and related resources. The ASAE provides standards used in industry for the manufacture of agricultural equipment. These standards are published by ANSI and

ASAE. For example, ANSI/ASAE S418-1988 is *Dimensions for Cylindrical Hydraulic Couplers for Lawn and Garden Tractors.*

The *Society of Automotive Engineers (SAE)* is a network of engineers, business executives, educators, and students from more than 80 countries who share the interest of advancing engineering of mobile systems. The SAE is a major source of technical information and expertise used in the design and manufacture of equipment powered by small engines. In addition, SAE provides a standardized method of testing engine output. Their standards are published by ANSI and SAE. For example, ANSI/SAE J1995-JAN90 is *The Engine Power Test Code – Spark Ignition and Compression Ignition – Gross Power Rating.*

ASTM International is the largest organization in the world devoted to developing and publishing voluntary, full-consensus standards. Standards developed by ASTM International include materials such as metals, paints, and plastics commonly used in the manufacture of small engines and equipment components. ASTM International publications include standard test methods, specifications, practices, guides, classifications, and terminology. Technical research or testing work is done voluntarily by qualified ASTM members throughout the world. More than 9100 ASTM standards are published each year in the 70 volumes of the *Annual Book of ASTM Standards.*

Private Organizations

Private organizations are organizations that develop standards from an accumulation of knowledge and experience with materials, methods, and practices. These standards are often modified from other standards to meet the needs of the organization. Private organizations often impose stricter standards than other organizations.

For example, *Underwriters Laboratories Inc. (UL®)* is an independent organization that tests equipment and products to verify conformance to national codes and standards. Equipment tested and approved by UL® carries the UL® label. UL®-approved equipment and products are listed in their annual publication, *UL® Standards for Safety.*

The *National Fire Protection Association (NFPA)* is a national organization that provides guidance in assessing hazards of the products of combustion. The NFPA publishes the *National Electrical Code® (NEC®).* The purpose of the NEC® is the practical safeguarding of persons and property from hazards arising from the use of electricity. The NFPA also develops informational material regarding hazardous materials. Although this material is not a regulatory standard, it provides information regarding properties of hazardous materials and required fire fighting procedures. Their standards are published by ANSI and NFPA. For example, ANSI/NFPA 49-1991 is *Hazardous Chemicals Data.*

Trade Associations

Trade associations are organizations that represent producers and distributors of specific products. For example, the *Outdoor Power Equipment Institute (OPEI)* is the national trade association representing manufacturers of consumer and commercial outdoor power equipment and their major components.

OPEI members include original equipment manufacturers (OEMs) that produce equipment and products such as walk-behind lawn mowers, lawn tractors, snow throwers, log splitters, commercial turf care products, golf course power equipment, and gasoline-powered hand-held equipment such as trimmers and chainsaws.

Most OPEI members are classified under the new North American Industry Classification System (NAICS) product codes 333112, "Lawn and Garden Equipment," and 333111J, "Commercial Turf and Grounds Care Equipment." These product codes replace the Standard Industrial Classification (SIC) codes 3524 and 3523F.

The *American Petroleum Institute (API)* is the primary trade association of the United States petroleum industry. The API provides public policy development and advocacy, research and technical services for advancing petroleum technology, and industry equipment and performance standards. See Figure 2-5. The API also addresses the enhancement of environmental, health, and safety performance concerns of the petroleum industry.

Training Organizations

The Equipment and Engine Training Council (EETC) is a nonprofit professional organization that promotes and supports the education and training of service technicians in outdoor power equipment, agriculture, and commercial and heavy equipment technology. The organization is made up of manufacturers and their service and training personnel, technical school instructors, equipment distributors, equipment dealers, associations, and other industry and educational leaders.

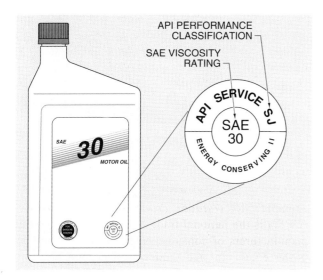

Figure 2-5. The API oil performance classification provides a standard for recommended usage.

The Outdoor Power Equipment and Engine Service Association (OPEESA) is an organization formed by the merging of the Engine Service Association (ESA) and the Outdoor Power Equipment Distributor Association (OPEDA). The OPEESA membership is comprised of OPE manufacturers and dealers, Central distributors representing air-cooled engine manufacturers, air- and water-cooled gas and diesel engine distributors, and original equipment services parts distributors and manufacturers. OPEESA assists its members through training, promoting the benefits of OPE technician certification, and other services and matters vital to the industry.

Associated Equipment Distributors (AED) is a membership association of 1200 independent distributors, manufacturers, and other organizations involved in the distribution of construction equipment and related products and services in North America and throughout the world. The Associated Equipment Distributors serves the industry by creating and providing high-quality products, services, and information, including industry education, training, and career development through the AED Foundation.

Student Organizations

The National FFA Organization (FFA) is dedicated to making a positive difference in the lives of young people through agricultural education. FFA prepares members for more than 300 careers in the science, business, and technology of agriculture. Curriculums addressed include agriscience, advanced agriscience and biotechnology, agricultural mechanics, horticulture, animal science, and environment-related agricultural science.

SkillsUSA-VICA is a national organization serving high school and college students and their instructors involved with the preparation for careers in technical, skilled, and service occupations including health occupations. SkillsUSA-VICA provides an opportunity for applying technical and leadership skills acquired in career and technical programs at the local, regional, state, and national level. The SkillsUSA Championships is the national-level competition that showcases over 4000 contestants in 75 separate events.

FIRE SAFETY

Small engine service facilities commonly have gasoline, cleaning solvents, and other combustible materials which require fire safety procedures. To start and sustain a fire, fuel, heat, and oxygen must be present. A fire is extinguished if any one of these is removed. The risk of fire in the small engine service facility may be reduced by following these general safety rules:

- Know the location of all fire extinguishers.
- Ventilate work areas to prevent accumulation of flammable vapors.
- Know the procedures in the emergency plan for the facility.
- Store gasoline, cleaning solvents, and all other flammable liquids in approved safety containers and cabinets.
- Do not perform electrical work near flammable materials.
- Label and store flammable materials for quick identification and access.
- Keep all emergency traffic paths and exits clear.

 Fire extinguishers must be installed by a qualified service agency.

Fire Extinguishers

Work areas must be equipped with the correct number and type of fire extinguishers. The number and type of fire extinguishers required are determined by the authority having jurisdiction (AHJ) based on how fast a fire may spread, potential heat intensity, and accessibility to the fire. Fire extinguishers are designed to extinguish fires quickly and safely. Fire extinguishers are rated for the class(es) of fire as Class A, B, C, or D based on the combustible material. See Figure 2-6.

Class A fires include combustibles such as paper, wood, cloth, rubber, plastics, refuse, and upholstery. Class B fires include combustible liquids such as gasoline, oil, grease, and paint. Class C fires include electrical equipment such as motors, appliances, wiring, fuse boxes, breaker panels, and transformers. Before extinguishing a Class C fire, power should be shut off as quickly as possible. Class D fires include combustible metals such as magnesium, potassium, sodium, titanium, and zirconium.

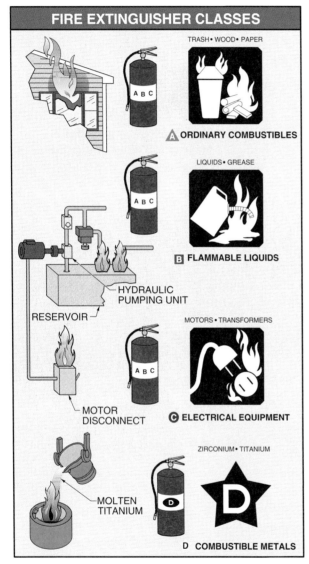

Figure 2-6. The class of fire determines the type of fire extinguisher and extinguishing agent used.

Common small engine service tasks require basic tools. Class A fire extinguishers are commonly filled with water. Water should never be used to extinguish a flammable liquid fire. Water disperses the flammable liquid and flames over a larger area. In addition, water is a conductor of electricity, and should never be used to extinguish an electrical fire. Class B-C fire extinguishers are commonly filled with carbon dioxide, potassium bicarbonate, or sodium bicarbonate. Class A-B-C fire extinguishers are commonly filled with monoammonium phosphate. Class D fire extinguishers are commonly filled with silica gel (mineral dust) or metal salt granules. Fire extinguishers used in a confined space may require respiratory protection.

The pressure gauge on top of the fire extinguisher must be checked periodically to verify the recommended pressure. Leaks in a fire extinguisher may cause a loss of pressure over time. State and local codes may require a scheduled inspection of fire extinguishers.

Flammable Liquids

A *flammable liquid* is a liquid that has a flash point below 100°F. *Flash point* is the lowest temperature at which a liquid gives off vapor sufficient to ignite when an ignition source is introduced. *Vapor pressure* is the pressure exerted by vapor above the surface of a liquid in a closed container. A *combustible liquid* is a liquid that has a flash point at or above 100°F.

Flammable liquids used in small engine service facilities must be stored in an approved safety can. A *safety can* is a UL®-approved container not exceeding 5 gal. that has a spring-loaded lid on the spout to prevent the escape of explosive vapors, yet allow relief of internal pressure. See Figure 2-7. A *flame arrestor* is a mesh or perforated metal insert within a safety can which protects its contents from external flame or ignition. When pouring a flammable liquid from one metal container to another, bonding is used to prevent possible sparks caused by static electricity. *Bonding* is the use of metal-to-metal contact or a wire between two containers to prevent possible ignition from static electricity sparks.

Safety cans should be stored in an approved fireproof metal safety cabinet. A *safety cabinet* is a double-walled steel cabinet specifically designed for storage of flammable liquid containers. Quantities stored should be limited to the amount required for immediate use. Fire codes may dictate the maximum amount of flammable liquid that may be stored in a work area at a given time. Labeling must comply with NFPA standards for easy identification in an emergency.

 In the outdoor power equipment industry, approximately 47% of dollars spent for components was for engines.

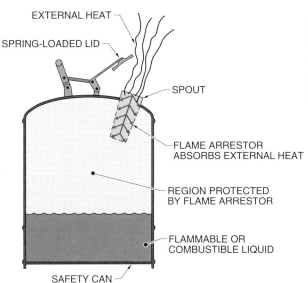

Figure 2-7. A safety can is used to safely store flammable and combustible liquids. *Eagle Manufacturing Company*

Service to an engine fuel system should be performed in a ventilated area or outdoors if possible. Gasoline drained from the fuel system must be collected and stored in an approved safety container prior to proper disposal. Shop rags used for removing grease, oil, gasoline, or solvent from parts and equipment must be stored in an approved oily waste can.

An *oily waste can* is a can designed for safe containment of rags and paper soiled with flammable materials. See Figure 2-8. The bottom is raised to provide a cooling space between the bottom and floor. A foot pedal opens the lid, and gravity causes the lid to close automatically. When closed, the lid provides a seal to keep out oxygen and prevent spontaneous combustion of the rags. *Spontaneous combustion* is self-ignition caused by chemical reaction and temperature buildup in waste material such as used oily rags.

Figure 2-8. An oily waste can seals out oxygen to prevent spontaneous combustion.

A horizontal shaft engine is used to power a pressure washer.

CARBON MONOXIDE

Carbon monoxide (CO) is a toxic (poisonous) gas produced by incomplete combustion of gasoline (HC-based fuels). Carbon monoxide is odorless and tasteless, providing no warning to its victims. It has a special affinity (attraction) for hemoglobin in the bloodstream. *Hemoglobin* is a blood component that transports oxygen. Carbon monoxide is easily absorbed by the normally oxygen-rich hemoglobin.

Once absorbed, carbon monoxide is released from hemoglobin at a much slower rate than oxygen. When the hemoglobin has absorbed a sufficient amount of carbon monoxide, it becomes unable to transport oxygen to vital organs and muscles of the body. The lack of oxygen may lead to dizziness, headaches, blurred vision, and possibly death.

Air normally contains 21% oxygen. Carbon monoxide is slightly lighter than air and rises as it is warmed in the combustion process. However, humid conditions may cause carbon monoxide to remain stationary, producing areas of high concentration. Carbon monoxide detectors are available for room installations. Carbon monoxide is measured in parts per million (ppm) and must be evacuated from the work area by an evacuation system and/or adequate ventilation to prevent health hazards. See Figure 2-9.

PERSONAL PROTECTIVE EQUIPMENT

Personal protective equipment (PPE) is safety equipment worn by a small engine service technician for protection against safety hazards in the work area. Personal protective equipment commonly used by a small engine service technician includes protective clothing, eye protection, ear protection, respiratory protection, hand protection, foot protection, and back protection.

 The National Safety Council estimates that 1,700,000 workers in the U.S. between the ages of 50 and 59 have compensable noise-induced hearing loss.

Protective Clothing

Protective clothing is worn by the small engine service technician to prevent injury and provide a professional appearance. Clothing such as coveralls made of durable material like denim offer protection from contact with sharp objects and hot equipment. Some tasks such as welding require special fire-resistant coveralls for additional protection. Clothing made of flammable synthetic materials should never be worn during engine service.

Protective clothing should be snug, but it also should allow ample freedom of movement. See Figure 2-10. Pockets should allow convenient access, but should not snag on tools or equipment. Soiled protective clothing should be washed as required to reduce flammability hazards.

Loose fitting clothing and long hair must be secured to prevent getting caught in rotating parts. Jewelry such as wrist watches, necklaces, and rings should be removed. Watches and rings are excellent conductors of electricity and may cause serious burns if contact is made with an electrical circuit.

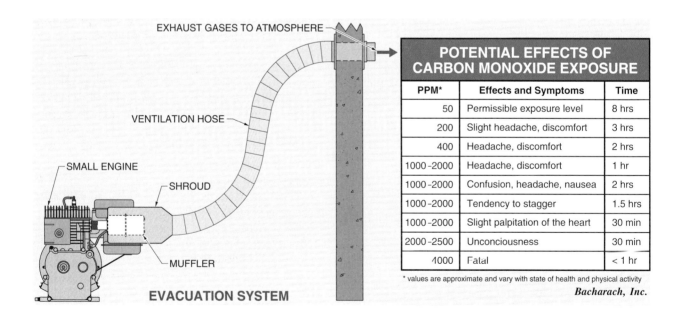

POTENTIAL EFFECTS OF CARBON MONOXIDE EXPOSURE		
PPM*	Effects and Symptoms	Time
50	Permissible exposure level	8 hrs
200	Slight headache, discomfort	3 hrs
400	Headache, discomfort	2 hrs
1000–2000	Headache, discomfort	1 hr
1000–2000	Confusion, headache, nausea	2 hrs
1000–2000	Tendency to stagger	1.5 hrs
1000–2000	Slight palpitation of the heart	30 min
2000–2500	Unconsciousness	30 min
4000	Fatal	< 1 hr

* values are approximate and vary with state of health and physical activity

Bacharach, Inc.

Figure 2-9. An engine operated in an enclosed space is connected to an evacuation system for safe removal of carbon monoxide.

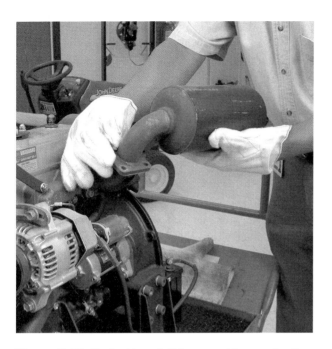

Figure 2-10. Protective clothing provides protection from sharp objects and hot equipment during engine service.

Eye Protection

Eye protection is required by OSHA when there is a reasonable probability of preventing injury to the eyes or face from flying particles, molten metal, chemical liquids or gases, radiant energy, or a combination of any or all of these. Eye protection must comply with OSHA 29 CFR Part 1910.133. Standards for eye protection are specified in ANSI Z87.1-1989, *Practice for Occupational and Educational Eye and Face Protection.*

Both moving parts and service procedures which involve chiseling, cutting, grinding, or welding may cause flying objects. Most eye injuries are preventable by wearing safety glasses, face shields, or goggles. See Figure 2-11. *Safety glasses* are glasses with impact-resistant lenses, reinforced frames, and side shields. The lenses are made of special glass or plastic to provide maximum protection against impact. The frame is designed to keep lenses secured in the frame during impact. Side shields provide additional protection from flying objects. Safety glasses are available with prescription lenses.

 A Bureau of Labor Statistics survey found that 60% of workers who suffered eye injuries on the job were not wearing eye protective equipment at the time of the injury.

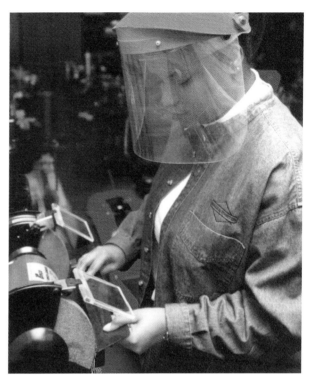

Figure 2-11. A face shield is worn when grinding to provide eye and face protection.

A *face shield* is an eye protection device that covers the entire face with a plastic shield. The face shield is used for eye and face protection from flying objects or splashing liquids. *Goggles* are an eye protection device secured on the face with an elastic headband that may be used over prescription glasses. Goggles provide full contact around the entire eye area for maximum protection from flying objects.

Goggles with clear lenses provide protection against flying objects or splashing liquids. Goggles with colored lenses also provide protection from harmful ultraviolet (UV) rays produced by heating and welding equipment. UV rays may cause injury to the small engine service technician and/or others in the vicinity. Different welding tasks require certain colored lenses to provide adequate protection. Refer to American Welding Society (AWS) standards for the recommended eye protection when welding.

Eye wash stations are used to safely flush out particles or contaminants from eyes. Safety lenses must be maintained to provide protection and clear visibility. Cleaners are available for specific lenses to prevent damage when cleaning. Pitted or scratched lenses reduce vision and may cause lens failure on impact.

Ear Protection

Engines and tools produce high noise levels. A small engine service technician subjected to high noise levels may develop hearing loss over a period of time. The severity of hearing loss depends on the intensity and duration of exposure. Sound intensity is expressed in decibels. A *decibel (dB)* is a unit of expressing relative intensity of sound on a scale from 0 dB (average least perceptible) to 140 dB (deafening). See Figure 2-12. Ear protection devices are worn to prevent hearing loss.

Ear protection devices can be broadly classified as earplugs and ear muffs. *Earplugs* are ear protection devices made of moldable rubber, foam, or plastic which are inserted into the ear canal. *Ear muffs* are ear protection devices worn over the ears. A tight seal around the ear muff is required for proper ear protection.

Ear protection devices are assigned a noise reduction rating (NRR) number based on the noise level reduced. For example, an NRR of 27 means that the noise level is reduced by 27 dB when tested at the factory. To determine approximate noise reduction in the field, 7 dB is subtracted from the NRR. An NRR of 27 provides a noise reduction of approximately 20 dB in the field.

SOUND LEVELS		
Decibel (dB)	**Loudness**	**Examples**
140	Deafening	Jet airplane taking off, air raid siren, locomotive horn
130	Pain threshold	
120	Feeling threshold	
110	Uncomfortable	
100	Very loud	Chain saw
90	Noisy	Shouting, auto horn
80	Moderately loud	Vacuum cleaner
70	Loud	Telephone ringing, loud talking
60	Moderate	Normal conversation
50	Quiet	Hair dryer
40	Moderately quiet	Refrigerator running
30	Very quiet	Quiet conversation, broadcast studio
20	Faint	Whispering
10	Barely audible	Rustling leaves, soundproof room, human breathing
0	Hearing threshold	Intolerably quiet

Figure 2-12. A decibel is a unit for expressing relative intensity of sound.

Respiratory Protection

Respiratory protection is required for protection against chemical hazards in the small engine service facility. A *chemical hazard* is a solid, liquid, gas, mist, dust, and/or vapor that is toxic when inhaled, absorbed, or ingested. Airborne chemical hazards exist as concentrations of mists, vapors, gases, fumes, or solids. Chemical hazards vary depending on how they are introduced into the body. For example, some chemicals are toxic through inhalation. Others are toxic by absorption through the skin or through ingestion. Some chemicals are toxic by all three routes and may be flammable as well. Small engine service technicians may be subjected to hazards such as vapors from aerosol cleaning solvents. The degree of risk from exposure to any given substance depends on the nature and potency of toxic effects and the magnitude and duration of exposure. Respiratory protection required is determined by the hazards of the chemical.

Hand Protection

Hand protection is required to prevent injuries to hands from burns, cuts, and the absorption of chemicals. Work activities of the small engine service technician determine the duration, frequency, and degree of hazards to hands. See Figure 2-13. Common tasks requiring hand protection include cleaning parts, welding, and working with hot and sharp engine components and equipment. Gloves made from nitrile (synthetic rubber material) or neoprene provide protection from chemicals, resist puncture, and allow good dexterity. Leather gloves provide protection from heat and molten metal in the welding process. Heavy fabric gloves provide protection from hot engine parts such as mufflers and cylinder heads. These gloves also provide protection from possible cuts and abrasions when working with blades, chains, and sharp engine components. Gloves should be snug fitting. Gloves that are too large can pose a safety hazard when working around moving parts.

Figure 2-13. Gloves worn when cleaning parts prevent the absorption of chemicals through the skin.

Foot Protection

According to the Bureau of Labor Statistics, a typical foot injury is caused by objects falling less than 4′ with the average weight approximately 65 lb. Small engine service technicians perform many tasks which require handling of similar objects. Safety shoes with reinforced steel toes provide protection against injuries caused by compression and impact. See Figure 2-14. Some safety shoes have protective metal insoles and metatarsal (toe) guards for additional protection. Oil-resistant soles and heels are not affected by gasoline and oil and provide improved traction. Protective footwear must comply with ANSI Z41-1991, *Personal Protection – Protective Footwear*.

Figure 2-14. Safety shoes protect against falling objects and have soles that are not affected by gasoline and oil.

Back Protection

Back injuries are one of the most common injuries resulting in lost time in the workplace. Most back injuries are the result of improper lifting procedures. Through proper planning and work procedures, these injuries may be prevented. Assistance should be sought when moving a heavy object. Before moving the object, prepare the work space to provide the required clearance. Check the balance of the load. With a firm grip, bend at the knees and lift the object using leg muscles, not back muscles.

Work should be planned to minimize distance and number of moves required. Sustained service work should be positioned at a comfortable working height to reduce back strain. A back support belt may be used to maintain stability and provide additional support when lifting.

HAZARDOUS MATERIALS

A *hazardous material* is a material capable of posing a risk to health, safety, or property. Many materials used in a small engine service facility, such as gasoline, battery electrolyte, paint, and cleaning solvents, are classified as hazardous materials. OSHA 29 CFR 1910.1200 details hazard communication requirements in the workplace. Hazard communication is based on the worker's right to know (RTK) hazards involved when working with certain materials.

Employers must develop, implement, and maintain a written, comprehensive hazard communication program that includes provisions for container labeling, chemical inventory, material safety data sheets, and an employee training program. This program must also contain a list of hazardous chemicals in each work area. Information must be provided in a language or manner that employees understand. The two major components of a hazard communication program affecting small engine service technicians are container labeling and material safety data sheets.

 There are currently 25 states which administer their own occupational and health programs through plans approved to meet minimum OSHA requirements.

> ### Smart-Fill™ Fuel Can
>
> *According to the EPA, Americans spill 17,000,000 gal. of gasoline annually while filling lawn mowers and other outdoor power equipment. A Smart-Fill™ Fuel Can is UL®-approved and designed for consumer use. It has a spill-proof spout that can reduce spillage of gas during refueling. Once the Smart-Fill™ Fuel Can is filled with gas, the can is sealed to contain evaporative emissions. To fill a tank, the nozzle is turned to release any pressure. The can is lifted with the nozzle pressed against the tank opening. A spring-mounted valve in the nozzle allows fuel to flow with the applied weight of the can. When the fuel level in the tank reaches the tip of the nozzle, fuel flow from the can is stopped. As the can is lifted away from the tank, the spring-loaded valve closes. The nozzle is then turned to seal the can for storage.*

Container Labeling

All hazardous material containers must have a label which should be examined before using the product. Specific hazards, precautions, and first-aid information are listed on the label. For example, the material may be corrosive and require the use of gloves. Hazardous material containers are labeled, tagged, or marked with appropriate hazard warnings per OSHA 29 CFR 1910.1200(f). Material stored in a different container than originally supplied by the manufacturer must also be labeled. Unlabeled containers pose a safety hazard because users are not provided with content information and warnings.

Container labeling varies with each manufacturer. However, all container labels must include basic RTK information. The NFPA Hazard Signal System and the *Hazardous Material Information Guide (HMIG)* may be used to provide information at a glance. See Figure 2-15.

The NFPA Hazard Signal System uses a four-color diamond sign to display basic information about hazardous materials. Colors and numbers identify potential health (blue), flammability (red), reactivity (yellow), and special hazards (no particular color). The degree of severity by number ranges from the number 4, indicating a severe hazard, to the number 0, indicating no hazard.

A *health hazard* is the likelihood of a material to cause, either directly or indirectly, temporary or permanent injury or incapacitation due to acute exposure by contact, inhalation, or ingestion. The degrees of health hazard are ranked by number according to the probable severity of the effects or exposure of hazardous material. The health hazard is indicated on a blue background in the left diamond located at the nine o'clock position. The degree of health hazard determines specialized protective and respiratory equipment required by emergency response and fire fighting teams.

A *flammability hazard* is the degree of susceptibility of materials to burning based on the form or condition of the material and its surrounding environment. The degree of flammability hazard is ranked by number according to the susceptibility of hazardous materials to burning. Flammability hazard is indicated on a red background in the top diamond located at the twelve o'clock position.

A *reactivity hazard* is the degree of susceptibility of materials to release energy by themselves or by exposure to certain conditions or substances. The degree of reactivity is ranked by number according to the ease, rate, and quantity of energy released. Reactivity hazard is indicated on a yellow background in the right diamond located at the three o'clock position.

A *specific hazard* is any extraordinary properties and hazards associated with a particular material. This information is particularly useful for identifying special techniques for emergency response and fire fighting teams. For example, a letter W with a horizontal line through the center is used to indicate unusual reactivity with water. The letters OX indicate materials that possess oxidizing properties. Specific hazard symbols are located in the fourth space at the six o'clock position, or immediately above or below the entire symbol. No specific color is required.

Material Safety Data Sheets

A *Material Safety Data Sheet (MSDS)* is printed material used to relay hazardous material information from the manufacturer, importer, or distributor to the employer and employees. The information is listed in English and provides precautionary information regarding proper handling, emergency, and first-aid procedures. All chemical products used in a facility must be inventoried and have a MSDS.

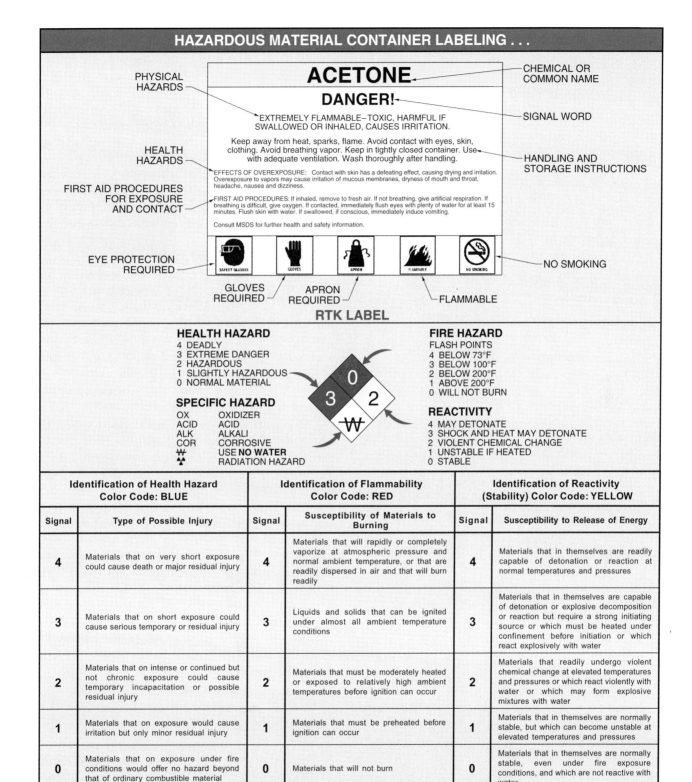

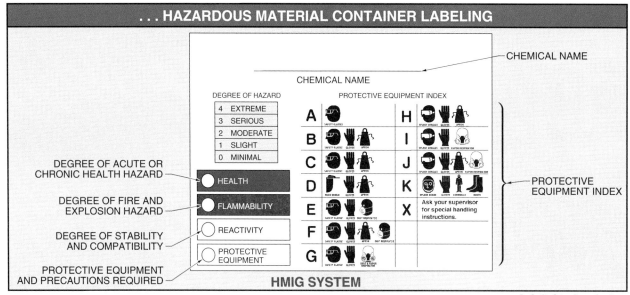

Figure 2-15. Container labeling must provide vital hazard information at a glance.

Chemical manufacturers, distributors, and importers must develop an MSDS for each hazardous material. If an MSDS is not provided, the employer must contact the manufacturer, distributor, or importer in order to obtain the missing MSDS. MSDS files must be kept up-to-date and readily available to employees. For example, when selecting a respirator, the MSDS is checked for potential hazards from the material or process.

Information may be filed according to product name, manufacturer, or a company-assigned number. If two or more MSDSs on the same material are found, the latest version is used. An MSDS has no prescribed format. Formats provided in ANSI Z400.1-1993, *Hazardous Industrial Chemicals – Material Safety Data Sheets – Preparation* may be used. All MSDSs include:

- Manufacturer's Information – Manufacturer's information provides the manufacturer's name, supplier of the hazardous material, emergency telephone numbers, preparation date, and manufacturer's address. Manufacturer's information is classified in nine sections. See Figure 2-16.
- Section I – Product Identification lists product number, name, and class.
- Section II – Hazardous Ingredients lists the chemical name or trade name and Chemical Abstracts Service (CAS) number. The CAS indexes information published in *Chemical Abstracts* by the American Chemical Society. These indexes assign CAS numbers to specific chemicals to allow quick access to information and a concise means of hazardous material identification. For example, the CAS number for acetone is 67-64-1. See Appendix.

The percent by weight column shows the approximate percentage of the hazardous ingredient as compared to the total weight or volume of the product. The *threshold limit value (TLV)* is an estimate of the average safe airborne concentration of a substance that represents conditions under which it is believed that nearly all workers may be exposed day after day without adverse effect. TLV values are published yearly by the American Conference of Governmental Industrial Hygienists (ACGIH). The acronym TLV is a trademark of ACGIH.

The *permissible exposure limit (PEL)* is the OSHA limit of employee exposure to chemicals. Exposure exceeding this limit requires the use of precautionary measures. Respiratory protection may be listed by PEL. For example, 10 × PEL means protection up to 10 times the permissible exposure limit. The TLV is listed as ppm or as milligrams of material per cubic meter of air (mg/m^3).

- Section III – Physical Data lists boiling point, specific gravity, evaporation rate, and other physical data.
- Section IV – Fire and Explosion Hazard Data lists flash point, extinguishing media required, special fire fighting procedures, and unusual fire and explosion hazards.

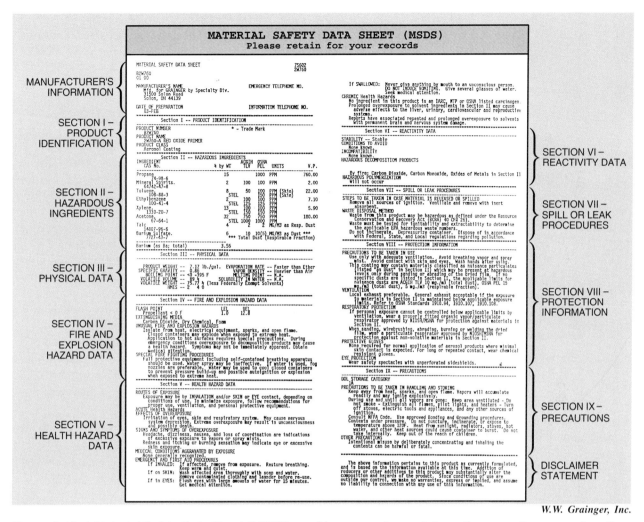

Figure 2-16. A Material Safety Data Sheet (MSDS) provides precautionary information regarding proper handling, emergency and first-aid procedures.

- Section V – Health Hazard Data provides information about the toxic effects of the material on health, including the routes of exposure, acute health hazards, effects of overexposure, signs and symptoms of overexposure, medical conditions aggravated by exposure, emergency and first aid procedures, and chronic health hazards.
- Section VI – Reactivity Data lists the stability of the material, other materials that could react with this material to create a hazardous condition, and conditions which create a violent reaction or toxic gas such as extreme temperatures, impact, or incorrect storage.
- Section VII – Spill or Leak Procedures details steps to take if there is a spill or leak, and proper waste disposal methods.
- Section VIII – Protection Information lists personal protective equipment required including respirator type, protective gloves, eye protection, and other protective equipment required.
- Section IX – Precautions lists any precautions to be taken in handling or storing the material. A disclaimer statement explains the limitations of the information listed on the MSDS.

Parts Cleaning

Parts cleaning during engine service is commonly performed in a cleaning tank. A *cleaning tank* is a tank used for cleaning parts in solvents with a lid that automatically closes to contain flames during a fire. A low-temperature solder joint on the safety prop is quickly melted from the heat of a fire. The prop breaks down, and gravity closes the lid. Most cleaning solvents used for parts cleaning are considered hazardous materials. See Figure 2-17.

38 SMALL ENGINES

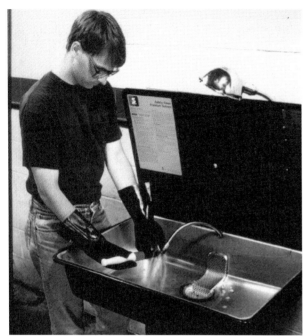

Safety-Kleen Corp.
Figure 2-17. A cleaning tank discharges cleaning solvent through a nozzle to flush any grease and dirt off parts.

In the cleaning tank, cleaning solvent is pumped through a hose and nozzle to flush grease and dirt off parts. Like aerosol spray cleaners, the cleaning solvent breaks down and washes away grease, oil, and sludge. A pump directs the cleaning solvent through a filter. The filter collects fine particles, and the heavier dirt particles collect at the tank bottom. Parts cleaning procedures must be performed in a well-ventilated area using proper eye and hand protection.

Hobart Brothers Company
A portable Welder/Generator can produce 140 A for DC welding or 3000 W auxiliary power.

 Sound measuring 74 dB is comparable to being 2′ from persons involved in a loud conversation.

Hazardous Material Disposal

Small engine service facilities commonly use and dispose of hazardous materials such as waste oil, gasoline, antifreeze, battery electrolyte, and cleaning solvents. When these materials are disposed of, they become hazardous waste. In 1986, Resource Conservation and Recovery Act (RCRA) regulations covering small quantity generators of hazardous waste went into effect. This Act makes removal and proper disposal of hazardous waste materials the responsibility of the small engine service facility or waste generator. Disposal options typically include transporting hazardous waste to an approved disposal site or contracting with a firm to pick up and dispose of the hazardous material. A manifest lists the content and quantity of hazardous material transported.

Hazardous waste is recycled or blended for safe burning to recover its heat value. For example, waste engine oil is typically processed into water (10%), light oil such as kerosene (12%), lubricating oil (65%), and asphalt bottoms used for roads (13%). If waste oil has been mixed with other hazardous materials, it is commonly burned above 2600°F in high-temperature cement kilns. These temperatures exceed the temperatures of commercial incinerators and completely destroy hazardous materials. For example, waste engine coolant (antifreeze) is commonly recycled to recover ethylene glycol for de-icing airplanes, to manufacture rug fiber polyesters, and to produce new antifreeze.

EMERGENCY PLANS

An *emergency plan* is a document that details the exact action to be taken in the event of an emergency. See Figure 2-18. The emergency plan lists procedures, exit routes, and assembly areas for company personnel. A designated company official is responsible for developing an emergency plan. Employees at the facility must be thoroughly familiar with procedures in the plan. In any emergency, the proper authorities must be notified immediately by calling 911. First aid and cardiopulmonary resuscitation (CPR) training prepares employees for assisting injured personnel until authorities arrive in an emergency.

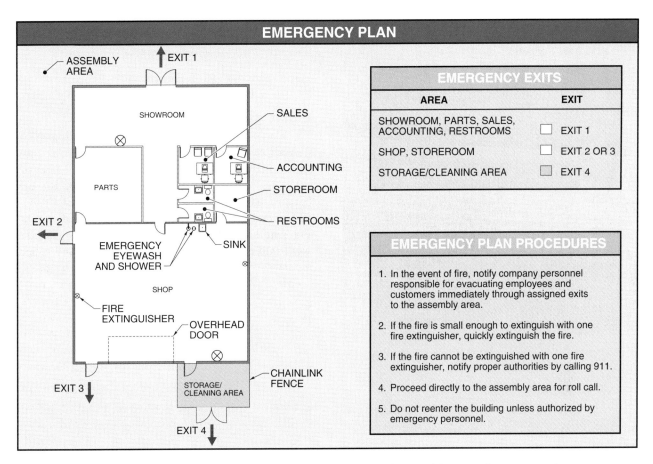

Figure 2-18. An emergency plan details procedures, exit routes, and assembly areas for personnel.

ACCIDENT REPORTS

An *accident report* is a document that details facts about an accident in the facility. See Figure 2-19. Accidents must be reported regardless of their nature. Accident report forms commonly include the name of the injured person, the date, time, and place of the accident, the name of the immediate supervisor, and the circumstances surrounding the accident. Accident reports are required for insurance claims and become a permanent part of company records. Information about the causes of the accident can be used to prevent future injuries.

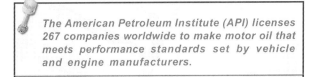
The American Petroleum Institute (API) licenses 267 companies worldwide to make motor oil that meets performance standards set by vehicle and engine manufacturers.

TOOLS

The small engine service technician uses a variety of tools to complete service and repair tasks. Tools are commonly classified by function as measurement, fastening, cutting, striking/driving, test tools, and specialty tools. Tools required vary, but some basic tools are frequently used for common service and repair tasks. See Figure 2-20. Most tools may be purchased from a variety of tool manufacturers. For specific tasks on certain engines, specialty tools are purchased from the manufacturer. See Appendix.

Larger small engine service facilities have equipment in-house to perform machine work such as engine cylinder boring. Smaller service facilities contract machine work out-of-house as required. Tools are also classified by the power supplied as hand tools and power tools. General tool safety rules which apply to both hand tools and power tools include:

- Wear proper eye protection, ear protection, and protective clothing.
- Secure hair and loose clothing.
- Work in areas with good lighting.
- Make sure all personnel are at a safe distance before using the tool.
- Keep tools free of oil, grease, and foreign matter.
- Use the tool for its designed use.
- Secure small work in a clamp or vise.
- Repair or replace damaged tools.
- Report any injuries to the supervisor.

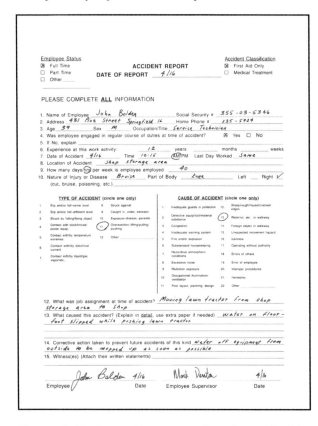

Figure 2-19. An accident report form is required for insurance claims.

Hand Tools

Hand tools are tools that are powered by hand. They are usually acquired and maintained by the small engine service technician. Hand tools used for measurement include tools such as rules, feeler gauges, and micrometers. A *micrometer* is a hand tool used to make very accurate inside, outside, or depth measurements. See Appendix. Hand tools used for fastening include tools such as wrenches, screwdrivers, and pliers. Hand tools used for cutting include tools such as hacksaws, files, and chisels. Hand tools used for striking/driving include tools such as hammers, mallets, and punches.

Hand Tool Safety. Accidents with hand tools may be reduced by following basic hand tool safety rules:

- Keep tools sharp and in proper working order. Look for wear that could cause an injury, such as a pitted hammer face, damaged insulation on a pliers, or splintered handle.
- Point cutting tools away from the body during use.
- Grind excess metal from mushroomed chisels.
- Organize tools to protect and conceal sharp cutting surfaces.
- Never use a hammer on another hammer. The impact of the hardened surfaces may cause the heads to shatter.
- Do not carry tools in a pocket. Transport sharp tools in a holder or with the blade pointed down.
- Remove fasteners by pulling the tool toward the body or pushing the tool away from the face.

Power Tools

Power tools are tools that are electrically, pneumatically, or hydraulically powered for greater efficiency when performing service and repair tasks. Power tools present in the facility vary with each company. Power tools can be portable or stationary. Portable power tools can be transported with the operator. Stationary power tools cannot be transported, and are commonly installed in a fixed position. Examples of power tools commonly used in a small engine service facility include grinders, drill motor, drill press, air chisels, and air impact wrenches. Small engine service facilities also commonly have blade grinders, chain saw sharpening equipment, battery chargers, wire wheels, and hydraulic presses. Power tools may cause serious injury if proper safety procedures are not followed.

 Crude oil from wells in the lower 48 states and over half of all refined petroleum products are shipped through 204,000 miles of pipelines.

TOOLS...

MEASURING

A. Rule
Measure linear distance

B. Feeler Gauges
Verify distance between parts

C. Micrometer
Measure thickness and diameter

D. Dial Indicator
Measure change in position of parts

E. Telescoping Gauge
Transfer distance from parts to measurement tool

F. Dial Caliper
Measure inside, outside, and depth of parts

G. Tachometer
Measure engine speed in rpm

FASTENING – WRENCHES

A. Open End
Loosen/tighten bolt or nut with contact on two flats

B. Box End
Loosen/tighten bolt or nut; less likely to slip off than open end wrench

C. Combination
Loosen/tighten bolt or nut with convenience of open end and box end on same wrench

D. Ratchet and Socket
Loosen/tighten bolt or nut in tight locations

E. Adjustable
Loosen/tighten different bolt or nut sizes

F. Hex
Loosen/tighten hex screws

G. Impact
Loosen/tighten bolt or nut with air or electric power with short, rapid impulses to socket

H. Torque
Tighten bolt or nut to specified torque

Safety and Tools 41

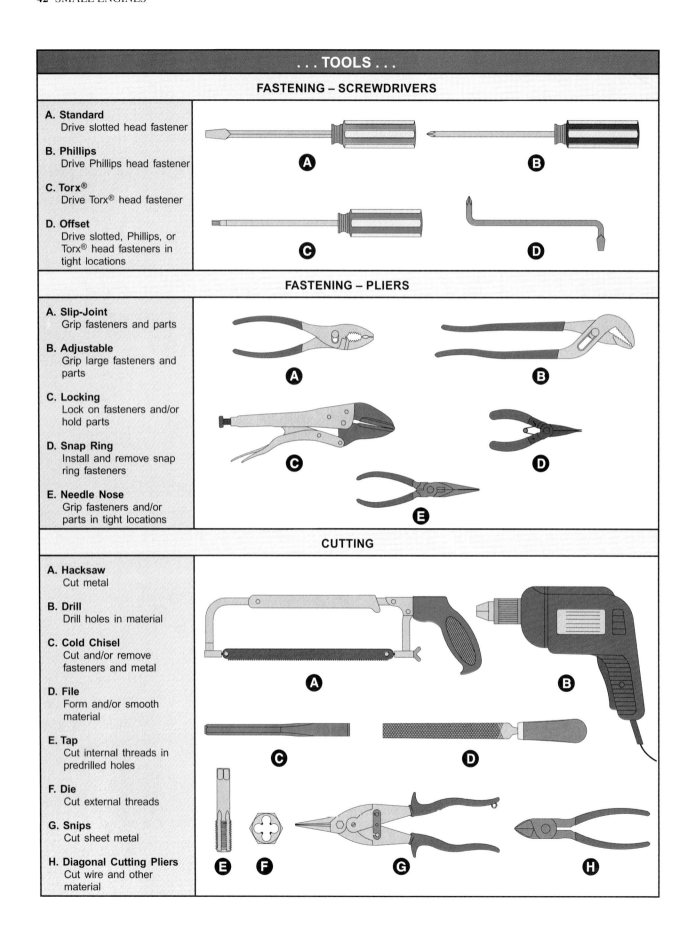

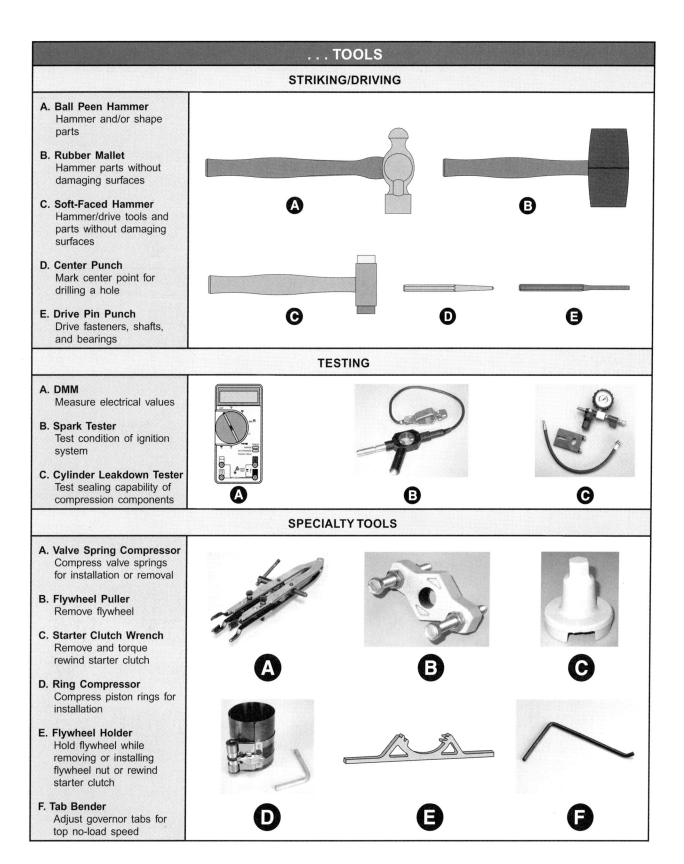

Figure 2-20. Tools commonly used by the small engine service technician can be classified by function as measuring, fastening, cutting, striking/driving, testing, and specialty tools.

Power Tool Safety. Safety risks when using power tools may be reduced by following basic power tool safety rules:

- Follow all manufacturer's recommended operating instructions.
- Use UL®- or CSA-approved power tools that are installed in compliance with the NEC®.
- Do not use electrical tools on or near a wet or damp area.
- Use power tools that are double-insulated or have a third conductor grounding terminal to provide a path for fault current.
- Ensure power switch is in OFF position before connecting to power source.
- Ensure that all safety guards are in place before starting.
- Arrange cords and hoses to prevent accidental tripping.
- Stand to one side when starting and using a grinder.
- Stand clear of operating power tools. Keep hands and arms away from moving parts.
- Use tools designed for compressed air service.
- Shut off, lock out, and tagout disconnect switches of power tools requiring service.

Test Tools

A *test tool* is a measurement tool used to test the condition or operation of an engine component or system. Different test tools can be used by the small engine service technician to troubleshoot and repair a malfunction or failure. Test tools required vary depending on the complexity of the engine problem. Test tools commonly used by the small engine service technician include the digital multimeter, spark tester, and cylinder leakdown tester.

Digital Multimeter. A *digital multimeter (DMM)* is a test tool used to measure two or more electrical values. The most common digital multimeter uses include testing for continuity, voltage, resistance, and current. Digital multimeters are commonly used by small engine service technicians for electrical troubleshooting tasks. See Figure 2-21.

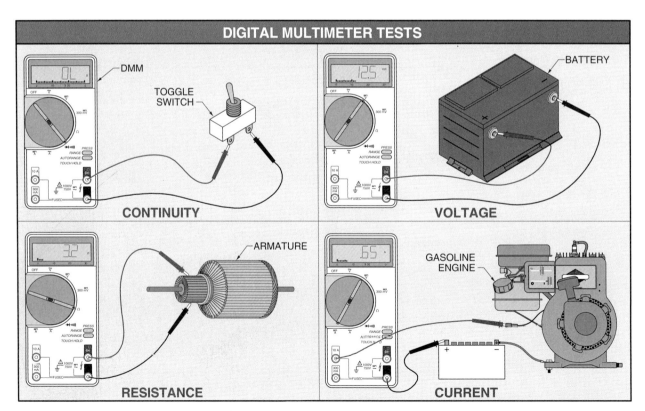

Figure 2-21. A digital multimeter (DMM) is used to perform electrical tests when troubleshooting.

Spark Tester. A *spark tester* is a test tool used to test the condition of the ignition system on a small engine. The spark plug wire is connected to the long terminal, with the spark tester grounded with an alligator clip. If a spark jumps across the gap in the spark tester when the starting system is actuated, the ignition system is considered acceptable. See Figure 2-22. If there is no spark, or if the spark is erratic, the ground lead is removed from the ignition armature and retested. On Magnetron® ignition armatures, if there is no spark, the ignition armature has failed. If there is a spark with the ground lead disconnected but no spark with the lead connected, there is a problem with a safety switch or the ground lead itself. Using a spark tester on a multiple-cylinder engine differs from using a spark tester on a single-cylinder engine. Refer to the service manual for proper test procedures.

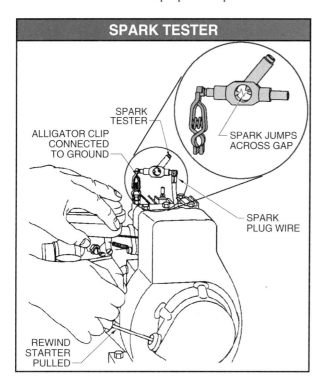

Figure 2-22. A spark tester is used to test the ignition system on a small engine.

Cylinder Leakdown Tester. A *cylinder leakdown tester* is a test tool designed to test the sealing capability of compression components of a small engine. Compression components include the cylinder, piston, rings, cylinder head, head gasket, valves, and valve seats. The leakdown tester design varies, but all are equipped with a spark plug fitting, outlet hose, outlet gauge, inlet gauge, and air pressure regulator. See Figure 2-23.

The overall condition of compression components is determined by the amount of leakage occurring after the combustion chamber is pressurized with compressed air. A clamping tool is attached to prevent the crankshaft from rotating when pressurizing the combustion chamber. A regulated amount of compressed air (70 psi minimum, 150 psi maximum) is used. To test compression components, the regulator adjustment knob is turned counterclockwise as far as possible. A compressed air hose fitting is connected to the air pressure regulator. The crankshaft must be locked to prevent movement before compressed air is introduced. The adjustment knob is then slowly turned clockwise until the needle reaches the set point to adequately pressurize the combustion chamber. The needle position on the outlet gauge indicates leakage from the pressurized combustion chamber.

An engine in good condition displays a reading in the green range on the leakdown gauge with a minimum of audible leakage. A reading in the red/green or red range, along with high audible leakage, indicates a problem with compression components. A small amount of air leakage is normal in all engines if the leakdown gauge remains in the green range. Audible air leaks should be used as an indicator of compression component condition rather than failure, providing the gauge is in green range.

Careful listening for air leaks when the combustion chamber is pressurized can isolate the specific component or components causing a compression leakage problem. However, a single component displaying audible air leakage could indicate a potential problem. For example, a slight audible air leak at the head gasket may not register on the leakdown gauge. A leaking head gasket has an adverse effect on engine performance and must be replaced.

Ariens Company
Lawn tractors are primarily designed for mowing applications.

46 SMALL ENGINES

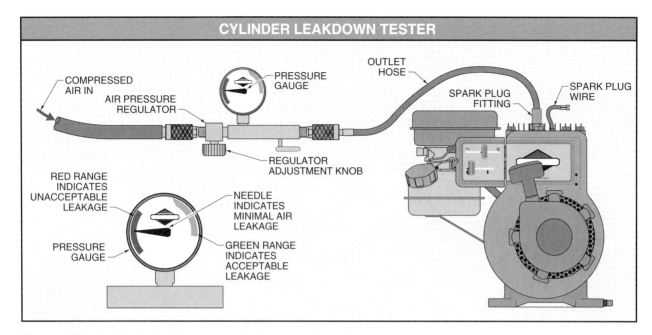

Figure 2-23. A cylinder leakdown tester indicates leakage from compression components when pressurized with compressed air.

Engine speed is measured using a digital tachometer.

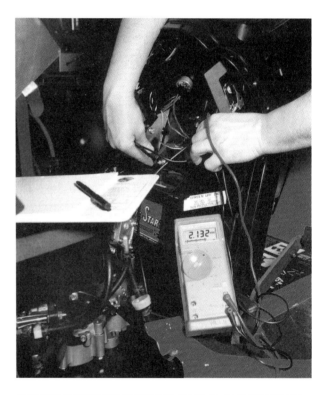

A digital multimeter is used to measure battery voltage.

ENGINE OPERATION

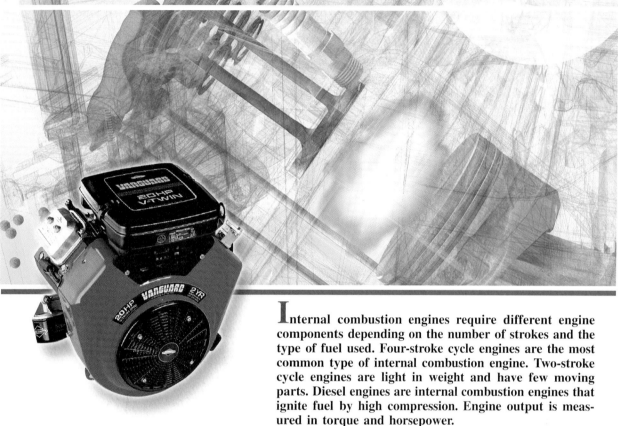

Internal combustion engines require different engine components depending on the number of strokes and the type of fuel used. Four-stroke cycle engines are the most common type of internal combustion engine. Two-stroke cycle engines are light in weight and have few moving parts. Diesel engines are internal combustion engines that ignite fuel by high compression. Engine output is measured in torque and horsepower.

ENGINE COMPONENTS

Internal combustion engines convert potential chemical energy in the form of heat derived from a fuel into mechanical energy. Approximately 30% of the energy released in an internal combustion engine is converted into work. The remaining energy is lost in the form of heat and friction in the engine. Engine components are designed to convert energy in an internal combustion engine for maximum efficiency. Material used for engine components must withstand heat and stress generated inside the engine during operation, and meet size and weight requirements. Engine components commonly required in reciprocating engines include the engine block, cylinder head, crankshaft, piston and piston rings, connecting rod, bearings, flywheel, and valve train.

Engine Block

The *engine block* is the main structure of an engine which supports and helps maintain alignment of internal and external components. The engine block consists of a cylinder block and a crankcase. See Figure 3-1. An engine block can be produced as a one-piece or two-piece unit. The *cylinder block* is the engine component which consists of the cylinder bore, cooling fins on air-cooled engines, and valve train components, depending on the engine design. The *cylinder bore* is a hole in an engine block that aligns and directs the piston during movement. The *bore* of an engine is the diameter of the cylinder bore. The *stroke* of an engine is the linear distance that a piston travels in the cylinder bore from top dead center (TDC) to bottom dead center (BDC).

48 SMALL ENGINES

Figure 3-1. The engine block is the main structure of the engine which helps maintain alignment of internal and external components.

Top dead center (TDC) is the point at which the piston is closest to the cylinder head. *Bottom dead center (BDC)* is the point at which the piston is farthest from the cylinder head. *Displacement (swept volume)* is the volume that a piston displaces in an engine when it travels from TDC to BDC during the same piston stroke. See Figure 3-2. When bore and stroke are known, the displacement of a single-cylinder engine is found by applying the formula:

$D = .7854 \times B^2 \times S$

where

D = displacement (in cu in.)

.7854 = constant

B^2 = bore squared (in in.)

S = stroke (in in.)

For example, what is the displacement of a single-cylinder engine that has a 2.5″ bore and a 2″ stroke?

$D = .7854 \times B^2 \times S$

$D = .7854 \times (2.5 \times 2.5) \times 2$

$D = .7854 \times 6.25 \times 2$

$D = 9.8175 =$ **9.82 cu in.**

Multiple-cylinder engine displacement is determined by multiplying the displacement of a single cylinder by the number of cylinders. Generally, the larger the displacement of the engine, the more power it can produce.

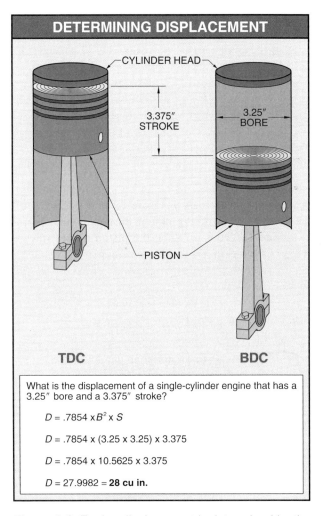

What is the displacement of a single-cylinder engine that has a 3.25″ bore and a 3.375″ stroke?

$D = .7854 \times B^2 \times S$

$D = .7854 \times (3.25 \times 3.25) \times 3.375$

$D = .7854 \times 10.5625 \times 3.375$

$D = 27.9982 =$ **28 cu in.**

Figure 3-2. Engine displacement is determined by the bore and stroke of the engine.

An air-cooled cylinder block has cooling fins on the exterior. A *cooling fin* is an integral thin cast strip designed to provide efficient air circulation and dissipation of heat away from the engine cylinder block into the air stream. Cooling fins increase the surface area of the cylinder block contacting ambient (existing) air for cooling efficiency. Cooling fins cast into or bolted onto the flywheel act as fan blades to provide air circulation around the cylinder block and head. Air circulation dissipates heat generated during combustion to maintain optimum engine temperatures. A *crankcase* is an engine component that houses and supports the crankshaft. In a four-stroke cycle engine, the crankcase also acts as an oil reservoir for lubrication of engine components. The crankcase may be a part of the engine block or a separate component. Some crankcases consist of multiple parts such as a sump or crankcase cover.

The *sump* is a removable part of the engine crankcase that serves as an oil reservoir and provides access to internal parts. The sump provides a bearing surface for vertical shaft engines and forms the lower section of the engine. Horizontal shaft engines do not have a sump. In horizontal shaft engines, the engine block and crankcase cover serve as an oil reservoir. The *crankcase cover* is an engine component that provides access to internal parts in the crankcase and supports the crankshaft.

The *crankcase breather* is an engine component that relieves crankcase pressure created by the reciprocating motion of the piston during engine operation. See Figure 3-3. When the piston moves toward TDC, volume in the crankcase increases, resulting in a lower than ambient (existing) pressure in the crankcase. When the piston moves toward BDC, the volume in the crankcase decreases, generating a higher than ambient pressure in the crankcase.

The crankcase breather functions as a check valve allowing more air to escape than can enter the crankcase. This maintains a crankcase pressure less than atmospheric pressure (14.7 psi at sea level). Crankcase gases, partially spent combustion gases, and other engine gases are then routed to the carburetor.

The crankcase breather also serves as an oil mist collector, preventing crankcase oil from escaping whenever the crankcase breather opens. The oil mist collector consists of a wire screen pack. The wire screen pack allows compression gases and other harmful vapors to pass through while collecting crankcase oil for return back to the crankcase. Crankcase breathers are available in different configurations and are required on most four-stroke cycle engines.

Cylinder blocks are cast from materials strong enough to withstand the heat and stress generated inside the engine during operation. The cylinder block must also meet size and weight requirements dictated by the specific application. Cylinder blocks are commonly constructed from cast aluminum alloy, cast iron, or cast aluminum alloy with cast iron cylinder sleeves.

Cast aluminum alloy cylinder blocks are lightweight and dissipate heat more rapidly than cast iron cylinder blocks. Cast iron cylinder blocks are heavier and more expensive, but are more resistant to wear and less prone to heat distortion than cast aluminum alloy cylinder blocks. Cast aluminum alloy cylinder blocks with cast iron cylinder sleeves combine the light weight of aluminum with the durability of cast iron. See Figure 3-4.

> *Outdoor power equipment sales in North America follow seasonal cycles. The highest monthly sales for walk-behind lawn mowers occur in May. The highest monthly sales for snow throwers occur in December.*

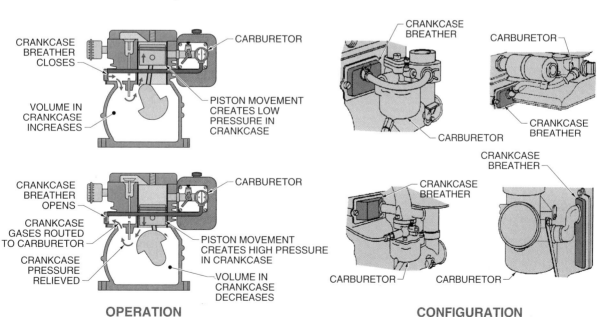

Figure 3-3. The crankcase breather functions as a check valve to maintain crankcase pressure and to route gases to the carburetor.

50 SMALL ENGINES

Figure 3-4. Cast aluminum alloy cylinder blocks with cast iron cylinder sleeves combine the light weight of aluminum with the durability of cast iron.

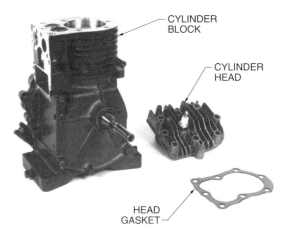

Figure 3-5. The head gasket is placed between the cylinder block and cylinder head to seal the combustion chamber and provide even heat distribution.

Cylinder Head

A *cylinder head* is a cast aluminum alloy or cast iron engine component fastened to the end of the cylinder block farthest from the crankshaft. A *head gasket* is the filler material placed between the cylinder block and cylinder head to seal the combustion chamber. Head gaskets are made from soft metals and graphite layered together. Head gaskets allow for even heat distribution between the cylinder block and cylinder head for efficient heat dissipation. See Figure 3-5.

Some two-stroke cycle engines combine the cylinder head and the cylinder block into a jug. A *jug* is an engine component in which the cylinder block and cylinder head are cast as a single unit. This provides maximum structural integrity and eliminates the potential for undesirable leaks in the combustion chamber.

Cylinder head design and components vary depending on whether the engine is an overhead valve engine or an L-head engine. An *overhead valve (OHV) engine* is an engine that has valves and related components located in the cylinder head. An *L-head engine* is an engine that has valves and related components located in the cylinder block.

GardenWay, Inc./Troy-Bilt®

A chipper/shredder is a machine that requires momentum created by high-inertia equipment components.

Crankshaft

The *crankshaft* is an engine component that converts the linear (reciprocating) motion of the piston into rotary motion. The crankshaft is the main rotating component of an engine and is commonly made of ductile iron. See Figure 3-6. Orientation of the crankshaft classifies the engine as a vertical shaft engine or a horizontal shaft engine.

Features of a crankshaft include the crankpin journal, throw, bearing journals, counterweights, crankgear, and a power take-off (PTO). A *crankpin journal* is a precision ground surface that provides a rotating pivot point to attach the connecting rod to the crankshaft. The *throw* is the measurement from the center of the crankshaft to the center of the crankpin journal, which is used to determine the stroke of an engine. The throw is equal to one-half the stroke. The longer the throw, the greater the stroke, or distance, a piston travels.

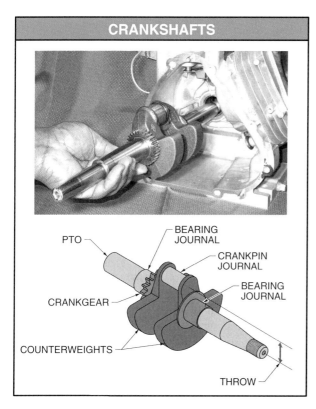

Figure 3-6. The crankshaft is the main rotating component of an engine and is commonly made of ductile iron.

A *bearing journal* is a precision ground surface within which the crankshaft rotates. Bearing journals mate with bearing surfaces in the cylinder block. Most bearing surfaces are machined integrally in the cylinder block. Some engines feature a low-friction bushing or a ball or tapered roller bearing. A *counterweight* is a protruding mass integrally cast into the crankshaft which partially balances the forces of a reciprocating piston and reduces the load on crankshaft bearing journals.

Some small engines have a separate counterweight shaft located in the crankcase that is driven by the crankshaft. The size of the counterweight is determined by the mass of the piston, the throw of the crankshaft, and the internal pressures and vibrations generated by the specific engine. The *crankgear* is a gear located on the crankshaft that is used to drive other parts of an engine. Crankgears are not required on all small engines. The *power take-off (PTO)* is an extension of the crankshaft that allows an engine to transmit power to an application. For example, the PTO shaft on a garden tractor can be fitted with a pulley for a drive belt to transmit power to the mower deck.

Murray Inc.
The cast iron cylinder sleeve in this lawn tractor extends engine life.

Piston and Piston Rings

A *piston* is a cylindrical engine component that slides back and forth in the cylinder bore by forces produced during the combustion process. The piston acts as the movable end of the combustion chamber. The stationary end of the combustion chamber is the cylinder head. Pistons are commonly made of a cast aluminum alloy for excellent and lightweight thermal conductivity. *Thermal conductivity* is the ability of a material to conduct and transfer heat. Aluminum expands when heated, and proper clearance must be provided to maintain free piston movement in the cylinder bore. Insufficient clearance can cause the piston to seize in the cylinder. Excessive clearance can cause a loss of compression and an increase in piston noise.

Piston features include the piston head, piston pin bore, piston pin, skirt, ring grooves, ring lands, and piston rings. The *piston head* is the top surface (closest to the cylinder head) of the piston which is subjected to tremendous forces and heat during normal engine operation. See Figure 3-7. The shape of the piston head is either flat or contoured, depending on engine design. Some engine designs use the piston head as an integral part of the combustion chamber. For example, a dished piston head shape creates a swirling effect to mix the air and fuel more completely as it enters the combustion chamber.

52 SMALL ENGINES

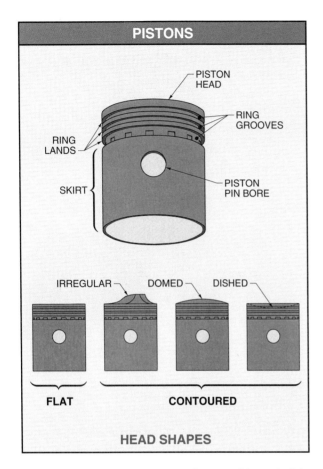

Figure 3-7. The piston acts as the movable end of the combustion chamber and is designed to utilize the forces and heat created during engine operation.

A *piston pin bore* is a through hole in the side of the piston perpendicular to piston travel that receives the piston pin. A *piston pin* is a hollow shaft that connects the small end of the connecting rod to the piston. The *skirt* of a piston is the portion of the piston closest to the crankshaft that helps align the piston as it moves in the cylinder bore. Some skirts have profiles cut into them to reduce piston mass and to provide clearance for the rotating crankshaft counterweights.

A *ring groove* is a recessed area located around the perimeter of the piston that is used to retain a piston ring. *Ring lands* are the two parallel surfaces of the ring groove which function as the sealing surface for the piston ring. A *piston ring* is an expandable split ring used to provide a seal between the piston and the cylinder wall. Piston rings are commonly made from cast iron. Cast iron retains the integrity of its original shape under heat, load, and other dynamic forces. Piston rings seal the combustion chamber, conduct heat from the piston to the cylinder wall, and return oil to the crankcase. Piston ring size and configuration vary depending on engine design and cylinder material.

Piston rings commonly used on small engines include the compression ring, wiper ring, and oil ring. A *compression ring* is the piston ring located in the ring groove closest to the piston head. The compression ring seals the combustion chamber from any leakage during the combustion process. See Figure 3-8.

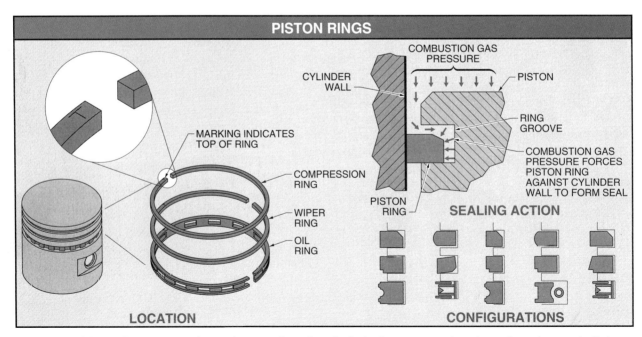

Figure 3-8. Piston rings commonly used on small engines include the compression ring, wiper ring, and oil ring.

When the air-fuel mixture is ignited, pressure from combustion gases is applied to the piston head, forcing the piston toward the crankshaft. The pressurized gases travel through the gap between the cylinder wall and the piston and into the piston ring groove. Combustion gas pressure forces the piston ring against the cylinder wall to form a seal. Pressure applied to the piston ring is approximately proportional to the combustion gas pressure.

A *wiper ring* is the piston ring with a tapered face located in the ring groove between the compression ring and oil ring. The wiper ring is used to further seal the combustion chamber and to wipe the cylinder wall clean of excess oil. Combustion gases that pass by the compression ring are stopped by the wiper ring.

An *oil ring* is the piston ring located in the ring groove closest to the crankcase. The oil ring is used to wipe excess oil from the cylinder wall during piston movement. Excess oil is returned through ring openings to the oil reservoir in the engine block. Two-stroke cycle engines do not require oil rings because lubrication is supplied by mixing oil in the gasoline, and an oil reservoir is not required.

Connecting Rod

A *connecting rod* is an engine component that transfers motion from the piston to the crankshaft and functions as a lever arm. Connecting rods are commonly made from cast aluminum alloy and are designed to withstand sudden impact stresses from combustion and piston movement. The small end of the connecting rod connects to the piston with a piston pin. See Figure 3-9. The *piston pin*, or wrist pin, provides a pivot point between the piston and connecting rod. Spring clips, or piston pin locks, are used to hold the piston pin in place.

The large end of the connecting rod connects to the crankpin journal to provide a pivot point on the crankshaft. Connecting rods are produced as one-piece or two-piece components. A *rod cap* is the removable section of a two-piece connecting rod that provides a bearing surface for the crankpin journal. The rod cap is attached to the connecting rod with two rod cap screws for installation and removal from the crankshaft.

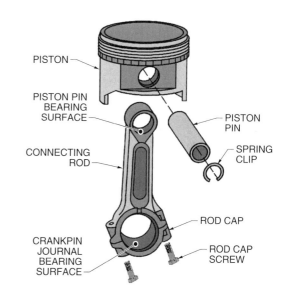

CONNECTING ROD

Figure 3-9. A connecting rod is designed to withstand sudden impact stresses from combustion and piston movement.

 Energy can be neither created nor destroyed. It can only be converted from one form to another.

Homelite, Inc.

Portable generators are rated in watts.

Bearings

A *bearing* is a component used to reduce friction and to maintain clearance between stationary and rotating components of an engine. Bearings, or bearing surfaces, are located on the crankshaft, connecting rod, and camshaft, and also in the cylinder block. Bearings can be subjected to radial, axial (thrust), or a combination of radial and axial loads. A *radial load* is a load applied perpendicular to the shaft. An *axial load* is a load applied parallel to the shaft. Bearings are classified as friction or antifriction bearings. See Figure 3-10.

A *friction bearing* consists of a fixed, non-moving bearing surface, such as machined metal or pressed-in bushing, that provides a low-friction support surface for rotating or sliding surfaces. Friction bearings commonly use lubricating oil to separate the moving component from the mated non-moving bearing surface. Friction bearing surfaces commonly consist of material that is softer than the supported component.

Friction bearings, because of their soft consistency, have the ability to embed foreign matter to prevent spreading in the engine. Friction bearings also have the ability to conform to slightly irregular mating surfaces. Friction bearings can be integrally machined, one-piece sleeve, split-sleeve for easy installation and removal, or a DU™. A *DU™* is a friction bearing that consists of a steel backing, a porous bronze innerstructure, and a polytetrafluoroethylene (PTFE) and lead overlay on the bearing surface. The steel backing provides high load-carrying capacity and excellent heat dissipation. The porous bronze innerstructure (approximately .010″) serves as a reservoir for the PTFE-lead mixture. The PTFE-lead overlay (approximately .001″) transfers an initial coating of oxide lubricant film to the mating surface. A DU™ bearing is a low-friction, self-lubricating bearing that can be used with or without lubrication.

An *antifriction bearing* is a bearing that contains moving elements to provide a low friction support surface for rotating or sliding surfaces. Antifriction bearings are commonly made with hardened rolling elements (balls and rollers) and races. A *race* is the bearing surface in an antifriction bearing that supports rolling elements during rotation. A *separator* is an antifriction bearing component used to maintain the position and alignment of rolling elements. Antifriction bearings reduce lubrication requirements and decrease starting and operating friction. Reduced friction results in less power required to rotate engine components and increases overall engine output.

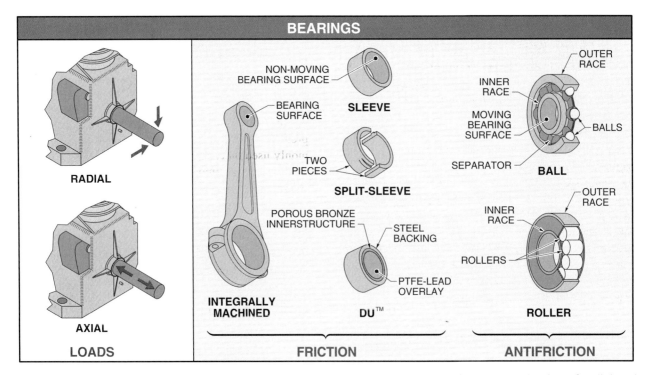

Figure 3-10. Bearings and bearing surfaces are subjected to radial, axial (thrust), or a combination of radial and axial loads.

The crankshaft is supported by main bearings. A *main bearing* is a bearing that supports and provides a low-friction bearing surface for the crankshaft. Small engines commonly have two main bearings, one at each end of the crankshaft. See Figure 3-11. Small engines with three or more cylinders may require more than two main bearings to provide additional support to the crankshaft. Main bearings are mounted in the crankcase and can be either friction or antifriction bearings. Antifriction bearings used for main bearings increase the radial and axial load capacity of the engine design.

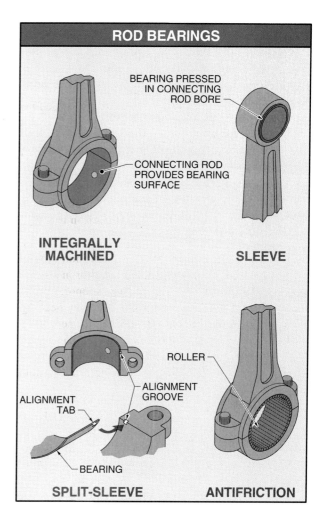

Figure 3-12. Rod bearings provide a low-friction pivot point between the connecting rod and the crankshaft and the connecting rod and piston.

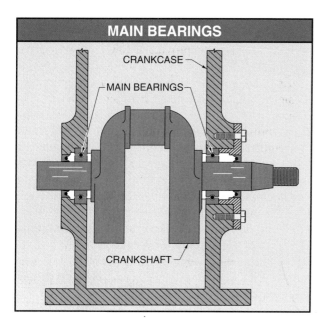

Figure 3-11. Small engines commonly have two main bearings to provide a low-friction bearing surface on each end of the crankshaft.

A *rod bearing* is a bearing that provides a low-friction pivot point between the connecting rod and the crankshaft and the connecting rod and piston. The large end of the connecting rod is connected to the crankpin journal. The small end of the connecting rod is connected to the piston pin. Rod bearings are friction bearings (integrally machined, sleeve, or split-sleeve) or antifriction bearings. Most connecting rods for small engines use integrally machined friction bearings. See Figure 3-12.

 The leading state for employment in manufacturing lawn and garden equipment is Tennessee, followed by Wisconsin.

Antifriction bearings used on connecting rods are precision ground from hardened steel and are commonly used on two-stroke cycle engines. Friction rod bearings are commonly made from nonferrous metals such as bronze, aluminum, and babbitt. A *nonferrous metal* is a metal that does not contain iron. *Bronze* is a nonferrous metal alloy that consists of brass and zinc. *Aluminum* is a nonferrous metal commonly alloyed with zinc or copper. *Babbitt* is a nonferrous metal alloy consisting of copper, lead, and tin or lead and tin. Babbitt is commonly used on split-sleeve bearings consisting of a steel backing coated with multiple thin layers of babbitt on the load bearing surface. Split-sleeve connecting rod bearing position in the large end of the connecting rod is maintained with an alignment tab. The alignment tab also prevents rotation of the bearing during engine operation.

Flywheel

The *flywheel* is a cast iron, aluminum, or zinc disk that is mounted at one end of the crankshaft to provide inertia for the engine. *Inertia* is the property of matter by which any physical body persists in its state of rest or uniform motion until acted upon by an external force. Inertia is not a force, it is a property of matter. During the operation of a reciprocating engine, combustion occurs at distinct intervals. The flywheel supplies the inertia required to prevent loss of engine speed and possible stoppage of crankshaft rotation between combustion intervals. See Figure 3-13.

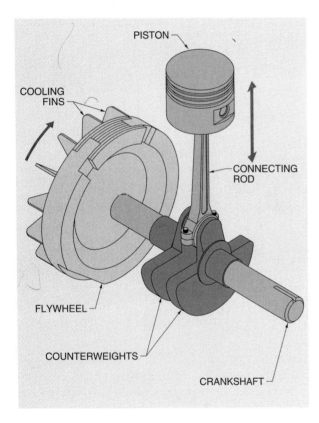

Figure 3-13. The flywheel supplies inertia to dampen acceleration forces caused by combustion intervals in an engine.

During each stroke of an internal combustion engine, the flywheel, crankshaft, and other engine components are affected by fluctuations in speed and force. During the power event in a four-stroke cycle engine, the crankshaft is accelerated rapidly by the sudden motion of the piston and connecting rod assembly. The flywheel absorbs some of the rpm and force deviation by its resistance to acceleration. The inertia of the flywheel provides a dampening effect on the engine as a whole to even out radical acceleration forces and rpm deviations produced in the engine.

Valve Train

The *valve train* of an internal combustion engine includes components required to control the flow of gases into and out of the combustion chamber. This includes valves and related components required to allow the air-fuel mixture to enter the combustion chamber, seal the combustion chamber during compression and combustion, and evacuate exhaust gases when combustion is complete. The type of valve train used for a reciprocating engine varies with a four-stroke cycle engine or two-stroke cycle engine.

FOUR-STROKE CYCLE ENGINES

A *four-stroke cycle engine* is an internal combustion engine that utilizes four distinct piston strokes (intake, compression, power, and exhaust) to complete one operating cycle. The piston makes two complete passes in the cylinder to complete one operating cycle. An operating cycle requires two revolutions (720°) of the crankshaft. The four-stroke cycle engine is the most common type of small engine. A four-stroke cycle engine completes five events in one operating cycle, including intake, compression, ignition, power, and exhaust events.

Intake Event

The *intake event* is an engine operation event in which the air-fuel mixture, or just air in diesel engines, is introduced to fill the combustion chamber. The intake event occurs when the piston moves from TDC to BDC and the intake valve is open. See Figure 3-14. The movement of the piston toward BDC creates a low pressure in the cylinder. Ambient atmospheric pressure forces the air-fuel mixture through the open intake valve into the cylinder to fill the low pressure area created by the piston movement. The cylinder continues to fill slightly past BDC as the air-fuel mixture continues to flow from motion and inertia while the piston begins to change direction. The intake valve remains open a few degrees of crankshaft rotation after BDC. The number of degrees the intake valve remains open after BDC depends on engine design. The intake valve then closes and the air-fuel mixture is sealed inside the cylinder.

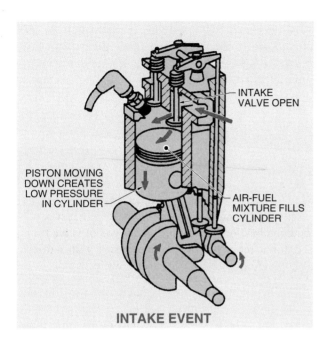

Figure 3-14. The intake event occurs when the air-fuel mixture is introduced into the combustion chamber as the piston moves from TDC to BDC.

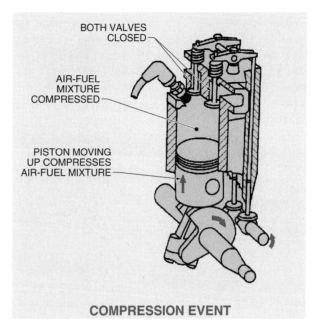

Figure 3-15. The compression event is an engine operation event in which the trapped air-fuel mixture is compressed to form the charge.

Compression Event

The *compression event* is an engine operation event in which the trapped air-fuel mixture, or just air in diesel engines, is compressed inside the cylinder. The combustion chamber is sealed to form the charge. The *charge* is the volume of compressed air-fuel mixture trapped inside the combustion chamber ready for ignition. Compressing the air-fuel mixture allows more energy to be released when the charge is ignited. Intake and exhaust valves must be closed to ensure that the cylinder is sealed to provide compression. *Compression* is the process of reducing or squeezing a charge from a large volume to a smaller volume in the combustion chamber. See Figure 3-15. The flywheel helps to maintain the momentum necessary to compress the charge.

When the piston of an engine compresses the charge, an increase in cohesive force supplied by work being done by the piston causes heat to be generated. The compression and heating of the air-fuel vapor in the charge results in an increase in charge temperature and an increase in fuel vaporization. The increase in charge temperature occurs uniformly throughout the combustion chamber to produce faster combustion (fuel oxidation) after ignition.

The increase in fuel vaporization occurs as small droplets of fuel become vaporized more completely from the heat generated. The increased droplet surface area exposed to the ignition flame allows more complete burning of the charge in the combustion chamber. Only gasoline vapor ignites. An increase in droplet surface area allows gasoline to release more vapor rather than remaining a liquid.

Whiteman Industries Inc.
Ride-on power trowels are used to finish concrete in large areas.

The tighter the charge vapor molecules are compressed, the more energy obtained from the combustion process. The energy needed to compress the charge is substantially less than the gain in force produced during the combustion process. For example, in a typical small engine, energy required to compress the charge is one-fourth the amount of gain in force produced during combustion. Gain in force during combustion from compression of the charge is considerably higher in diesel engines.

Compression Ratio. The *compression ratio* of an engine is a comparison of the volume of the combustion chamber with the piston at BDC to the volume of the combustion chamber with the piston at TDC. See Figure 3-16. This area, combined with the design and style of combustion chamber, determines the compression ratio. Gasoline engines commonly have a compression ratio ranging from 6:1 – 8.5:1. Diesel engines commonly have a compression ratio ranging from 14:1 – 25:1. The higher the compression ratio, the more fuel-efficient the engine. A higher compression ratio normally provides a substantial gain in combustion pressure or force on the piston. However, higher compression ratios increase operator effort required to start the engine. Some small engines feature a system to relieve pressure during the compression stroke to reduce operator effort required when starting the engine.

Sun-and-planet gearing was the forerunner to the crankshaft.

Ignition Event

The *ignition (combustion) event* is an engine operation event in which the charge is ignited and rapidly oxidized through a chemical reaction to release heat energy. See Figure 3-17. *Combustion* is the rapid, oxidizing chemical reaction in which a fuel chemically combines with oxygen in the atmosphere and releases energy in the form of heat.

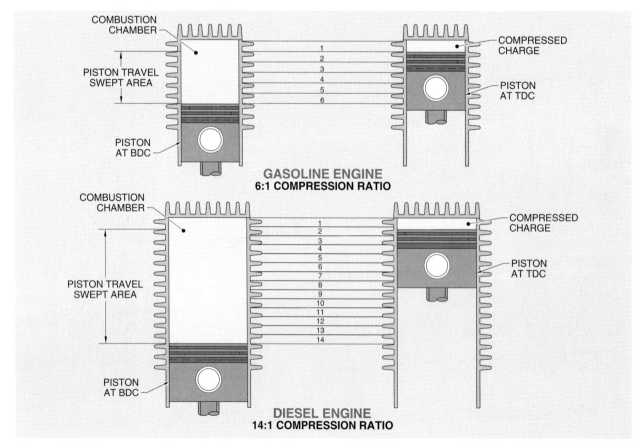

Figure 3-16. The compression ratio of an engine is a comparison of the volume of the combustion chamber with the piston at BDC and TDC.

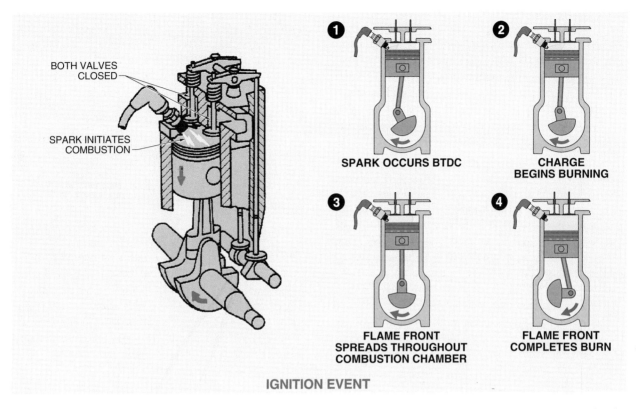

Figure 3-17. During the ignition event, atmospheric oxygen and fuel vapor in the charge are consumed by the progressing flame front.

Proper combustion involves a short but finite time to spread a flame throughout the combustion chamber. The spark at the spark plug initiates combustion at approximately 20° of crankshaft rotation before TDC (BTDC). The atmospheric oxygen and fuel vapor are consumed by a progressing flame front. A *flame front* is the boundary wall that separates the charge from the combustion by-products. The flame front progresses across the combustion chamber until the entire charge has burned.

Power Event

The *power event* is an engine operation event in which hot expanding gases force the piston head away from the cylinder head. Piston force and subsequent motion are transferred through the connecting rod to apply torque to the crankshaft. The torque applied initiates crankshaft rotation. See Figure 3-18. The amount of torque produced is determined by the pressure on the piston, the size of the piston, and the throw of the engine. During the power event, both valves must be closed.

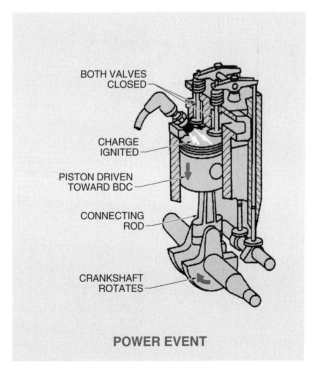

Figure 3-18. During the power event, hot expanding gases force the piston head away from the cylinder head.

Exhaust Event

The *exhaust event* is an engine operation event in which spent gases are removed from the combustion chamber and released to the atmosphere. The exhaust event is the final event and occurs when the exhaust valve is open and the intake valve is closed. Piston movement evacuates exhaust gases to the atmosphere.

As the piston reaches BDC during the power event, combustion is complete and the cylinder is filled with exhaust gases. See Figure 3-19. The exhaust valve opens, and inertia of the flywheel and other moving parts push the piston back to TDC, forcing the exhaust gases past the open exhaust valve. At the end of the exhaust stroke, the piston is at TDC and one operating cycle has been completed.

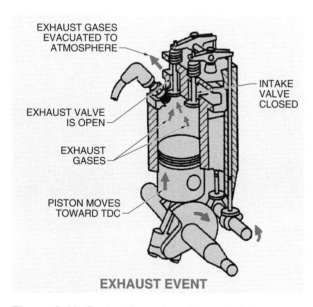

Figure 3-19. During the exhaust event, piston movement evacuates exhaust gases to the atmosphere.

Valve Overlap

Valve overlap is the period during engine operation when both intake and exhaust valves are open at the same time. Valve overlap occurs when the piston nears TDC between the exhaust event and the intake event. Duration of valve overlap is between 10° – 20° of crankshaft rotation, depending on the engine design. See Figure 3-20. The intake valve is opened during the exhaust event just before TDC, initiating the flow of a new charge into the combustion chamber.

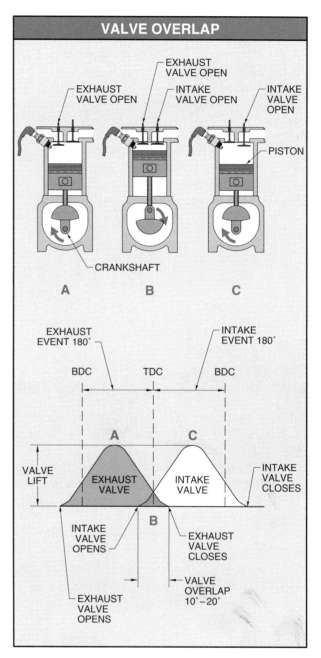

Figure 3-20. Valve overlap is the period between the exhaust event and the intake event when the piston nears TDC.

As the exhaust gases are evacuated from the combustion chamber, a small but distinct low-pressure area is created on the surface of the piston head. By opening the intake valve earlier than TDC, the charge begins to fill this low-pressure area while exhaust gases exit. The low-pressure area on the head of the piston assists the fresh charge in filling the combustion chamber to its maximum capacity.

Valve overlap is designed into the engine and is most useful at higher speeds. At higher speeds, the head start in filling the combustion chamber provides a substantial increase in available power. The amount of time that both valves are open is directly related to engine rpm. The higher the engine rpm, the shorter the amount of time that both valves are open. The degrees of crankshaft rotation when both valves are open do not change. Only the amount of time both valves are open varies with the engine rpm. For example, at idle, the amount of time both valves are open is relatively long compared to top no-load speed.

TWO-STROKE CYCLE ENGINES

A *two-stroke cycle engine* is an internal combustion engine that utilizes two distinct piston strokes to complete one operating cycle of the engine. The crankshaft turns only one revolution for each complete operating cycle, providing twice as many power strokes in the same number of crankshaft rotations as a four-stroke cycle engine. The valving system in a two-stroke cycle engine requires fewer parts, making it lighter in weight than a four-stroke cycle engine. Weight reduction is especially desirable in applications such as chainsaws, leaf blowers, and other hand-held outdoor power equipment. Like a four-stroke cycle engine, a two-stroke cycle engine completes five events in one operating cycle. Some events occur concurrently, such as ignition/power and exhaust/intake. See Figure 3-21.

The ignition/power event occurs when the piston moves toward TDC and the compressed charge in the cylinder is ignited. During this time, the crankcase has already filled with a fresh air-fuel mixture. When the charge is ignited, expansion of hot combustion gases force the piston toward BDC. Piston motion is transferred from the piston through the connecting rod, causing the crankshaft to rotate. When the piston moves toward BDC, the exhaust port is uncovered and exhaust gases are discharged through the side of the cylinder.

The exhaust/intake event occurs as the piston continues moving toward BDC. The intake port opens and air-fuel mixture is routed into the cylinder. The shape of the piston head helps to divert the incoming air-fuel mixture to the top of the cylinder. This prevents incoming air-fuel mixture from passing across the top of the piston and out the exhaust port without burning.

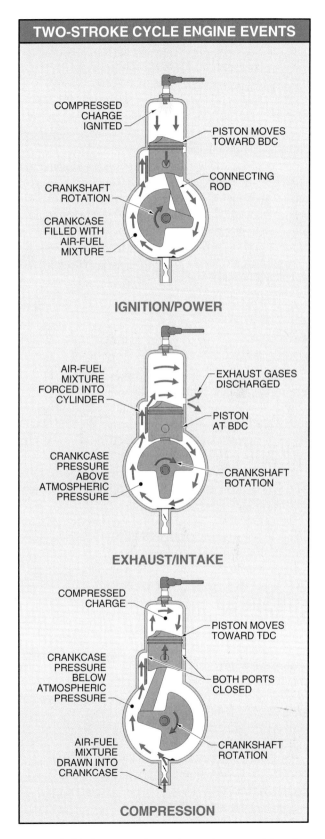

Figure 3-21. A two-stroke cycle engine completes five events in one operating cycle.

Diversion of the air-fuel mixture also helps when discharging or scavenging exhaust gases. *Scavenging* is the process of using the introduction of the fresh air-fuel mixture to help remove exhaust gases from the cylinder in a two-stroke cycle engine.

The compression event occurs when the piston is at BDC, exhaust gases have been discharged, and the cylinder is filled with a new charge. The compression stroke begins as the piston starts moving toward TDC. The piston closes the intake and exhaust ports, trapping the charge in the cylinder. The piston functions as a slide valve, exposing the intake and exhaust ports as it moves in the cylinder. The charge is compressed as the piston continues to move toward TDC. Piston movement toward TDC causes more air-fuel mixture to be drawn into the crankcase. When the piston reaches TDC, the engine has completed one full operating cycle.

Two-Stroke Cycle Engine Applications

Two-stroke cycle engines are widely used in outdoor power equipment applications such as chain saws, trimmers, and leaf blowers. Two-stroke cycle engines have specific characteristics that provide advantages and disadvantages for different applications. See Figure 3-22. Two-stroke cycle engines ignite the air-fuel mixture once every revolution of the crankshaft, while four-stroke cycle engines ignite the air-fuel mixture once every other revolution.

The power-to-weight ratio for two-stroke cycle engines is greater than for four-stroke cycle engines. Weight of the engine is a particularly important factor for equipment that is hand-held or mounted for portable use. Two-stroke engines do not have valves, which simplifies their construction and also reduces overall weight. Additionally, a two-stroke cycle engine cost less to manufacture because there are fewer moving parts. Two-stroke engines can work in several positions. Lubrication is continuously provided by the oil mixed into the gasoline. Proper operation in several positions is important on equipment such as a chainsaw. A four-stroke cycle engine used in this application would require special modification to the lubrication system.

Two-stroke cycle engines operate at higher speed and temperatures, and typically do not have as long a service life as four-stroke cycle engines with a sophisticated lubrication systems. Fuel consumption is greater in two-stroke cycle engines, and the cost of oil required (approximately 3 to 4 oz per gal.) also adds to overall operating expense.

TWO-STROKE CYCLE ENGINE CHARACTERISTICS*

- Fewer moving parts
- Less weight for comparable output
- Higher fuel consumption
- More noise
- Higher operating speed and temperature
- Smaller size for comparable output
- Greater exhaust emissions

* compared with four-stroke cycle engines

Figure 3-22. Two-stroke engines are widely used in the outdoor power equipment industry for hand-held equipment applications such as chain saws, trimmers, and leaf blowers.

Two-Stroke Cycle Engine Emissions. Two-stroke cycle engines use a mixture of oil and gasoline. The combustion of the oil causes all two-stroke engines to produce a smokier exhaust. In older engines with worn parts, this problem is even more evident. The design and operation of a two-stroke cycle engine also creates emission problems from the leakage of the air/fuel mixture through the exhaust port. Each time a new charge of air/fuel is introduced into the combustion chamber, a portion leaks out through the exhaust port. The hydrocarbons from the unburned fuel mixture are discharged into the environment as a pollutant.

In general, two-stroke engines have primarily been used where the power-to-weight ratio is critical, and in applications where the engine is not used continuously for long periods. However, pollution standards have mandated increasingly more stringent controls of two-stroke engine emissions. Manufacturers have been working on better methods of regulating the air-fuel mixture and combustion process for greater efficiency and fewer emissions. Manufacturers are also working to reduce the size and weight of four-stroke cycle engines to match the benefits of a two-stroke cycle engine.

VALVING SYSTEMS

The *valving system* is the system on an internal combustion engine that controls the flow of gases into and out of the combustion chamber. The three primary functions of a valving system are to allow air-fuel mixture to enter the combustion chamber, to seal the combustion chamber during compression and combustion, and to remove exhaust gases from the cylinder when combustion is complete. The valving system used depends on whether the engine is a four-stroke cycle engine or two-stroke cycle engine.

Four-Stroke Cycle Engine Valving Systems

Four-stroke cycle engines control the flow of gases in the cylinder with valves. A *valve* is an engine component that opens or closes at precise times to allow the flow of air-fuel mixture into the cylinder and to allow the flow of exhaust gases from the cylinder. Valve features include the valve head, margin, valve face, valve seat, valve stem, valve neck, and retainer groove. See Figure 3-23. The *valve head* is the large end of the valve that contains the margin and the valve face. The *margin* is the surface of a valve joining the valve face and the top surface of the valve head. The *valve face* is the machined surface of a valve that mates with the valve seat to seal the combustion chamber.

The *valve seat* is the machined stationary surface that mates with the valve face to seal the combustion chamber. The *valve stem* is the long part of a valve that aligns the valve. The *valve neck* is the part of a valve joining the valve head and valve stem. The *retainer groove* is the part of a valve that is recessed for mounting the valve spring retainer.

Valve overlap and engine performance

Valve overlap is used in high-performance automobile engines to increase performance. These engines are equipped with a high-performance camshaft, which provides long periods of valve overlap. Increasing the valve overlap results in maximum performance at a higher rpm than a standard passenger car engine. However, the long periods of valve overlap tend to cause high-performance engines to idle roughly. The rough idle is caused by a misdirection of the air-fuel mixture and exhaust gases at very low idle speeds. When the rpm of a high-performance engine is increased, engine roughness dissipates, and power increases.

Some engines reduce valve and valve seat wear with a valve rotator. A *valve rotator* is a mechanical device on a valve used to rotate the valve each time it opens to provide even wear and distribution of heat. A *valve spring* is a compression spring that closes the valve and holds it tightly against the valve seat. A *valve spring retainer* is an engine component that compresses and secures the valve spring on the valve stem. Valve spring retainer designs vary with different engine models. Common valve spring retainers include the pin, collar, or keyhole retainer.

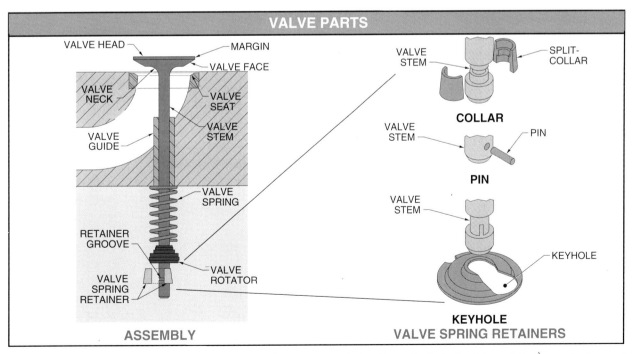

Figure 3-23. Valves seal the combustion chamber to control the flow of air-fuel mixture into the cylinder and exhaust gases out of the cylinder.

Small engines commonly have two valves (intake and exhaust) per cylinder. The intake valve allows the air-fuel mixture to flow into the cylinder. The exhaust valve allows exhaust gases to flow out of the cylinder. Both valves must withstand a peak temperature of approximately 3000°F during combustion and a sustained operating temperature of approximately 1200°F. Contact between the valve face and the valve seat provides a seal and cools the valve by transferring heat from the valve to the cylinder block or head.

Exhaust valves are subjected to greater thermal and structural stress than intake valves due to direct contact with hot exhaust gases. Intake valves are cooled by the passing air-fuel mixture. Exhaust valves are commonly made from austenitic steel. *Austenitic steel* is a heat-resistive metal alloy consisting of cobalt, tungsten, and chromium. Some valve seats are machined directly into the cast iron cylinder block. On some aluminum heads and blocks, the valve seat is a hardened metal insert to provide additional wear resistance.

Valve Location. The location of valves determines the type of head design and the necessary components for the valve train. Valves are located in the cylinder block or in the cylinder head. In an L-head engine, the valves are located in the cylinder block to one side of the cylinder. Small engines are commonly L-head engines. See Figure 3-24.

Valves in L-head engines are opened by the movement of a tappet. A *tappet* is an engine component that rides on the camshaft and pushes the bottom of the valve stem to open the valve. A *cam lobe* is an egg-shaped protrusion on the camshaft that moves a tappet to open a valve. Cam lobes change rotary motion into linear motion for actuating valve movement. Cam lobe orientation on the cam shaft is different for intake and exhaust valves. Cam lobe dimensions control the lift and duration of the opening and closing of the valve.

Overhead valve engines provide greater efficiency and are increasing in popularity. In an overhead valve engine, the valve head is positioned upside down in relation to the cylinder when compared to the L-head configuration. As the cam rotates, the lobe moves the tappet that moves the pushrod to open the valve. A *pushrod* is an engine component that transfers motion from the tappet to one side of the rocker arm in overhead valve engines. A *rocker arm* is an engine component that acts as a pivoting device for opening and closing overhead valves. A *valve guide* is an engine component that aligns a valve stem in a linear path. The valve guide allows rotation of the valve during the opening and closing to increase valve life.

The opening and closing of the valve is precisely timed by the camshaft to control the air-fuel mixture entering the cylinder and the exhaust gases leaving the cylinder. The *camshaft* is an engine component that includes the cam gear, cam lobes, and bearing surfaces. The *cam gear* is the portion of the camshaft that meshes with the crankgear. The camshaft is driven by the crankgear and is usually constructed as one unit. See Figure 3-25.

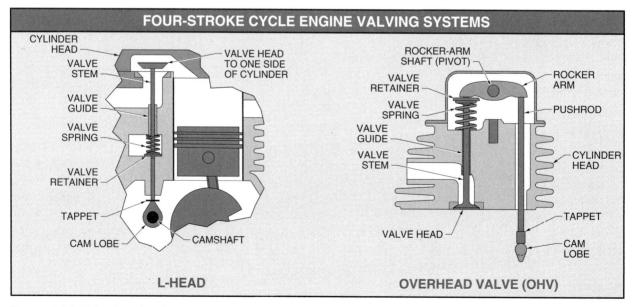

Figure 3-24. Valve location determines whether an engine is an L-head or OHV engine.

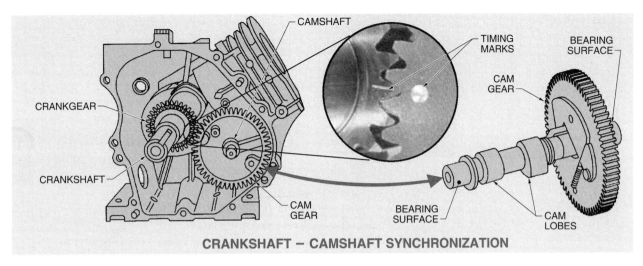

Figure 3-25. Timing marks on the cam gear and crankgear indicate the proper gear teeth mesh required to prevent damage to engine components.

The camshaft and crankshaft must be perfectly synchronized to prevent damage to engine components. Timing marks on the cam gear and crankgear or counterweights indicate the gear teeth mesh required for proper synchronization. On four-stroke cycle engines, the camshaft rotates at one-half the speed of the crankshaft. For example, in an engine operating at 3600 rpm, the camshaft rotates at 1800 rpm. At 3600 rpm, valves open and close 1800 times per min (30 times per sec).

Two-Stroke Cycle Engine Valving Systems

Two-stroke cycle engine valving systems require fewer parts than four-stroke cycle engine valving systems. Engine components such as the camshaft and tappets are not required. This results in less engine size and weight. Engine components used for controlling the air-fuel mixture and exhaust gases are also less complicated. Two-stroke cycle engines commonly use a reed valve valving system, a three-port valving system, or a rotary valve valving system. See Figure 3-26.

A *reed valve valving system* is a two-stroke cycle engine system that uses a valve made from thin spring steel that opens and closes with pressure changes in the crankcase to control air-fuel mixture flow. A reed valve valving system is the most common type of valving system used for two-stroke cycle engines. Reed valves are located in the crankcase on the back side of the carburetor. As the piston moves toward TDC, pressure in the crankcase drops and the reed valve is forced open by ambient atmospheric pressure. This allows the air-fuel mixture to enter the crankcase. When the piston moves toward BDC, pressure in the crankcase increases and the reed valve is forced closed.

A *three-port valving system* is a two-stroke cycle engine system that uses three ports to control the flow of air-fuel mixture and exhaust gases. This system has no moving parts other than the piston, and the air-fuel mixture is introduced by pressure differences in the crankcase and combustion chamber created by piston movement. The air-fuel mixture enters the combustion chamber through the transfer port.

A *rotary valve valving system* is a two-stroke cycle engine system that uses a rotating flat disk with a section removed to control air-fuel mixture flow. The rotary valve is attached to the crankshaft and seals the crankcase until the piston starts to move toward TDC. The combination of pressure drop in the crankcase with the rotary valve in open position draws air-fuel mixture into the crankcase.

Exhaust gases are discharged from the cylinder of a two-stroke cycle engine similarly regardless of the valving system used. When combustion occurs, the piston is driven toward BDC, uncovering the exhaust port to allow exhaust gases to flow out of the cylinder and into the atmosphere.

DIESEL ENGINES

A *diesel engine* is a reciprocating internal combustion engine that ignites fuel by high compression. The high compression of the diesel engine increases fuel vaporization, resulting in more complete combustion. Diesel engines require high compression ratios, have a longer stroke, and produce more torque than gasoline engines with the same displacement. Diesel engines can be either four-stroke cycle or two-stroke cycle. Four-stroke cycle diesel engines are more common and are used in small engine applications. The primary difference between diesel and gasoline engines is how fuel is ignited.

66 SMALL ENGINES

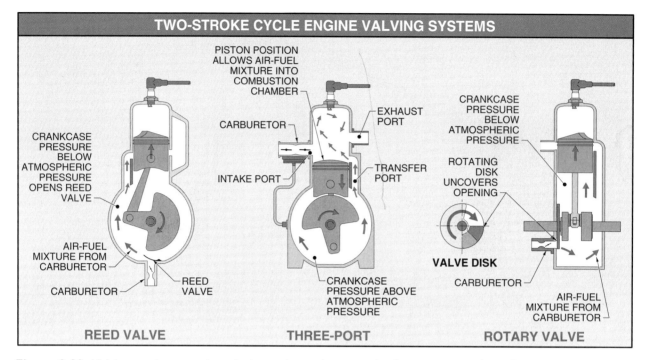

Figure 3-26. Valving systems on two-stroke cycle engines require fewer parts and are less complicated than four-stroke cycle engine valving systems.

Diesel Engine Components

The components of a diesel engine are similar to the components of a gasoline engine. Diesel engines utilize a piston with rings, valves, crankshaft, camshaft, connecting rod, piston pin, engine block, and cylinder head. Some components require additional strength to accommodate the higher compression ratios. Components specific to diesel engines and required for operation include the injection pump, injector, and glow plug.

Injection Pump. The *injection pump* is a diesel engine component that provides pressurized fuel to the cylinder at precise intervals. See Figure 3-27. The injection pump is similar to the distributor cap used on multiple cylinder spark-ignition engines. Instead of providing a spark to a specific cylinder at the precise time, the injection pump provides high-pressure fuel to the cylinder at the precise time. One injection pump may supply multiple cylinders, or a separate injection pump may be required for each cylinder. To be fed into the cylinder, fuel pressure must exceed the pressurized air in the cylinder. The injection pump is mechanically timed to piston movement and is usually driven by the crankshaft or camshaft.

Injector. The *injector* is a diesel engine component that functions as an ON/OFF valve to introduce fuel into the cylinder. The injector is located in the cylinder head.

See Figure 3-28. Engines that have multiple cylinders require an injector for each cylinder, similar to spark plugs for multiple cylinder gasoline engines.

Figure 3-27. Diesel engines use an injection pump to deliver pressurized fuel to the cylinder at precise intervals.

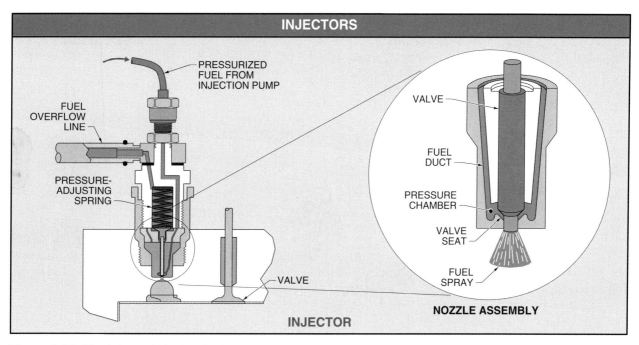

Figure 3-28. The injector is hydraulically activated by the pressurized fuel delivered from the injection pump.

The injection pump pressurizes the fuel to be pumped to the injector. The injector is hydraulically activated by the pressurized fuel delivered from the injection pump. Fuel pressure builds up in the pressure chamber until it overcomes the pressure-adjusting spring in the injector. The valve in the nozzle assembly releases the fuel from the fuel duct, and it is injected into the precombustion chamber. The fuel instantly ignites from the pressurized, superheated air. Excess fuel in the injector is routed back through the fuel overflow line to the injector pump.

Glow Plug. A *glow plug* is a diesel engine component that preheats air inside the combustion chamber to facilitate ignition of the charge. Diesel fuel is more difficult to ignite at lower temperatures than gasoline. The glow plug is located in the cylinder head. Heat is created by electrical resistance to current passed through a heating coil. The heating coil begins to glow and heat the air inside the cylinder. See Figure 3-29.

> The first engine patented by inventor Rudolph Diesel exploded as he tried to start it. In 1897, Diesel produced a successful 25 HP single-cylinder engine. A year later, the first commercial diesel engine, a 60 HP two-cylinder engine, was placed in service.

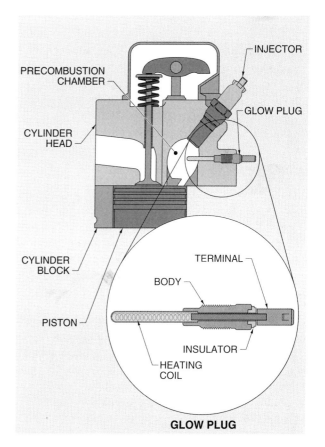

Figure 3-29. Heat in the glow plug is created by resistance to current passed through a heating coil.

Four-Stroke Cycle Diesel Engine Operation

Four-stroke cycle diesel engine operation includes the five events of intake, compression, ignition, power, and exhaust for one cycle to be completed as in a four-stroke cycle gasoline engine. During the intake event, air is forced into the cylinder by atmospheric pressure. As the piston moves toward BDC, volume inside the cylinder increases. A pressure lower than atmospheric pressure causes air to flow by the open intake valve into the cylinder. Only air is drawn into the cylinder during the intake stroke. When the piston is at BDC, the intake event is complete and the cylinder is filled with air.

The compression event occurs when the piston moves from BDC to TDC and both intake and exhaust valves are closed. Air in the cylinder is compressed as the piston moves to TDC. Compression superheats the air trapped inside the cylinder. Fuel must not be present during the compression stroke or it may preignite and cause severe engine damage. The ignition event occurs as the piston approaches TDC and fuel is injected into the combustion chamber, which is filled with compressed, superheated air. The fuel ignites on contact with the superheated air, and a flame front progresses across the combustion chamber.

The power event occurs when the hot expanding gases force the piston toward BDC. All valves are closed during the power event to ensure that all combustion pressure force is applied to the piston head. The force on the piston is transferred through the connecting rod to the crankshaft. The exhaust event occurs when the piston reaches BDC and the cylinder is filled with exhaust gases. The exhaust valve opens, and piston movement toward TDC discharges exhaust gases out of the cylinder.

Diesel Engine Turbocharging

Turbocharging is the process of using a compressor to raise the pressure and the density of the air entering an engine. Energy for compressing the intake air is scavenged from the exhaust. An exhaust-driven turbine drives a small compressor on the same shaft to pressurize the intake air. This increases the total mass of air introduced per cycle. The increased air mass increases the power output, improves engine thermal efficiency (fuel economy), and can reduce overall emissions. Turbocharging is the primary means for reducing initial cost and weight per horsepower in diesel engines.

Turbochargers must operate at high temperatures and high rotational speeds for maximum benefit. With some turbochargers, there can be a turbo lag. Turbo lag is the time period between when the engine operator demands more power and the time when the intake air reaches maximum pressure for the desired power. Turbo lag has been minimized in most turbocharged diesel engines by improved designs.

Diesel Fuel

Diesel fuel is a generic term for any fuel commercially made for use in a diesel (compression ignition) engine. In the United States, the most common diesel fuel is Grade No. 2-D diesel fuel. Two other grades, Grade No. 1-D and Grade No. 4-D, are also used commercially. Diesel fuel grade designations are assigned by ASTM International in order of increasing density and viscosity, with No. 1-D the lightest and No. 4-D the heaviest. Heating value is the amount of heat produced by a fuel measured in Btu per gallon. Diesel fuel has a heating value of approximately 138,000 Btu compared with approximately 115,000 Btu for gasoline.

Diesel fuel is rated by cetane number. A cetane number is the measure of how easily diesel fuel will ignite. Cetane numbers range from a low of 33 to a high of 64. The higher the number, the easier the fuel will ignite. If the cetane number is too low, the engine may be difficult to start. An engine requires a higher cetane number fuel to start easily in colder ambient temperatures. When a cold diesel engine is started, the heat of compression is the sole energy source for heating the combustion chamber to a temperature that will cause combustion of the fuel (about 750° F).

In a cold diesel engine, the walls of the combustion chamber are at ambient temperature and act as a heat sink rather than a heat source. The cranking RPM is slower than operating RPM, and the compression process in the cylinder is also slower, which allows more time for the compressed air to lose heat to the chamber walls. A glow plug can be used to provide an additional source of heat for preheating the air in the combustion chamber to aid starting. After starting, heat from the operating temperature of the combustion chamber facilitates efficient combustion.

Diesel fuel requires special handling compared to gasoline. Both diesel fuel and gasoline accumulate water due to condensation. However, water will readily mix with diesel fuel. Additionally, microorganisms can grow and accumulate in diesel fuel. Fresh diesel fuel should always be used. Diesel fuel in bulk quantities should be properly stored in approved, sealed containers. Special fuel conditioners may be required to reduce the possibility of fuel separation for maximum performance in cold climates.

ROTARY ENGINES

A *rotary engine* is an internal combustion engine that operates using the rotating motion of a rotor. Like a four-stroke cycle reciprocating engine, the rotary engine performs the same five engine operation events of intake, compression, ignition, power, and exhaust. Unlike reciprocating engines that change linear motion into rotary motion, the rotary engine has a circular motion. This requires fewer moving parts than the four-stroke cycle engine and the two-stroke cycle engine. See Figure 3-30.

During intake, the air-fuel mixture is sealed in the recessed area of a rotor face inside the rotor housing. The rotor rotates in the rotor housing and the air-fuel mixture is compressed. The air-fuel mixture is further compressed until the rotor face is located at the spark plug. Ignition occurs and hot expanding gases drive the rotor face. As the gases continue to expand, the rotor rotates to discharge exhaust gases. The three rotor faces and their simultaneous operation are equivalent to a three-cylinder reciprocating engine. At any given time, there are three of the five engine operation events (intake, compression, ignition, power, and exhaust) taking place.

ENGINE OUTPUT

Engine output is measured by torque and horsepower. Torque is measured in pound-feet (lb-ft) or newton-meters (Nm). Torque is the measure of the ability of an engine to do work. Power is the measure of the rate at which work is done. Horsepower (HP) is the most common measurement of power. For example, the amount of torque determines if an engine can drive a lawn mower through tall grass. The amount of power determines how quickly the mower progresses through the grass.

Engines have different torque requirements at different speeds, and engine design is based on compromising for the best overall operating performance. The primary difference between long- and short-stroke engines is the profile of the torque curve. Long-stroke engines develop maximum torque at lower rpm. This has a profound effect on certain engine applications requiring maximum torque at high rpm, such as a chipper/shredder or some generators.

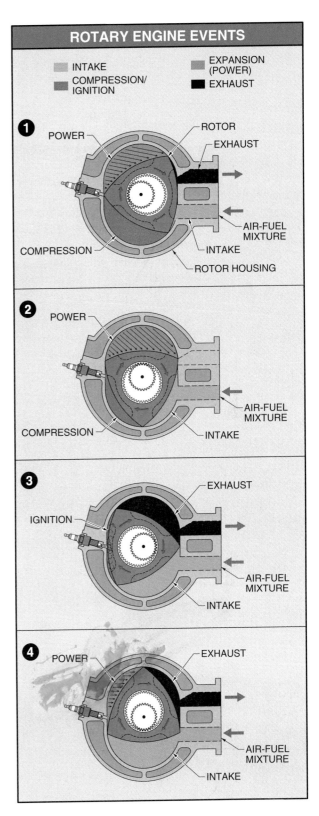

Figure 3-30. A rotary engine performs the same five engine operation events of intake, compression, ignition, power, and exhaust as a four-stroke cycle engine.

Horsepower is a timed rate of doing work and is a calculation of the various conditions of a given engine operating under various loads and speeds. Horsepower can be classified as brake, friction, or indicated horsepower.

Brake Horsepower

Brake horsepower (BHP), or shaft horsepower, is the amount of usable power taken from an engine. Brake horsepower is the amount of power possible, including any loss from internal friction. The term brake refers to the device that absorbs power to measure engine output. Engine output power is absorbed by applying friction against the rim of a moving component, such as the engine flywheel.

Friction Horsepower

Friction horsepower (FHP) is the amount of power required to overcome the internal friction of engine moving parts. Power produced by expanding gases in the cylinder is reduced as actual output is transferred through moving parts. Moving parts, such as the piston, bearings, and valve train components, create friction. Friction results in a loss of power and is compounded by each moving part in the engine.

Indicated Horsepower

Indicated horsepower (IHP) is the power produced inside the engine cylinder and is the sum of the usable power (BHP) plus the power used to drive the engine (FHP). Indicated horsepower accounts for heat energy converted into mechanical energy, but does not account for any energy (power) required to move the engine parts. Indicated horsepower is always greater than BHP.

Measuring Horsepower

Torque is generated in an engine by the combustion pressure pushing the piston to the bottom of the cylinder, forcing the crankshaft of the engine to turn. Torque from the crankshaft is then applied to the application, such as a lawn mower, an electric generator, or the drive shaft of a vehicle. Torque is directly related to the power of an engine, and determining torque and speed allows the calculation of horsepower. Knowing force and distance allows the calculation of torque, or how much turning effort the crankshaft applies to a load.

Engine output in torque or horsepower can be measured by a dynamometer. The *dynamometer* is a device that applies a load to an operating engine and measures torque, load, speed, or horsepower. Dynamometers are commonly classified as a water dynamometer, electric dynamometer, eddy current dynamometer, or prony brake dynamometer.

Water Dynamometer. The *water dynamometer* is a dynamometer used to measure engine torque using load produced by a water pump. See Figure 3-31. The function of a water dynamometer is similar to the function of a common water pump. The output shaft of the engine drives an impeller located inside the water dynamometer. Load is increased or decreased by adding or removing water from the impeller housing. As water is added, it moves to the outside surfaces of the impeller housing, increasing the load. A load cell measures the twisting force exerted on the outside of the impeller housing. Water dynamometers are most commonly used in industry.

Figure 3-31. Load is increased or decreased by adding or removing water from the impeller housing of a water dynamometer.

Electric Dynamometer. An *electric dynamometer* is a dynamometer used to measure brake horsepower by converting mechanical energy into electrical energy. See Figure 3-32. The output shaft of the engine drives an electric generator that is connected to a load bank. Load is increased until the engine reaches wide open throttle (WOT) at the rated rpm. The electric power produced is consumed by the load bank and measured in watts (W).

 The first 50,000,000 Briggs & Stratton engines were produced between 1924 and 1967. The second 50,000,000 engines were produced between 1968 and 1975.

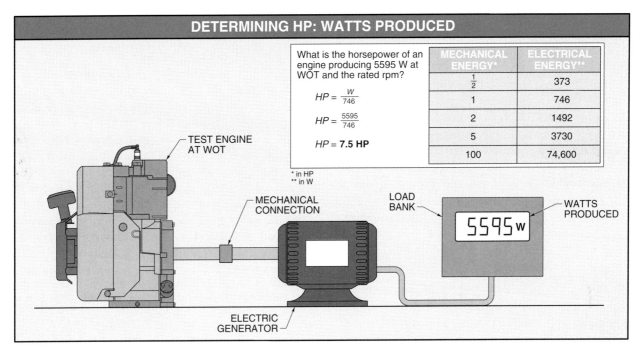

Figure 3-32. The electric dynamometer measures brake horsepower by converting mechanical energy into electrical energy.

When converting mechanical energy into electrical energy, 1 HP is equal to 746 W. When watts produced are known, engine horsepower is found by applying the formula:

$$HP = \frac{W}{746}$$

where

HP = horsepower

W = watts

746 = constant

For example, what is the horsepower of an engine producing 8952 W at WOT and the rated rpm?

$$HP = \frac{W}{746}$$

$$HP = \frac{8952}{746}$$

$$HP = \mathbf{12\ HP}$$

Electric dynamometers are the most accurate and expensive dynamometers available. A disadvantage of an electric dynamometer is that the efficiency of the generator must be accounted for. Some electric dynamometers use a load cell that measures the torque placed on the generator housing when a load is applied.

Long stroke versus short stroke engines

An engine with a longer stroke produces more torque than an engine with a shorter stroke having the same displacement. If a longer stroke produces more torque, why not manufacture all engines with a long stroke? Manufacturers try to find a combination of bore and stroke specifications which best meets overall performance requirements. In some applications, maximum torque must occur at higher rpm.

For example, an engine driving a portable generator in North America commonly operates at 3600 rpm to generate the proper frequency of electricity. A long-stroke engine develops maximum torque at an rpm much lower than 3600 rpm. Torque produced by the engine at 3600 rpm may be less than the torque required to drive the generator. Consequently, a larger engine or an engine with a shorter stroke would be required.

In the small engine industry, a near balance of bore and stroke specifications is referred to as unity. Unity, in theory and practice, is the best overall ratio for bore to stroke dimensions. Unity occurs at a 1:1 ratio of bore to stroke dimensions to produce the best overall performance throughout the usable horsepower range.

Eddy Current Dynamometer. The *eddy current dynamometer* is a dynamometer used to measure engine torque using load produced by a magnetic field. See Figure 3-33. The eddy current dynamometer consists of a disk that is driven by the engine being tested. The disk turns in a magnetic field controlled by varying the current through a series of coils located on both sides of the disk. Increasing the current in the coils causes an increased magnetic field, resulting in a greater load on the engine. The increased load on the engine results in more torque and pressure applied to the load cell.

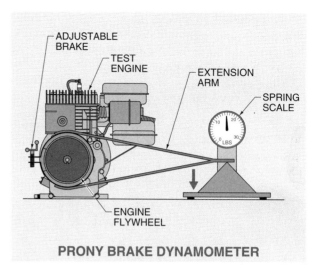

PRONY BRAKE DYNAMOMETER

Figure 3-34. The prony brake dynamometer measures engine torque using an adjustable brake that exerts pressure on a spring scale.

Factors Affecting Engine Output

Factors affecting engine output include engine displacement, volumetric efficiency, thermal efficiency, and air density. An increase in engine displacement increases the volume of air-fuel mixture in the cylinder. With a larger charge, more power is produced. *Efficiency* is the ratio of fuel energy supplied to work produced. An engine that operates with an efficiency of 100% has all fuel energy used to produce work. No engine is 100% efficient. Gasoline engines commonly have an efficiency of 25% to 32%, compared to diesel engines, which have an efficiency of 32% to 38%.

Volumetric efficiency is the ratio of volume available in the engine to the actual volume filled during operation. Volumetric efficiency pertains to how well an engine breathes. As an engine increases speed, less time is available for the cylinder to be filled. Reducing time to fill the cylinder reduces the amount of charge and energy released during combustion. This decreases volumetric efficiency.

Thermal efficiency is a measurement that compares the amount of chemical energy available in fuel converted into heat energy used to produce useful work. It is expressed as a percentage of heat energy available in the fuel converted into torque at the crankshaft.

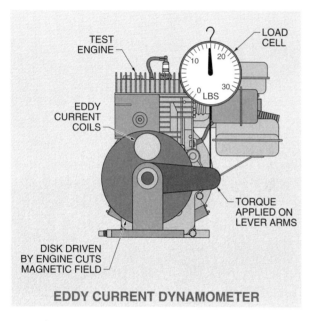

EDDY CURRENT DYNAMOMETER

Figure 3-33. The eddy current dynamometer measures engine torque using load from the magnetic field produced by current in eddy current coils.

Prony Brake Dynamometer. The *prony brake dynamometer* is a dynamometer for measuring engine torque using a brake that exerts pressure on a spring scale. An adjustable brake with an extension arm is mounted in contact with the flywheel. The end of the extension arm is connected to a spring scale. The engine is operated at a set speed while the adjustable brake is tightened to increase the load. As the load increases, the governor system keeps the engine at a constant speed by opening the throttle. Load is increased until reaching WOT. At WOT a reading is taken from the spring scale and is used to determine torque produced. See Figure 3-34. The prony brake dynamometer has been largely replaced by the electric dynamometer and water dynamometer.

 In 1953, Briggs & Stratton introduced the first die-cast aluminum engine for consumer use.

Brake mean effective pressure (BMEP) is the average effective pressure placed on a piston during one complete operating cycle. Pressures used to draw in the air-fuel mixture, compress it, expel burnt gases, and overcome internal friction are subtracted from the power delivered during the power stroke.

Air density affects engine output. Denser air allows for a greater charge, resulting in greater engine output. Air density is most affected by altitude, temperature, and humidity. Higher altitudes have less oxygen in the air or thinner air. Thinner air results in a smaller charge inside the cylinder. Engine horsepower decreases $3\frac{1}{2}\%$ for each 1000′ above sea level. See Figure 3-35. For example, a 10 HP engine operating at 10,000′ produces only 6.5 HP. See Appendix.

$$\frac{10,000}{1000} = 10$$

$$10 \times 3.5\% = 35\%$$

$$10 \times 35\% = 3.5$$

$$10 - 3.5 = \textbf{6.5 HP}$$

Air density decreases as ambient (existing) temperature increases. A decrease in air density results in a smaller charge and less energy released after ignition. Engine horsepower decreases by 1% for each 10°F above standard temperature of 60°F. For example, a 10 HP engine operating in a temperature of 100°F produces only 9.6 HP.

$$100 - 60 = 40$$

$$\frac{40}{10} = 4$$

$$10 \times 4\% = .4$$

$$10 - .4 = \textbf{9.6 HP}$$

Humidity affects engine output as dry air is less dense than moist air. The effect of humidity is less significant than that of temperature or altitude.

 The unit watt, which is used to define power rating and consumption, was named after James Watt.

ALTITUDE AND ENGINE HORSEPOWER											
Altitude*	3.0 HP	3.5 HP	4.0 HP	5.0 HP	7.0 HP	8.0 HP	10.0 HP	12.0 HP	14.0 HP	16.0 HP	20.0 HP
0	3.0	3.5	4.0	5.0	7.0	8.0	10.0	12.0	14.0	16.0	20.0
1000	2.9	3.4	3.9	4.8	6.75	7.7	9.65	11.6	13.5	15.4	19.3
2000	2.8	3.25	3.7	4.65	6.5	7.4	9.3	11.1	13.0	14.9	18.6
3000	2.7	3.1	3.6	4.5	6.3	7.1	8.95	10.7	12.5	14.3	17.9
4000	2.6	3.0	3.4	4.3	6.0	6.9	8.6	10.3	12.0	13.8	17.2
5000	2.5	2.9	3.3	4.1	5.8	6.6	8.25	9.9	11.55	13.2	16.5
6000	2.4	2.8	3.2	3.95	5.5	6.3	7.9	9.5	11.0	12.6	15.8
7000	2.3	2.6	3.0	3.8	5.3	6.0	7.55	9.0	10.6	12.0	15.1
8000	2.2	2.5	2.9	3.6	5.0	5.8	7.2	8.6	10.0	11.5	14.4
9000	2.05	2.4	2.7	3.4	4.8	5.5	6.85	8.2	9.6	11.0	13.7
10,000	1.95	2.3	2.6	3.25	4.55	5.2	6.5	7.8	9.1	10.4	13.0

* in ft above sea level

Figure 3-35. Engine horsepower decreases $3\frac{1}{2}\%$ for each 1000′ above sea level.

In 1952, the Briggs & Stratton Corporation West Plant assembly line produced 1,000,000 engines in a single year.

GardenWay, Inc./Troy-Bilt®

A chipper/vac has a high-inertia chipper assembly.

Wacker Corporation

This vibratory plate, powered by a 5 HP horizontal shaft engine, produces 5800 vibrations per minute.

COMPRESSION SYSTEM
CHAPTER 4

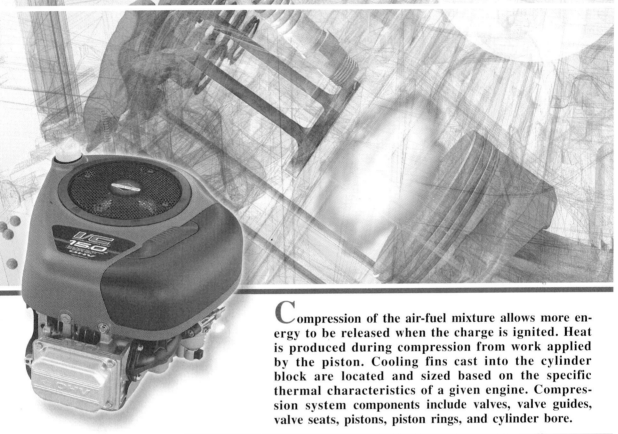

Compression of the air-fuel mixture allows more energy to be released when the charge is ignited. Heat is produced during compression from work applied by the piston. Cooling fins cast into the cylinder block are located and sized based on the specific thermal characteristics of a given engine. Compression system components include valves, valve guides, valve seats, pistons, piston rings, and cylinder bore.

COMPRESSION

Compression is required in a small gasoline engine to prepare the charge for ignition. Compressing the air-fuel mixture allows more energy to be released when the charge is ignited. Compression system components direct, contain, and compress the air-fuel mixture, and discharge exhaust gases. The compression ratio of most small gasoline engines ranges between 6 : 1 and 8.5 : 1. For example, with a compression ratio of 8 : 1, the charge is compressed into a space ⅛ the original volume of air-fuel mixture before the compression event. Compression of the charge in an internal combustion engine is an adiabatic process.

Adiabatic Process

An *adiabatic process* is a process in which heat is derived from the process itself. During compression, heat is produced from the work applied by the piston. Heat is not introduced from an external source. The increase in temperature of the charge is directly proportional to the quantity of work applied to compress the charge. See Figure 4-1.

Heating of the charge occurs in hundredths of a second. As the piston moves toward TDC, the volume of the combustion chamber is reduced. The charge in the combustion chamber is compressed in the smaller space. As the compression ratio increases, charge temperature increases proportionately from the work applied to the charge by the piston. There is little time for any heat energy to leave or to be transferred from the combustion chamber to the cylinder block.

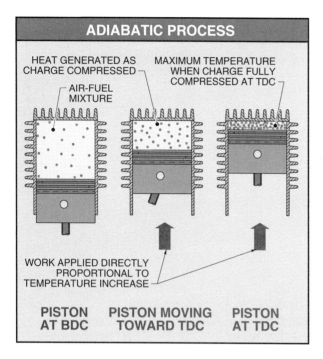

Figure 4-1. Compression of the air-fuel mixture is an adiabatic process in which heat is derived from the process itself and does not come from an external source.

Changes in the Charge

The charge enters the combustion chamber in a gaseous state. In a gaseous state at the molecular level, atoms and molecules are as far apart from each other as possible, yet held together by cohesion. *Cohesion* is the molecular attraction by which atoms and molecules are united throughout the mass. As distance increases between the molecules, cohesion force is proportionately weaker. When the piston compresses the charge, an increase in cohesive force causes heat to be generated.

As the temperature of the charge is raised, gasoline molecules become more active. The increased activity from speed and molecular vibration and from being forced together by compression causes additional and increasing molecular collisions. This results in the addition of internal energy to the charge. As the process continues, the temperature of the charge increases hundreds of degrees.

When gasoline is heated, it also changes rapidly from a liquid to a vapor. In a vapor state, small droplets of gasoline liquid are suspended in the gaseous medium of the charge. The vapor is suspended, similar to the suspension of water in a rain cloud. The rate at which the gasoline becomes a vapor is a function of temperature, and the propensity of the gasoline to become a vapor. As the temperature increases, gasoline droplets release more vapor.

In addition to an increase in vapor, larger droplets tend to break apart at higher temperatures. The smaller droplets increase the total surface area and expose more of the liquid gasoline to the air. This reduction in size and increase in surface area of the droplets provides additional volatile vapor for the combustion process.

The increase in the temperature of the charge reduces the energy needed to maintain the combustion process. The energy required to initiate combustion is provided by the spark jumping across the gap in the electrode of the spark plug. The energy required to compress the charge before combustion is typically 25% of the energy released during combustion, or a 1 : 4 ratio.

Compression Problems

Compression problems occur from inadequate or excessive compression. Inadequate compression is commonly caused by leaks. Any compression leaks in the system cause an exponential decrease in performance, efficiency, and available power, and a possible increase in exhaust gas temperature. If a compression component such as an exhaust valve leaks during the compression process, there is a decrease in the overall pressure of the charge and a dilution in the concentration of the charge by maverick air. See Figure 4-2. *Maverick air* is undesirable, unaccounted for air entering the engine through leaks caused by worn, loose, or failed engine components. An exhaust valve leak can result in maverick air entering the combustion chamber through the muffler.

Under ideal conditions, air normally picks up fuel vapor in the carburetor. Maverick air bypasses the carburetor and causes a lean air-fuel mixture that can result in a dramatic increase of exhaust gas temperature in the combustion chamber. This accelerates the wear and degradation of all compression components. Like loss of compression, excessive compression can also cause engine performance problems, and can lead to detonation and preignition.

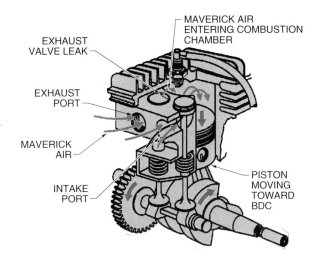

Figure 4-2. A leaking exhaust valve results in a loss of compression and allows maverick air to dilute the charge during the intake stroke.

Detonation. *Detonation* is an undesirable engine condition in which there is spontaneous combustion of a significant portion of the charge before the spark-induced flame front reaches it. See Figure 4-3. Detonation is sometimes called knocking, spark knock, or pinging. The release of energy during detonation occurs at a much faster rate than by normal spark-induced ignition. This causes a pressure differential in the combustion chamber, which produces shock waves causing a knocking or pinging noise. The noise itself is caused by the radical vibration of oxidizing combustion gases.

One of the major determining factors of detonation is the compression ratio. Compression ratio is increased when a large volume of combustion chamber deposits accumulate in the engine. With the increased compression ratio, the accumulation of carbon deposits slows the movement of the charge into and out of the combustion chamber. Increased heating of the valve area occurs. In addition, the carbon deposits act as an insulator and reduce the heat transferred to cool the engine parts.

The extreme pressure and temperature fluctuations caused by detonation can cause severe engine damage. Engine damage from detonation includes piston failures (melting or breakage), piston skirt damage, connecting rod breakage, and occasional crankshaft failures.

Preignition. *Preignition* is an undesirable engine condition which occurs when a small portion of a combustion chamber component or a particle in the combustion chamber becomes excessively heated and ignites the charge as it enters the combustion chamber. This hot deposit creates a separate and distinct flame front before the spark-induced flame front. Preignition decreases performance and results in an audible pinging or knocking sound in the engine. See Figure 4-4. Preignition increases the peak combustion pressures of the engine as well as the internal temperatures. Preignition also causes peak combustion pressures earlier in the compression cycle and opposes piston motion. Like detonation, this can lead to engine damage from increased loads and stresses such as broken pistons, connecting rods, and crankshafts.

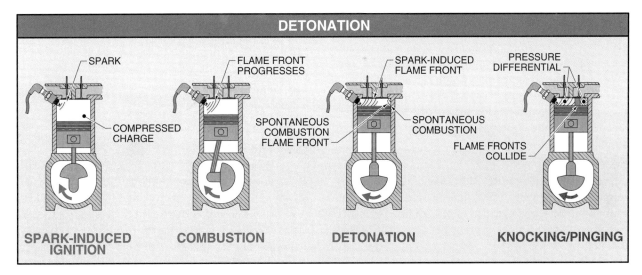

Figure 4-3. The release of energy from spontaneous combustion during detonation causes a pressure differential in the combustion chamber, resulting in audible knocking and pinging.

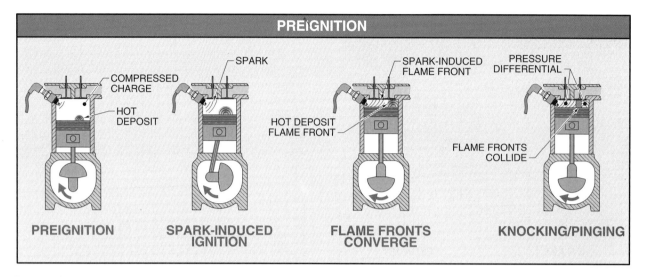

Figure 4-4. Preignition causes two separate flame fronts which increase internal pressure and temperature to reduce combustion efficiency and engine performance.

The ability to properly seal the combustion chamber for proper compression greatly affects the overall performance of the engine. The position, condition, and operation of the compression system components affect the compression problems present in the engine. Compression system components include valves, valve guides, valve seats, pistons, piston rings, and cylinder bore.

VALVES

Valves allow the flow of air-fuel mixture into the cylinder, and the flow of exhaust gases from the cylinder. Most exhaust valves used in Briggs & Stratton engines are made from austenitic steel. *Austenitic steel* is a heat-resistive metal alloy consisting of cobalt, tungsten, and chromium. Austenitic steel has properties similar to stainless steel. *Stainless steel* is a ferrous alloy primarily consisting of chromium or nickel. Austenitic steel used for valves offers similar heat and corrosion resistance at a lower cost than stainless steel.

Valves are exposed to various chemical, mechanical, and thermal stresses during operation. Valves must maintain their basic shape and dimension throughout the expected life of the engine. In addition, the integrity of the sealing surface of the valve and mating valve seat is critical to durability and performance. Engineers determine the valve material, shape, specifications, and surface coatings to match the specific engine family, expected operating environment, and projected length of service. Valves commonly used in small engines are classified as one-piece, projection-tip welded, or two-piece-stem welded valves. See Figure 4-5.

One-Piece Valve

A *one-piece valve* is a valve that is constructed from a single piece of austenitic steel. One-piece valves are commonly used on engines that drive lighter loads, have occasional use, or where economy is the main consideration. Ample durability is achieved using austenitic steel alloys that are hard enough to provide long life and reliable performance. One-piece valves have been used successfully for decades in all types of internal combustion engines. Technology and manufacturing processes to produce better valve construction at an acceptable price have only recently been developed. One-piece valves are not normally recommended for engines used in high-load industrial/commercial applications.

Projection-Tip Welded Valve

A *projection-tip welded valve* is a valve that is constructed from austenitic steel that has approximately .09″ of hardened steel welded on the end of the valve stem. The hardened end of the valve stem provides better wear resistance against the scrubbing action of the rocker arm or tappet during normal operation. Projection-tip welded valves are commonly used as exhaust valves, and on some OHV engines.

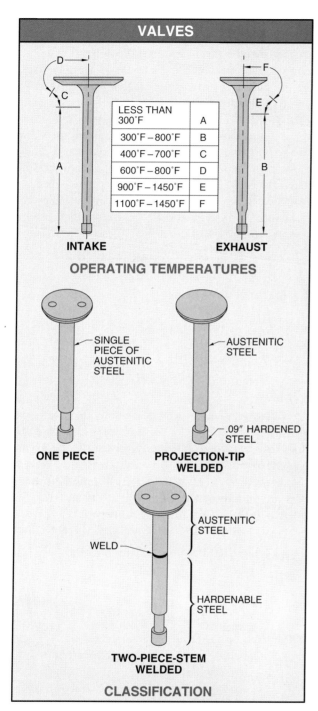

Figure 4-5. The valve material, shape specifications, and surface coatings match the expected operating environment and projected length of service.

A greater than 25% difference between the compression readings on a two-cylinder engine indicates a loss of compression in the cylinder with lower pressure.

Two-Piece-Stem Welded Valve

A *two-piece-stem welded valve* is a valve that is constructed from an austenitic steel valve head welded to a hardened valve stem. The two pieces are joined by friction welding. *Friction welding* is a metal joining process in which heat and pressure cause fusion as one or both pieces are rotated and pressed against each other. The different metals used to fabricate the valve provide the desired characteristics required by different valve parts. The harder valve stem offers maximum wear resistance at the valve spring retainer groove areas and running surface at the valve guide. A *running surface* is the portion of an engine component which interacts with a lubricated mating engine bearing surface during operation. The weld on the valve remains in the valve guide during engine operation. The hardened stem portion of the valve is usually .12″ shorter in length than the total valve lift dimension. This tolerance ensures that the hardened end of the valve has minimum exposure to hot, caustic exhaust gases exiting the combustion chamber.

Valve Hardfacing

Some valves have special coatings (hardfacing) to improve durability and service life. *Hardfacing* is the application of material to an engine component to improve wear resistance from load, heat, and chemical corrosion. Although austenitic steel exhaust valves provide ample resistance to corrosion and wear, some valves and valve seats used in heavy service applications or premium engines require hardfacing to extend valve life.

In the past, alloys known commercially as Stellite® were the predominant alloys used for valve hardfacing. These alloys contained a high percentage of the element cobalt. The rising cost and strategic value of cobalt required alternative materials of equal quality. One of the most successful of these newly developed coatings for use in small engines is the iron-based hardfacing alloy marketed by Briggs & Stratton known as Cobalite™. Cobalite™ has now become widely used in the internal combustion engine market.

Stellite® hardfacing alloys were deposited on the valve head using an oxyacetylene welding (OAW) process. Cobalite™ alloys are deposited on the valve head using the plasma spraying process. The *plasma spraying process (PSP)* is a thermal process in which hardfacing metal in powder form is molten with an electric arc and propelled to coat the desired surface. The plasma spraying process provides an even hardfacing coating and has become the industry standard for depositing Cobalite™ and other iron-based, nickel-based, and Stellite® hardfacing alloys. This has resulted in improved valve quality, better performance, and lower production costs. Hardfacing thicknesses on valves vary with each engine type and anticipated operating environment. Most hardfacing deposits on valves are approximately .03″ thick on the sealing surface or face of the valve.

Properties of a compressed gas

The expansion or compression of a gas results in predictable changes in properties. For example, when a gas is released from a pressurized container such as an aerosol spray can, the compressed gas releases heat as it expands and feels cool. This principle can be demonstrated with a simple experiment. Hold your hand 4″ from your mouth. Open your lips and blow some air across your hand. The air feels warm across your hand. Continue the experiment by blowing some air across your hand again, but this time with your lips almost closed, similar to an attempt to whistle. This produces a cool volume of air crossing the hand.

Air leaving the lungs enters the mouth and is compressed by the partially-closed lips. As the air expands when leaving the closed lips, it releases heat and becomes cooler. The change in temperature is proportional to the volume of air expanding from the lips. The greater the volume of air leaving the lips, the cooler the air temperature.

Valve Stem Surface Treatment. In the past, large clearances between the valve guide running surface and the valve stem allowed ample lubrication to reduce friction and wear. However, the lubrication provided also increased the hydrocarbon (HC) emissions of the engine. To comply with more stringent federal emission control standards, valve stem to valve guide clearances had to be reduced. This challenged engineers to find an alternative method to reduce friction and valve stem wear from the valve guide wall surface.

Engineers developed valve stem surface treatments that allowed a reduction in valve stem to valve guide clearance and reduced friction-related wear. A reduction in friction reduces the amount of heat generated on the running surface of a valve stem and valve guide. Reducing the valve stem to valve guide clearance also provides better control of valve movement for a more consistent valve seal between the valve and valve seat. This sealing capability has become more important, with the effects of increased temperatures in the combustion chamber from leaner air-fuel mixtures required by stricter exhaust emissions standards. Valve stem surface treatment can be as important to the overall durability of the valve as hardfacing of the valve head.

Valve Head Design

Valve head designs have changed over the past several decades. Many of the design changes affected the valve face and corresponding valve seat angles. For optimum performance and durability, a 45° exhaust valve face angle is the industry standard. A 45° valve face angle provides a smooth transitional flow from the carburetor to the combustion chamber and an unimpeded path during exhaust gas evacuation. In addition to gas flow characteristics, the 45° exhaust valve face angle provides sufficient valve seat pressure to crush and void small combustion deposit particles. This improves the performance and longevity of the entire valve train.

All Briggs & Stratton small engines use a 45° exhaust valve face angle and a 45° or 30° intake valve face angle. Valve face angle is selected as a compromise between the needs of engine performance and durability. For example, a 30° intake valve face angle requires a 60° intake valve seat angle. The flatter angle allows slightly better flow into the combustion chamber. In addition, a 30° valve face angle provides increased wear resistance as well as some improvement in air flow. However, the greater valve face angle diminishes the ability to crush combustion deposit particles between the valve face and valve seat.

Each valve and corresponding valve seat has a slight deviation in the mating angles. For example, a 45° valve face angle is expected to have a true 45° angle, but it is commonly manufactured with an interference angle. An *interference angle* is the intentional deviation from a specification of two mating machine components to improve seating quality after a sufficient break-in period. See Figure 4-6. The valve face is machined with a 15 minute (15′) deviation from 45° with an actual measurement of 44°-45′. The valve seat also deviates from 45° with an actual measurement of 45°-15′. The deviation from the desired 45° anticipates a small amount of initial wear at the valve interface. The *valve interface* is the point of contact between the valve face and the valve seat. After a short break-in period, the valve face and valve seat each wear slightly to seat at a perfect 45° for an improved seal.

Valve systems equipped with the keyhole valve spring retainer produce nominal rotation. The motion of the valve is inhibited by the constant contact between the retainer and the valve spring. Each time the valve is lifted from the seat by the tappet, the retainer and valve spring apply pressure to the valve stem retainer groove. The valve spring itself influences some rotation of the valve with a keyhole valve spring retainer. The wound spiral wire of the valve spring initiates a small torque through the retainer into the valve stem, which causes some rotation. See Figure 4-7.

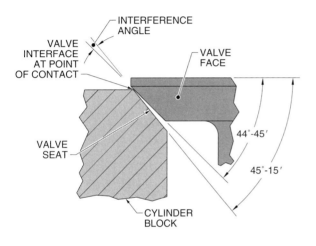

Figure 4-6. An interference angle anticipates a small amount of initial wear at the valve interface to produce a perfect 45° for an improved seal.

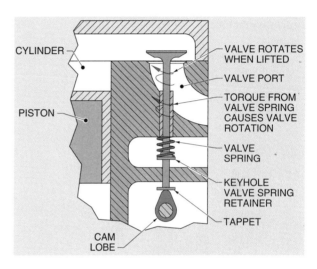

Figure 4-7. A keyhole valve spring retainer produces nominal valve rotation, with the valve spring initiating a small torque.

Valve Dynamics

Each valve design utilizes components in the compression system to maintain proper valve and valve spring position in the cylinder block. A valve spring retainer maintains spring contact with the valve stem and helps to align the valve spring. Although valve spring retainers come in different styles, all provide the essential requirements of maintaining valve spring contact and spring alignment. The primary difference in valve spring retainer styles is most evident when considering valve rotation.

Most valves in an operating small engine rotate about the valve stem axis at varying rates. Valve rotation has an overall positive effect on valve life. Rotation provides improved temperature distribution in the valve head and a mild scraping action that cleans the valve interface of any crushed combustion deposits.

The Toro Company
Engine design and operating characteristics must allow maximum performance in all climate conditions.

Another engine component that helps initiate valve rotation is the camshaft. The orientation of the camshaft to the valve tappets in a Briggs & Stratton engine is intentionally offset, initiating a rotation of the tappet. This transfers rotation of the tappet to the valve stem, which also causes the valve to rotate. Although small, this offset results in a relatively constant valve rotation speed.

Some valves on small engines use a two-piece automotive-style valve spring retainer. This consists of a valve rotator with a tapered hole in the center and a two-piece tapered collar with a projection. The collar holds the valve in contact with the valve spring by locking against the groove in the valve stem. See Figure 4-8.

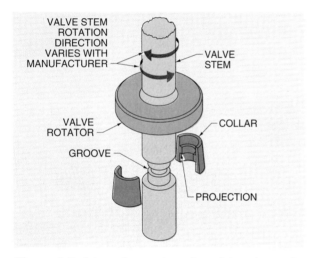

Figure 4-8. A two-piece automotive-style valve spring retainer has minimal pressure applied at the valve stem groove, allowing the valve stem to rotate freely.

Valve rotation is very important in some engine families and in heavy service engine applications. A valve rotator may be added to initiate and maintain a steady rate of rotation. Valve rotators used in some Briggs & Stratton engines contain springs which impart a small torque on the valve stem. This causes an incremental rotation of the valve with each opening and closing cycle. With a valve rotator, when the valve is lifted off its seat, the combination of rotation of the tappet, vibration from the engine, and coiled nature of the valve spring design contributes to the rotation of the valve.

Valve rotation can also be caused by the complex harmonic vibration of the operating engine. Some engines exhibit more valve rotation based on their vibration frequency and profile. Excessive rotation can cause surface wear between the valve face and valve seat. The signature of excessive rotation is a wiping or smearing of metal on the valve face or valve seat. Another common signature of excessive rotation is a circular wear pattern on the end of the valve stem. The circular pattern looks similar to the annular rings of a tree. See Figure 4-9.

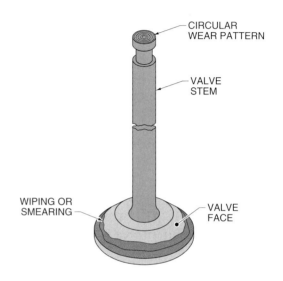

Figure 4-9. Excessive rotation is indicated by a wiping or smearing on the valve face and/or a circular wear pattern on the end of the valve stem.

Valve rotation is not recommended on engines that use LP gas or propane. These fuels do not provide the inherent lubricating qualities of gasoline. Without lubrication, the valve interface wears prematurely if significant valve rotation is allowed to occur. Engines using LP gas or propane typically operate hotter than comparable gasoline engines. The increase in combustion gas temperatures increases the surface temperatures in the valve interface. This can lead to metal transfer between seat and valve and premature wear leading to failure.

VALVE GUIDES

The valve guide is the foundation of the valve system and provides a consistent path for the valve stem to ensure maximum engine power, valve performance, and overall valve train durability. Proper valve guide design, material, and clearances allow the valve to operate freely throughout the operating temperature range of the engine.

The valve guide features a machined cylinder that has specific size and thermal conductivity properties. *Thermal conductivity* is the ability of a material to conduct and transfer heat. The ability of the valve guide material to conduct and transfer heat is important in maintaining a proper clearance between the valve guide and the valve stem.

Valve Guide Design

In small engines, a valve guide is susceptible to dramatic fluctuations in temperature, chemical corrosion, and ingestion of foreign material. The intake valve guide is not exposed to the high exhaust gas temperatures at the exhaust valve guide. The exhaust valve guide is also exposed to caustic elements in the exhaust gas stream each time the valve opens.

The valve guide must be designed to provide a predictable and consistent clearance between itself and the valve stem regardless of operating conditions. The valve guide length to valve stem diameter ratio is typically 7:1. This ratio provides a sufficient running surface area for an oil film layer to provide lubrication for the sliding valve stem. The valve guide length also provides sufficient protection from side loading of the valve stem from the camshaft, rocker arm, and/or tappet combination. *Side loading* is the application of an undesirable unilateral (one-sided) force to an engine component or components. Side loading of the valve stem reduces the normal bearing surface of the valve stem as one side of the valve guide and valve stem absorbs a greater amount of the force.

Valve guides allow a specific clearance between the valve stem and the valve guide wall. For example, the clearance specification on Briggs & Stratton L-head engines may range from .002″ – .003″ on the intake valve to .004″ – .005″ on the exhaust valve (cold). The valve stem to valve guide clearance specification is based on the location of the valve, cylinder block material, and thermal expansion of the valve and valve guide material. *Thermal expansion* is the expansion of a material when it is subjected to heat. There are no standard clearance specifications for all engines. Each engine is tested with clearances specified to match engine component design and materials, cylinder block material, load, anticipated operating environment, and fuel used. Valve guides can be aluminum, brass, or sintered iron. *Sintered iron* is a powdered iron compound that is heated and compressed to form the desired shape.

Aluminum Valve Guides. An *aluminum valve guide* is an integrally machined part of an aluminum cylinder block and is not a separate engine component. Aluminum valve guides are the most common valve guides used in small engines. See Figure 4-10. Aluminum valve guides offer a predictable thermal expansion rate, good performance, and low manufacturing cost. With proper clearances and lubrication, aluminum valve guides can be used successfully for most light- to medium-duty consumer applications for intake and exhaust valves.

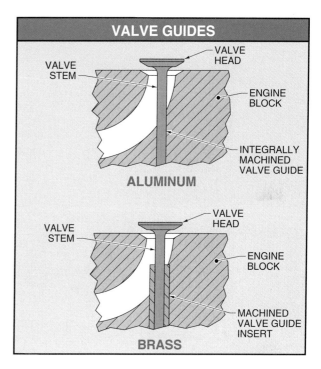

Figure 4-10. An aluminum valve guide is integrally machined into an aluminum cylinder block and is the most common valve guide used in small engines.

Brass Valve Guides. A *brass valve guide* is a separate machined valve guide insert manufactured from brass alloy. They are commonly used on many types of small engine cylinder blocks. Brass valve guides offer greater resistance to wear than aluminum valve guides, increasing valve train longevity. Brass is much harder than aluminum and provides better resistance to abrasive foreign particles, which can cause premature wear. Brass valve guides also have the ability to embed abrasive foreign particles within the valve guide for removal from the lubrication system.

 In 1996, over 5,355,000 walk-behind lawn mowers were shipped by U.S. manufacturers.

In addition to resistance to abrasive particles, brass has a lower coefficient of thermal expansion than aluminum. *Coefficient of thermal expansion* is the unit change in dimension of a material by changing the temperature 1°F. Brass has a .00001 coefficient of thermal expansion. Aluminum has a .00001244 coefficient of thermal expansion. See Figure 4-11.

Thermal expansion occurs in all dimensions when metals are exposed to heat. To find the amount of thermal expansion or the new length of a metal when heated using the coefficient of thermal expansion, apply the formula:

$L_n = L_o + (C \times \Delta T \times L_o)$

where
L_n = new length
L_o = original length
C = coefficient of thermal expansion
DT = temperature difference

For example, what is the new length of a 2″ brass rod that has a .00001 coefficient of thermal expansion when the temperature is increased 300°F?

$L_n = L_o + (C \times \Delta T \times L_o)$
$L_n = 2 + (.00001 \times 300 \times 2)$
$L_n = 2 + .00600$
$L_n = \mathbf{2.00600″}$

Brass valve guides expand at a much slower rate than the surrounding aluminum. This provides a more consistent clearance between the valve stem and the valve guide regardless of temperature fluctuations in the engine. Brass valve guides are commonly used on both the intake and exhaust sides of the engine on heavy service applications.

Sintered Iron Valve Guides. A *sintered iron valve guide* is a separate machined valve guide insert manufactured from a powdered iron compound that is heated and compressed to form the desired shape. It has excellent lubricating capabilities and hardness to provide resistance to abrasive wear. These features make the sintered iron valve guide the highest quality valve guide in the industry and the valve guide of choice for premium engines.

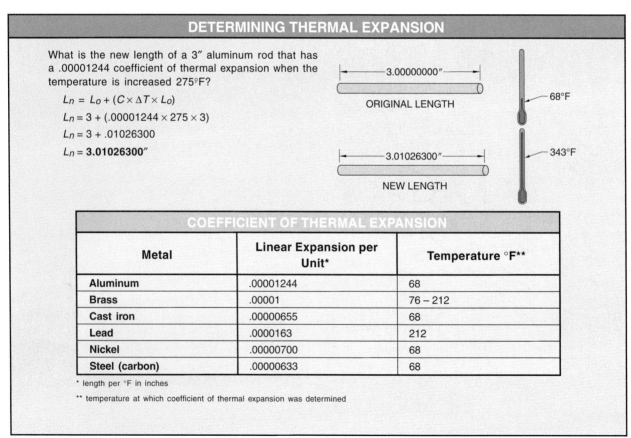

Figure 4-11. Thermal expansion which occurs when metals are exposed to heat is determined by using the coefficient of thermal expansion of the specific metal.

Sintered iron valve guides have been used in the automotive and small engine industry for many years. The lubricating capabilities can be identified by viewing the valve guide running surface under a microscope. Sintered iron has microscopic pockets that provide an additional reservoir of oil to help maintain the proper amount of lubrication. Most engines with sintered iron valve guides include valve stem oil seals to help control oil consumption by wiping the valve stem during each valve motion. The oil seal leaves a small amount of oil on the surface of the stem to provide adequate lubrication.

VALVE SEATS

The valve seat mates with the valve face to seal the combustion chamber. In addition to the sealing function, the valve seat also removes a significant amount of the heat away from the valves. Two common types of valve seats are the integrally machined valve seat and the valve seat insert. See Figure 4-12.

An *integrally machined valve seat* is a machined portion of the cylinder block that provides the sealing surface for a valve. This seat is machined directly into a cast iron cylinder block and is commonly used for intake valves. Because the intake valve and valve seat operate at a substantially lower temperature, an integrally machined valve seat in a cast iron cylinder block provides the required service and durability.

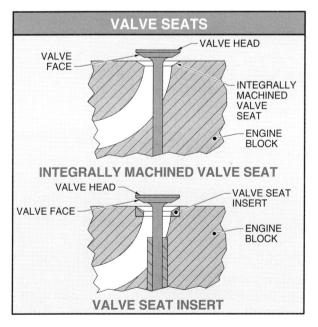

Figure 4-12. Valve seat inserts are commonly used for exhaust valves because of their resistance to heat and durability.

A *valve seat insert* is a separate machined engine component pressed into the cylinder block that provides the sealing surface for a valve. Valve seat inserts are used for both intake and exhaust valves to provide additional durability and accommodate any increased temperatures. Most valve seat inserts are made from powdered metal materials in a process similar to that used for sintered iron valve guides. Valve seats used in Briggs & Stratton engines are commonly hardfaced with Cobalite™.

Valve seat insert materials are selected based on a similar coefficient of thermal expansion as the cylinder block. Many different valve seat insert materials have been used over the years. Recently, a sintered iron material has been developed that more closely matches expansion characteristics of the cast aluminum alloy used for Briggs & Stratton cylinder blocks. This has virtually eliminated valve seat expansion failures due to manufacturing tolerance and process quality problems.

Valve seat inserts are installed in the cylinder block using a press fit. A *press fit* is a method of attaching two mating components where one component is pressed into a machined hole having a slightly smaller diameter than the component inserted. Valve seat inserts are commonly .003″ – .005″ larger than the machined hole in the cylinder block. The pressed fit retains the valve seat insert in position with a constant force. This force is necessary as small engines have significant structural responses to heat generated in the combustion process. Both integrally machined valve seats and valve seat inserts can be re-surfaced during engine service.

PISTONS

The piston acts as the movable end of the combustion chamber and must withstand pressure fluctuations, thermal stress, and mechanical load. Piston material and design contribute to the overall durability and performance of an engine. Most pistons are made from die- or gravity-cast aluminum alloy. Cast aluminum alloy is lightweight and has good structural integrity and low manufacturing costs. The light weight of aluminum reduces the overall mass and force necessary to initiate and maintain acceleration of the piston. This allows the piston to utilize more of the force produced by combustion to power the application. Piston designs are based on benefits and compromises for optimum overall engine performance.

Piston Design

Pistons are designed with features which perform specific functions during engine operation. The piston head receives the majority of the initial pressure and force caused by the combustion process. The piston pin area is exposed to a significant amount of force due to rapid directional changes. It is also subjected to thermal expansion caused by the transfer of heat from the head to the body of the piston. The piston pin area is subject to more thermal expansion than other areas of the piston. This occurs from the thermal expansion properties of cast aluminum alloy and the mass in the piston pin area.

Some pistons are cast and machined at the factory into a cam ground (elliptical shape). An *elliptical shape* is an oval shape in which one-half is a mirror image of the other half. These piston shapes provide an advantage in conforming to the ever-changing dimensions of the cylinder bore. The piston is designed to be an elliptical shape when cold. See Figure 4-13. As the engine reaches operating temperature, the piston pin bore area expands more than other thin areas of the piston. At operating temperature, the piston shape becomes a circular shape. The circular shape matches the cylinder bore for combustion efficiency.

Some pistons are designed with a taper, with the smallest diameter of the taper at the piston head. The taper shape compensates for thermal expansion and thermal growth. *Thermal growth* is the increase in size of a material when heated, with little or no change back to original dimensions. The taper design allows the piston to move freely in the cylinder bore regardless of the heat applied to the piston head.

Some Briggs & Stratton engines use a barrel-shaped piston skirt. The barrel shape provides a smoother transition during directional changes of the piston. The piston rolls into the cylinder wall when changing direction at the end of a stroke. This reduces noise, spreads the force of the directional change across a greater surface, and reduces side loading on the piston skirt.

Some piston designs have the piston pin offset from center in the piston. The proper orientation of the piston pin offset is marked by a notch or an arrow on the piston head. The mark on all Briggs & Stratton pistons should be facing or closest to the flywheel on all one- and two-cylinder engines. See Figure 4-14. The offset piston pin design offers a quieter running engine by reducing piston wobble and related noise. This results in truer linear movement of the piston in the cylinder bore.

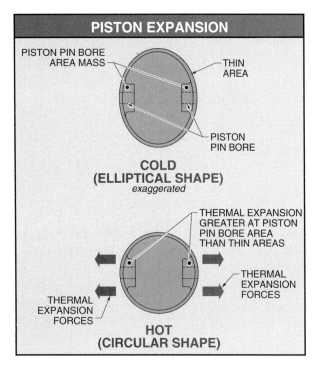

Figure 4-13. An elliptical piston shape conforms to the cylinder bore by changing to a circular shape when reaching operating temperature.

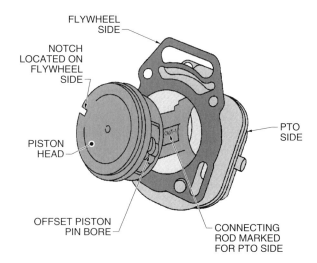

Figure 4-14. A mark on the piston head indicates the side facing or closest to the flywheel for proper orientation of the piston pin offset.

In an engine operating at 3600 rpm for 1000 hours, the crankshaft completes 216,000,000 revolutions.

Each piston design must have a provision for returning oil to the oil reservoir and the crankcase. During operation, a significant amount of oil is accumulated in the piston oil ring groove. This oil is returned to the reservoir through piston windows or through a machined channel near the piston pin.

Piston windows are a series of small holes machined into the oil ring groove surface of the piston. The oil ring collects excess oil from the cylinder bore. Piston windows allow oil in the oil ring groove to drain into the oil reservoir. See Figure 4-15.

Another common method used to return oil to the oil reservoir is through a machined channel near the piston pin. Oil collects in the rear of the oil ring groove and is routed back to the oil reservoir through the channel ending at the piston pin. This provides a path for oil to return to the oil reservoir along the outside surface of the piston when the machined channel is exposed to the oil reservoir at BDC.

Ring Grooves

Ring grooves are machined to specific tolerances into the piston, creating a three-sided machined surface. *Ring lands* are the two parallel surfaces of the ring groove which function as the sealing surface for the piston ring. Ring lands also function as the running surface for the side of the corresponding piston ring.

The clearance between the piston ring and the ring lands is critical for proper rotation, flexing, sealing of piston rings, and routing of combustion gases. See Figure 4-16. The depth of the ring groove must provide sufficient clearance for the piston ring to compress into the ring groove without contacting the piston.

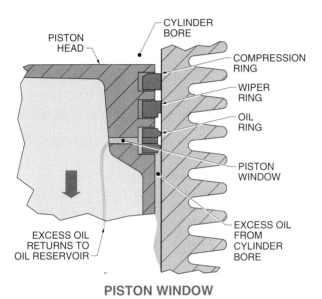

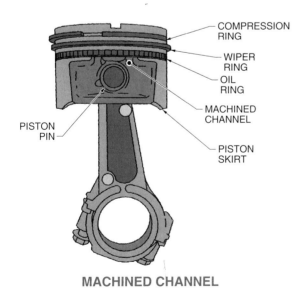

Figure 4-15. Excess oil from the cylinder bore is routed back to the oil reservoir through piston windows or a machined channel near the piston pin.

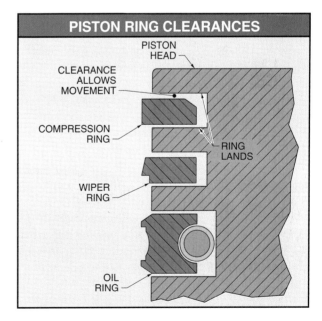

Figure 4-16. The clearance between the piston ring and ring lands allows the required piston ring movement to seal the combustion chamber and to route combustion gases.

Piston Dynamics

The piston head is exposed to over 500 psi when the engine is operating under load. The force differentials caused by the expanding combustion gases and the flame front crossing the piston head can reach two to three times this force. The piston is also exposed to relatively high temperature fluctuations during operation. The temperature of the initial flame front during combustion exceeds 3000°F. Although the piston is subjected to this temperature for a minute amount of time, the thermal stress and expansion of the piston head are significant.

In addition to the forces and thermal fluctuations incurred by the piston, the piston changes direction in the cylinder bore 120 times per sec at 3600 rpm. The changing of direction, with its inherent acceleration of mass from a static state, causes variable forces at the piston pin connection. The design, material selection, and manufacturing of a piston considers these operating conditions.

The piston is cooled by the contact of crankcase oil to the underside of the piston head and by contact with the piston rings and cylinder wall. The underside of the piston head is designed to remain open, allowing crankcase oil to contact the piston to remove combustion heat. Piston ring contact with the cylinder wall also transfers heat from the piston. The heat is then dissipated from the engine through cooling fins on the outside of the engine block.

Piston Surface Treatment

The surface material of a piston affects the overall performance and durability of an internal combustion engine. In early internal combustion engines, all cylinder bores were made from some variation of cast iron. Aluminum alloy used for the piston offered compatibility, durability, and a high strength to weight ratio. Cast iron and aluminum alloy were dissimilar metals, and the aluminum alloy piston became the standard of most engine manufacturers.

When Briggs & Stratton introduced the first aluminum alloy engine for consumer use in 1953, one of the major engineering challenges was material selection for the piston. Use of an aluminum alloy piston in an aluminum alloy cylinder bore posed a problem, as metals with similar hardness and other properties tend to adhere to each other when thermally stressed. This can cause severe scuffing, metal transfer, and eventual engine seizure. Plating of aluminum alloy pistons for use in aluminum alloy cylinder bores eliminated the problem. Aluminum alloy pistons are commonly plated with a chromium alloy to provide a hardened piston running surface. An unplated aluminum alloy piston can provide adequate performance in cast iron cylinder bores. A newer process, iron plating, creates the dissimilar metal requirements of the piston and cylinder bore. Iron plating greatly reduces the environmental concerns in the chromium alloy process. See Figure 14-17.

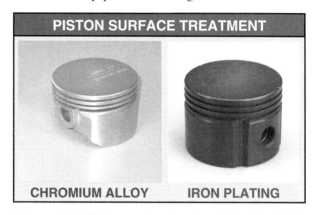

Figure 4-17. Piston surface treatment is selected for engine performance and economy.

Piston Rings

Piston rings seal the combustion chamber, transferring heat to the cylinder wall and controlling oil consumption. A piston ring seals the combustion chamber through inherent and applied pressure. *Inherent pressure* is the internal spring force that expands a piston ring based on the design and properties of the material used. Inherent pressure requires a significant force needed to compress a piston ring to a smaller diameter. Inherent pressure is determined by the uncompressed or free piston ring gap. *Free piston ring gap* is the distance between the two ends of a piston ring in an uncompressed state. See Figure 4-18. Typically, the greater the free piston ring gap, the more force the piston ring applies when compressed in the cylinder bore.

A piston ring must provide a predictable and positive radial fit between the cylinder wall and the running surface of the piston ring for an efficient seal. The radial fit is achieved by the inherent pressure of the piston ring. The piston ring must also maintain a seal on the piston ring lands.

In addition to inherent pressure, a piston ring seals the combustion chamber through applied pressure. *Applied pressure* is pressure applied from combustion gases to the piston ring, causing it to expand. Some piston rings have a chamfered edge opposite the running surface. This chamfered edge causes the piston ring to twist when not affected by combustion gas pressures.

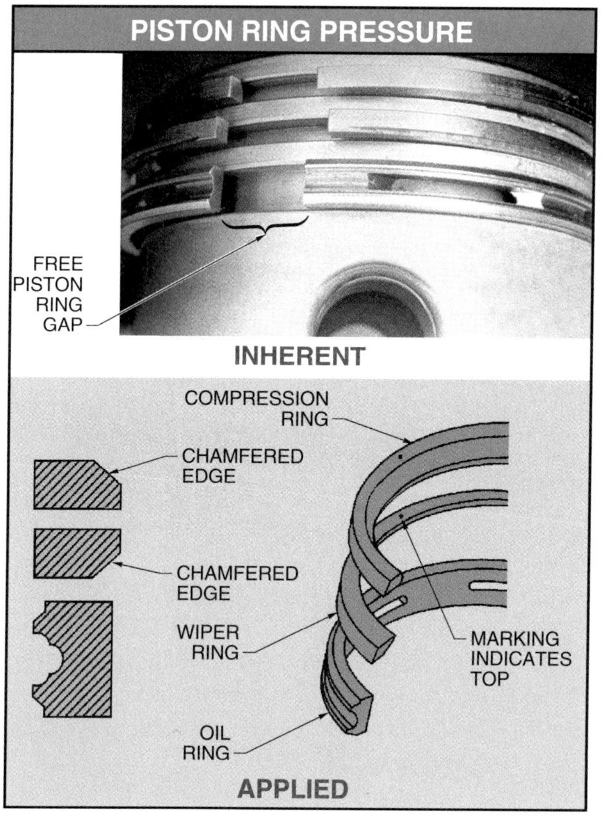

transfers 70% of the combustion chamber heat from the piston to the cylinder wall. Most Briggs & Stratton engines use either taper-faced or barrel-faced compression rings. A *taper-faced compression ring* is a piston ring that has approximately a 1° taper angle on the running surface. This taper provides a mild wiping action to prevent any excess oil from reaching the combustion chamber.

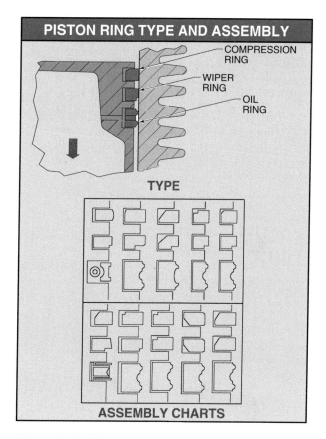

Figure 4-18. Piston ring pressure is the result of internal spring force (inherent) and combustion gasses (applied).

Another piston ring design consideration is cylinder wall contact pressure. This pressure is usually dependent on the elasticity of the piston ring material, free piston ring gap, and exposure to combustion gases. All piston rings used by Briggs & Stratton engines are made of cast iron. Cast iron easily conforms to the cylinder wall. In addition, cast iron is easily coated with other materials to enhance its durability. Care must be exercised when handling piston rings, as cast iron is easily distorted. Piston rings commonly used on small engines include the compression ring, wiper ring, and oil ring.

Currently, most piston ring design improvements are dictated by stricter emission control standards. Each piston ring design and ring set are specifically designed to maximize efficiency and reduce oil consumption. Engine manufacturers provide piston ring assembly charts to ensure the proper ring type and position. See Figure 4-19.

Figure 4-19. Piston rings must be installed per manufacturer specifications for proper performance.

A *barrel-faced compression ring* is a piston ring that has a curved running surface to provide consistent lubrication of the piston ring and cylinder wall. This also provides a wedge effect to optimize oil distribution throughout the full stroke of the piston. In addition, the curved running surface reduces the possibility of an oil film breakdown due to excess pressure at the ring edge or excessive piston tilt during operation.

Compression Ring. The compression ring is the top or closest ring to combustion gases and is exposed to the greatest amount of chemical corrosion and the highest operating temperature. The compression ring

Wiper Ring. The wiper ring, sometimes called the scraper ring, Napier ring, or back-up compression ring, is the next ring away from the cylinder head on the piston. The wiper ring provides a consistent thickness of oil film to lubricate the running surface of the compression ring. Most wiper rings in Briggs

& Stratton engines have a tapered angle face. The tapered angle is positioned toward the oil reservoir and provides a wiping action as the piston moves toward the crankshaft.

The taper angle provides contact that routes excess oil on the cylinder wall to the oil ring for return to the oil reservoir. A wiper ring incorrectly installed with the tapered angle closest to the compression ring results in excessive oil consumption. This is caused by the wiper ring wiping excess oil toward the combustion chamber.

Oil Ring. An oil ring includes two thin rails or running surfaces. Holes or slots cut into the radial center of the ring allow the flow of excess oil back to the oil reservoir. Oil rings are commonly one piece, incorporating all of these features. Some one-piece oil rings utilize a spring expander to apply additional radial pressure to the piston ring. This increases the unit (measured amount of force and running surface size) pressure applied at the cylinder wall.

The oil ring has the highest inherent pressure of the three rings on the piston. Some Briggs & Stratton engines use a three-piece oil ring consisting of two rails and an expander. The oil rings are located on each side of the expander. The expander usually contains multiple slots or windows to return oil to the piston ring groove. The oil ring uses inherent piston ring pressure, expander pressure, and the high unit pressure provided by the small running surface of the thin rails.

Piston Ring Dynamics

Proper ring dynamics requires the presence of an oil film between the piston ring running surface and the cylinder wall. During normal operation, piston rings should not touch the cylinder wall. A small section of the piston ring running surface actually contacts the cylinder wall only during engine break-in. *Break-in* is the process that causes the running surfaces of piston rings and the surface of the cylinder bore to conform to each other. Friction created by piston ring to cylinder wall contact helps remove small imperfections to develop a uniform seal.

Piston Ring Inertia. Piston rings have mass and therefore have inertia. When the piston changes direction, piston rings move within the ring groove and resist piston motion. As the piston moves away from the crankshaft, piston rings are forced against the ring land surface closest to the piston pin. As the piston moves in the opposite direction, the piston rings are forced against the ring land surface closest to the piston head. This action provides additional sealing capabilities between the piston rings and the ring land surfaces.

The inertia qualities of the piston rings cause a positive flat seal against the corresponding ring land surfaces to ensure that neither the compressed charge nor combustion gases can escape between the piston ring and ring groove. A combination of inertia, radial pressure, and conformability of piston rings ensures the required seal for combustion efficiency.

Piston Ring Twist. Some Briggs & Stratton piston rings incorporate positive or negative twists featuring a tapered running surface in the piston ring structure. These twists cause the piston ring to provide a seal against the ring land during non-combustion strokes of the piston. The axial position of the piston ring within the piston ring lands is determined by combustion gas pressure and inertia and friction forces of the piston ring. The position of the piston ring can alternate between the ring land surfaces depending on the motion of the piston and combustion gas pressure. See Figure 4-20.

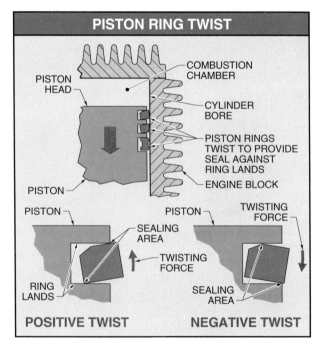

Figure 4-20. Piston ring twist causes piston rings to provide a seal against the ring land during non-combustion strokes of the piston.

The tapered leading edge of the running surface prevents any compression gases or oil from entering the ring land. This helps to maintain acceptable exhaust gas emissions on the engine. The tapered leading edge of the piston ring also provides a mild wiping action during piston motion in the direction of the crankshaft to reduce oil consumption.

All twisted piston rings have a small but necessary tapered running surface. The tapered running surface angle ranges from 15′ to 1°-15′. This angle is usually not discernible by the naked eye. For proper installation, markings on the piston rings indicate the top. The twist is added to the piston ring so when in motion, it twists to force the angle of the running surface firmly against the cylinder wall. The wiping action of the leading edge of the piston ring provides additional oil control and increases the unit pressure at the piston ring wall interface.

A piston ring with a negative twist causes the piston ring to rock slightly in the ring lands, forcing the inside top corner of the piston ring to contact the ring land surface closest to the piston head. The same action causes the outside lower corner of the piston ring to contact and seal against the ring land surface closest to the piston pin. This movement provides better seal integrity on non-combustion piston strokes.

A piston ring with a positive twist performs in approximately the same fashion with the twist in the reverse direction. The bottom inside corner of the piston ring contacts and seals on the ring land surface closest to the piston pin and the top outside corner of the piston ring contacts and seals on the ring land surface closest to the piston head.

Piston Ring Rotation. In four-stroke cycle small engines, all piston rings rotate during operation. Piston ring rotation has long been associated with the cross-hatched pattern in a cylinder bore after a standard honing operation. *Honing* is the process of using a hone with rigid Carborundum stones rotated in the cylinder bore to remove small surface irregularities and any glazing. See Figure 4-21. Piston ring rotation has also been a concern of the small engine service technician because of the ramifications of having all the piston ring end gaps aligned on the piston. Piston ring rotation is not a direct result of cross-hatched pattern, but is caused by a combination of variables including vibration, piston ring land clearance, and engine condition.

Vibration in the engine is the primary cause of piston ring rotation. Oscillating vibratory waves pass through the cylinder in varying strengths as the engine operates. Just as a loose bolt rotates in its threaded hole until it falls out, the same vibrations cause the unattached piston rings to rotate around the piston.

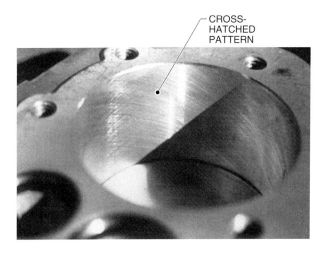

Figure 4-21. The cross-hatched pattern in a cylinder bore has no direct effect on piston ring rotation during engine operation.

Along with vibration, piston ring land clearance is instrumental in assisting piston ring rotation by the tilting of the piston during peak combustion pressures. The tilting of the piston is caused by the initial force applied to the piston head and by the cylinder wall to piston clearance. The slight piston tilting and ring land clearance provide room for the piston rings to inch around the piston.

Piston ring rotation is critical to engine durability. Without rotation, inherent gas pressures and inertia can cause piston ring ends to wear into the piston ring lands. This leads to rapid piston and piston ring failure due to uneven heating of the piston ring.

If piston ring end gaps are aligned on the piston, it is usually caused by the deformation of the cylinder bore or simply by chance. Piston rings are designed to have a slightly higher radial pressure at the piston ring end gap. A cylinder bore commonly deforms into an asymmetrical shape from excessive heat or load. In an asymmetrical cylinder bore, piston rings rotate until they reach the greatest cylinder bore diameter. The piston rings then remain in the same location in the piston ring lands.

Piston ring alignment and poor performance are often caused by cylinder bore deformation and the resulting localized heat rather than piston ring alignment alone. Normal, acceptable leakage of compression gases is measurably greater by volume and pressure than leakage caused by any alignment of (unworn) piston rings. Alignment of the piston ring end gaps under normal conditions does not cause an immediate decrease in performance or an increased oil consumption. Excessive (heat-induced) deformation of the cylinder bore can exceed conformability characteristics of the piston rings, leading to cylinder bore hot spots.

The Briggs & Stratton Model F was produced in 1921 and 1922 and featured a cooling fan and spiral cooling fins.

Piston Ring Materials

Piston ring material is selected on the basis of the operating environment and cylinder material thermal conductivity. Piston rings are commonly made of gray or ductile cast iron alloy. Piston ring material characteristics can be altered with surface treatment by the addition of a different material added to the entire surface of the piston ring. This provides better break-in performance and inhibits corrosion during storage. The break-in period can range from fast (1 hour – 8 hours) to slow (8 hours – 25 hours). Surface coatings commonly used on piston rings include chromium plating, phosphating and ferroxiding, and nitriding.

Chromium Plating. *Chromium plating* is a piston ring surface treatment that adds a layer of chromium to increase hardness and durability. Chromium used on piston rings is much harder than the cast iron piston ring material but softer than the chromium used on an automobile bumper. All chromium-plated piston rings used in Briggs & Stratton engines are barrel-faced. Chromium plating and the increased contact pressures of barrel-shaped piston rings provide enhanced oil control for the engine.

Under a microscope, chromium-plated surfaces reveal multiple cracks or fissures that act as oil reservoirs. These oil reservoirs increase lubrication on the piston ring surface to provide a cooler cylinder bore during operation. In addition to providing lubricating oil, the cracks and fissures provide a spot for cylinder wall debris to accumulate. By providing a place for debris to accumulate, overall engine durability is enhanced. Fissures that remain unfilled continue to provide lubrication to the cylinder bore. This permits a shorter break-in period. However, chromium-plated rings still require a relatively long break-in period. Some cylinder wall debris is normal for an engine, but accumulation can lead to premature cylinder wall wear.

The total surface area of the piston ring is important to the overall wear patterns between cylinder bore and piston ring. With no abrasive ingestion, the harder material (chromium) wears more slowly than the cylinder bore. *Abrasive ingestion* is a cause of engine failure through the undesirable introduction of abrasive particles into a small engine. Under normal operating conditions, the wear ratio between the cylinder bore and the piston ring should be equal. A *wear ratio* is a comparison of the rate of material loss from the piston ring and the cylinder bore over an extended period of time. The preferred wear ratio on a Briggs & Stratton engine is 1 : 1. This means that in an engine exhibiting no abrasive ingestion, piston rings wear at the same rate as the cylinder bore. The total piston ring running surface area is appreciably less with a chromium-plated barrel-faced piston ring. This smaller area, with its increased hardness from the chromium plating, results in a 1 : 1 wear ratio for the engine.

The piston ring manufacturer does not suggest or approve the use of chromium-plated rings in cylinders that exhibit wear due to abrasives. By installing chromium-plated rings into an abrasive-filled cylinder bore, the desired wear ratio is altered and the chromium-plated rings may show signs of premature wear.

Briggs & Stratton does not recommend glaze breaking. *Glaze breaking* is the process of using a flexible hone consisting of small Carborundum stones rotated in the cylinder bore to remove glazing and to obtain the desired surface appearance. The small Carborundum stones on flexible mounts can cause increased cylinder bore distortion by following existing cylinder bore surface irregularities. In addition, the process can leave residual abrasives in the cylinder bore.

Installation of a cross-hatched pattern when installing chromium-plated piston rings in a worn cylinder bore is generally not recommended. However, the process can be acceptable if the cylinder bore is honed with rigid Carborundum stones and thoroughly cleaned with hot water and detergent only. Insufficient cleaning of the cylinder bore before assembly can leave residual abrasives and Carborundum from the hone. This causes a dramatic increase in the volume of debris on the cylinder wall, which rapidly fills the fissures in the piston ring surface critical for proper engine operation.

Phosphating and Ferroxiding. Phosphating and ferroxiding are surface treatment processes for piston rings applied using a thermochemical process. *Phosphating* is a piston ring surface treatment process that changes the outer surface of the piston ring to phosphate crystals. Phosphate crystals are much softer than the cast iron base material of the piston ring and provide a faster break-in period. *Ferroxiding* is a piston ring treatment process that changes the outer surface of the piston ring to iron oxide. Iron oxide is much harder than the cast iron base material of the piston ring and assists break-in by acting as an abrasive on harder cylinder bore surface materials. Both phosphating and ferroxiding are surface treatment processes and peripheral coatings. A *peripheral coating* is a coating that is applied only to the wear surface of the piston ring. It is used to lengthen the service life and to provide maximum running surface protection to the piston ring and cylinder bore.

Nitriding. *Nitriding* is a piston ring surface treatment process that uses a thermal process in which nitrogen and some carbon are absorbed into the piston ring surface. This provides additional wear resistance on the piston ring. The nitriding process is only used on piston rings with a base material of high chromium steel (13% – 18%) which increases the cost. In some cases, nitriding piston rings is a less costly process than manufacturing a chromium-plated piston ring. In most cases, nitriding provides similar performance and durability as chromium-plated piston rings. However, a nitrided piston ring can only be used in conjunction with a cast iron cylinder bore.

 A valve spring only assists in the closing of a valve. Force on the valve head from combustion pressure adds significantly to the sealing properties of the valve.

Piston Ring Break-In

During the break-in period of a piston ring, the piston ring and cylinder bore wear at an accelerated rate and conform to a mutual shape and size. *Break-in period* is the period of time required for the running surfaces of piston rings and the surface of the cylinder bore to conform to one another after initial startup. In the past, the engine break-in period was very important to the overall life and durability of the engine. The break-in period required has changed over the years with improved piston ring materials and designs. The break-in period now is short in comparison with that of engines of the past. Aluminum cylinder bore engine piston rings break-in faster than those used on cast iron cylinder bores.

During the break-in period, the piston rings and cylinder bore wear rapidly to remove any rough edges on the piston ring running surface and cylinder wall. In general, there is no special engine operation procedure required during the break-in period. Break-in is accomplished at any speed above idle and may occur faster if the engine is operated at varying loads and speeds. However, break-in occurs at an acceptable rate if the engine is operated at slightly less than top no-load speed with or without a moderate load. Combustion pressures at this speed are sufficiently high to cause piston rings to conform to the cylinder wall. It is recommended that an engine not be operated continuously at full load during the initial hours of operation. This can lead to permanent deformation of the cylinder bore.

Piston Ring Installation

Proper installation of a piston ring is critical for optimum engine performance. Any improper stressing or bending of the ring during installation can cause serious problems. All piston rings should be installed using a ring expander. See Figure 4-22. A *ring expander* is a tool that expands the piston ring uniformly and causes no permanent distortion to the piston ring. Use of hands to install pistons rings is not recommended. The piston ring should be expanded only enough to slide over the head of the piston. A *ring compressor* is a tool that is used to compress piston rings for installation in the cylinder bore. Distortion of the piston ring dimensions can remove the intentional positive or negative twist causing increased oil consumption, scuffing, and/or eventual failure.

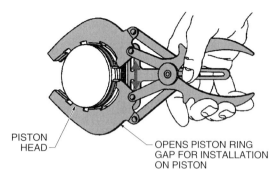

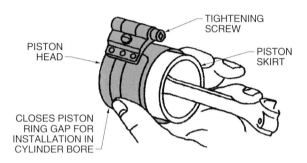

Figure 4-22. A piston ring expander prevents improper stressing or bending of the piston ring during installation. A piston ring compressor is used to install the piston and piston rings into the cylinder bore.

CYLINDER BORE

Cylinder bores on Briggs & Stratton engines are cast aluminum alloy or cast iron. Aluminum alloy is the most common material used for engine construction as well as cylinder wall material. Aluminum alloy is a malleable and durable material that has high heat dissipation characteristics. Although aluminum alloy also exhibits some thermal expansion, it most commonly returns to its nominal size when cooled.

The surface of an aluminum cylinder bore contains microscopic convolutions. A *convolution* is an irregularly-shaped pocket that acts as a small reservoir for lubricating oil. In addition to lubricating piston rings when passing through the cylinder bore, these convolutions tend to capture some of the small particles of debris that may be ingested by the engine during operation. Cylinder wall finish quality is required to control oil consumption and maintain optimum operating performance.

A cast iron cylinder block or cylinder sleeve provides the harder and more durable cylinder surface required in more demanding applications. Cast iron offers less heat dissipation and a lower coefficient of thermal expansion than aluminum. Cast iron cylinder blocks provide greater durability under load because the structural integrity of the cylinder is less prone to asymmetrical expansion. The surface of a cast iron cylinder bore is more porous and has more convolutions for lubrication than an aluminum alloy cylinder bore.

Cylinder Bore Design

The cylinder bore design selected is based on maintaining the integrity of cylinder dimensions during engine operation. The cylinder block and bore design accounts for the probability of asymmetrical or unilateral thermal expansion. Cooling fins cast into the cylinder block are located and sized based on the specific thermal characteristics of a given engine.

Cylinder block and bore temperatures vary greatly based on load and the location of the temperature measurement. In most cases, cooling fins are located in the areas of the cylinder block where temperature measurements are most extreme. The more cooling fins in a specific area, the greater the likelihood of localized heating in the engine block design. Cooling fin location and number is also based on the location of various components, such as valves and exhaust systems. Cooling fins increase heat radiation surfaces of the engine to efficiently and quickly remove the heat of combustion from the engine. See Figure 4-23.

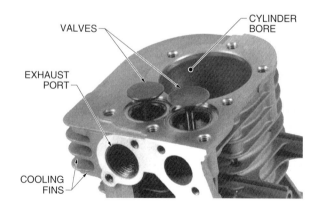

Figure 4-23. The size and quantity of cooling fins located on the cylinder block is determined by localized heating and engine components, such as valves and exhaust systems.

Some engines incorporate special asymmetrical cooling fins and head bolt locations in the cylinder block design. As thermal expansion cannot be completely eliminated, the asymmetrical location of holes and cooling fins helps control where cylinder block expansion occurs. Proper cooling fin location promotes uniform cylinder block and bore expansion to minimize distortion of the sealing components in the compression system.

Cylinder Bore Finish

Cylinder bores commonly have a cross-hatched pattern from the machining process used on a new cylinder bore. As cast aluminum alloy or cast iron is machined through the honing process, the cutting action of the tool tends to smear metal as it is removed from the surface. This smearing action distributes metal throughout the cylinder bore and partially or completely occludes (closes up) the small convolutions desired in the cylinder bore surface. The last .001″ is removed from the surface using slower strokes of the hone. The result is a cleaning action on the surface of the cylinder bore to remove any metal from occluded convolutions. The cross-hatched pattern left behind on the cylinder bore surface is a series of linear, curved scratches or asperities approximately .0005″ deep that intersect at approximately a 45° angle.

Many Briggs & Stratton small engines are manufactured with a Diamond Bore™ cylinder bore machining process. A mirror finish is left on the cylinder bore surface after sizing and shaping. There is no cross-hatched pattern left in the cylinder wall. Piston rings installed in a Diamond Bore™ cylinder bore rotate the same as piston rings in a conventionally-honed cylinder.

CRANKCASE BREATHER SYSTEM

The crankcase breather system maintains pressure in the crankcase at less than ambient pressure to assist in the control of oil consumption. The crankcase breather system is often not included as a component of a compression system. However, a compression loss related to the cylinder bore or to piston ring failure causes the crankcase breather to be rendered useless. Escaping combustion gases entering the crankcase eliminate the existing negative pressure and cause excessive oil consumption. In addition, as combustion gases escape the combustion chamber by the piston rings or cylinder wall, oil temperature in the oil reservoir increases dramatically. Overheating of oil commonly results in viscosity loss and increased oil loss by vaporization as hot combustion gases enter the crankcase.

COMPRESSION RELEASE SYSTEM

Some small engines incorporate a compression release system to decrease operator effort when pulling a rewind starter. A *compression release system* is a system that relieves excess pressure during the compression event by allowing a small amount of compressed gas to be released through the muffler or carburetor. The compression release system lifts either the exhaust or the intake valve slightly off its seat during the compression event. This releases pressure and reduces the force required to pull the starter rope or load on the starter motor. A compression release is used on most Briggs & Stratton engines and does not affect engine performance above engine starting rpm.

Compression release systems commonly used on Briggs & Stratton engines include the Easy Spin® compression release system and a mechanical compression release system. The Easy Spin® system is the most common compression release system. It has a small projection machined into the camshaft lobe. The small projection raises the intake or exhaust valve .003″ – .005″ off the valve seat just before maximum pressure of the compression event is reached. This releases a small amount of the compression pressure to reduce effort when starting the engine. See Figure 4-24.

After the engine is started, the Easy Spin® system has little effect on engine performance. The short lift of the valve results in a limited amount of time the valve is actually open. As engine speed increases, the amount of time the valve is open decreases proportionately. The minor power loss from the Easy Spin® compression release system is undetectable to the operator.

A simple check for compression can be made by spinning the flywheel counterclockwise (flywheel side) against the compression stroke. A sharp rebound indicates satisfactory compression. Slight or no rebound indicates poor compression.

96 SMALL ENGINES

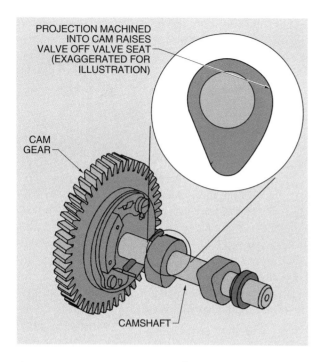

Figure 4-24. The Easy Spin® compression release system uses a projection machined into the camshaft lobe to lift the intake or exhaust valve off the valve seat just before the maximum compression pressure is reached.

A *mechanical compression release system* is a compression release system that incorporates a weighted lever or arm attached to the camshaft or cam gear. During starting rpm or when the engine is at rest, the arm on the camshaft or cam gear lifts the valve off its seat at approximately the same time as the Easy Spin® system. The main difference between the systems is that when the engine using the mechanical compression release system reaches approximately 800 rpm, the centrifugal force on the arm forces it to move away from the center of rotation. See Figure 4-25. When this occurs, the portion of the arm that actuates the valve is moved to a position where it no longer touches the valve stem or tappet. As engine speed increases, the mechanical compression release does not affect engine operation.

In the future, stricter emission control regulations for small engines are going to result in the phasing out of the Easy Spin® system. Although this system has been used for many years, the small amount of slightly compressed and unburned air and gasoline vapor released from the combustion chamber affects overall engine emissions.

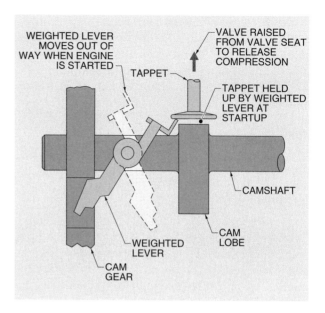

Figure 4-25. A mechanical compression release system incorporates a weighted lever that lifts the valve off the valve seat during engine startup.

VALVE RESURFACING SERVICE PROCEDURES

Although the practice of resurfacing valve faces is common in the automobile industry, resurfacing valve faces on Briggs & Stratton engines is strongly discouraged. The valve head margin on a new Briggs & Stratton valve is $1/32''$. Removal of material from the face of the valve decreases the valve head margin. This can leave the valve head with insufficient material to withstand thermal stresses of an operating engine. In addition, the decrease in the material in the valve head can cause an overheated valve and result in exceeding the yield point of the material. The *yield point* of a material is the limit at which the material can be exposed to thermal and/or mechanical stress and still return to its original size and chemical composition. Exceeding the yield point can lead to a distortion of the sealing face of the valve.

Approximately .03″ of Cobalite™ alloy is deposited on valve heads. Resurfacing removes this coating and decreases efficiency and durability of the valve interface.

 When manufacturing a cast aluminum alloy cylinder block with a cast iron cylinder sleeve, molten aluminum is poured around the cylinder sleeve.

FUEL SYSTEM

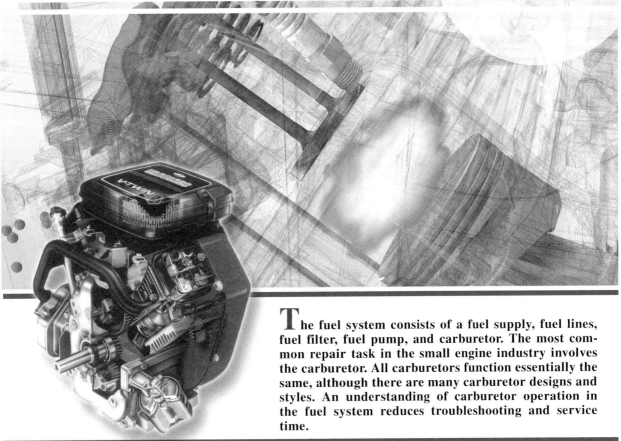

The fuel system consists of a fuel supply, fuel lines, fuel filter, fuel pump, and carburetor. The most common repair task in the small engine industry involves the carburetor. All carburetors function essentially the same, although there are many carburetor designs and styles. An understanding of carburetor operation in the fuel system reduces troubleshooting and service time.

FUEL

The most common fuel used in small engines is gasoline. *Gasoline* is a liquid fossil fuel derivative that primarily consists of the elements hydrogen (H) and carbon (C). A *fossil fuel* is a fuel derived from previously living things that have been preserved in a mineralized or petrified state. Gasoline is the most common fuel used for internal combustion engines and was originally an undesirable by-product of the crude oil refining process. The flammable vapors from gasoline posed safety concerns to early refiners. Before the advent of the internal combustion engine, gasoline was primarily used as lamp fuel or as a cleaning agent. In the early 1900s, demand for the internal combustion engine made the refining of crude oil into gasoline a growing, profitable venture.

A fuel can designed for consumer use may have a spring-mounted valve in the nozzle that allows fuel to flow with the applied weight of the can.

97

Although gasoline is commonly available, it is not composed of a single substance and varies greatly in content. Gasoline is a complex blend of chemicals mixed together to provide a predictable performance when used in specific applications. Gasoline is a hydrocarbon (HC) based fuel that is derived from the elements hydrogen and carbon. A *hydrocarbon (HC) molecule* is a molecule held together by a loose bond between hydrogen and carbon atoms that occurs naturally in all fossil fuels. See Figure 5-1. The looser the bond, the easier it is for energy to be released from an HC molecule. The most efficient method for releasing potential chemical energy in gasoline is through burning in the combustion process.

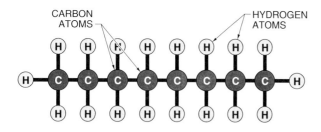

HYDROCARBON MOLECULE (GASOLINE)

Figure 5-1. Gasoline is derived from a hydrocarbon molecule formed by a loose bond between hydrogen and carbon atoms.

Combustion and energy release

The combustion process is similar to a rock rolling down a hill. The rock releases energy while rolling down the hill toward the stronger gravitational force of the earth. Energy in gasoline is released as heat during combustion in a similar manner. The exchange of electrons during the combustion process releases energy as the molecule moves from a weak bond (similar to the rock on the top of the hill) to a stronger bond (similar to the rock at the bottom of the hill) resulting in the release of energy. The heated expanding gases from combustion are used to move the piston.

Combustion Chemistry

Combustion is the rapid, oxidizing chemical reaction in which a fuel chemically combines with oxygen in the atmosphere and releases energy in the form of heat. Combustion produces force that can be used to do work. Most gasoline molecules contain more elements than hydrogen and carbon. For simplicity, a generic gasoline molecule consisting of these elements is identified by the chemical formula C_8H_{18}. The formula indicates that a generic gasoline molecule contains 8 atoms of carbon (C_8) and 18 atoms of hydrogen (H_{18}). See Figure 5-2.

Ignition of the charge causes an exchange of elements which releases heat energy. To release energy during combustion, weaker chemical bonds must be broken before new bonds can be formed. As the HC molecule fragments from intense heat during ignition, oxygen (O_2) atoms bond with H atoms from the fragmented HC molecule.

The joining of these atoms releases energy in the form of heat and new chemical compounds. When C_8H_{18} is combined with O_2 in a perfect ratio, only water (H_2O) and carbon dioxide (CO_2) are produced. To produce maximum heat energy, sufficient oxygen must be present during the combustion of the charge. The combustion chamber must also provide ample opportunity for the flame originating from the ignition to burn the charge completely. The energy released and rapid gas expansion from the chemical reaction provides the force to move the piston to produce engine torque.

 A 3 HP engine operating at 3600 rpm uses approximately 390 cu ft of air per hour, which enters the engine at a rate of 24 mph.

Stoichiometric Ratio

Stoichiometric ratio is the specific air-fuel ratio (by weight) of atmospheric air to fuel at which the most efficient and complete combustion occurs. Although the stoichiometric ratio is theoretically the most efficient air-fuel ratio, it is not the air-fuel ratio which produces maximum engine power. A power decrease occurs when an engine is operated with a stoichiometric ratio compared to the same engine operated with an slightly richer air-fuel mixture. A power decrease also occurs from inefficient ignition of the charge and engine design limitations at the stoichiometric ratio.

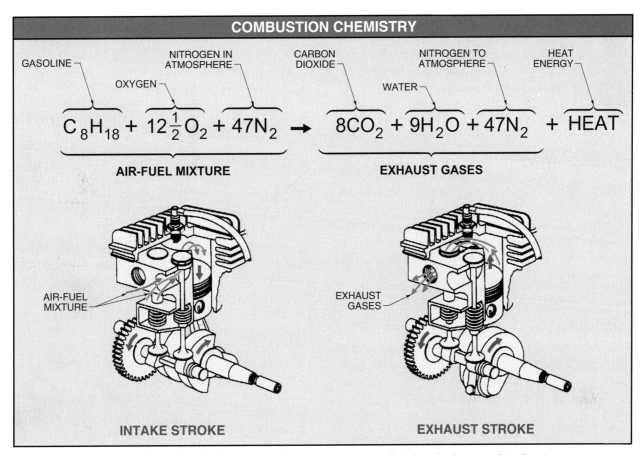

Figure 5-2. Combustion is a chemical process which converts potential chemical energy into heat energy.

At the stoichiometric ratio, the charge is more difficult to ignite and may not ignite until the engine has rotated several degrees past the optimum point of ignition. Engine design limitations influence the ability to provide a smooth transition area for the flame front to spread throughout the combustion chamber. A charge with an air-fuel ratio slightly richer in fuel than the stoichiometric ratio is easier to ignite and burns efficiently to provide ample power.

The stoichiometric ratio for a specific fuel or combination of fuels varies and is expressed as the Lambda (λ) excess air factor. The *Lambda excess air factor (λ factor)* is a numerical value assigned to represent the stoichiometric ratio of atmospheric air to any hydrocarbon fuel. A 1.0 λ factor is the theoretically perfect ratio for atmospheric air and a specific hydrocarbon fuel at sea level. For example, a 1.0 λ factor for gasoline is 14.7 : 1. The most efficient and complete combustion occurs when there are 14.7 parts atmospheric air for every 1 part fuel. Lambda factors vary for different fuels. See Figure 5-3.

| LAMBDA (λ) EXCESS AIR FACTOR ||
Fuel	Air-Fuel Ratio at 1.0 λ
Alcohol	9.0 : 1
Butane	14.3 : 1
Gasoline	14.7 : 1
Isobutane	15.4 : 1
Methane	17.2 : 1
Propane	15.7 : 1

Figure 5-3. A 1.0 Lambda (λ) factor is the theoretically perfect air-fuel ratio for the most efficient and complete combustion.

Small engines are designed to operate at less than a 1.0 λ factor with typical λ factors ranging from .6 to .8. A small engine operated at or near a 1.0 λ factor overheats from the intense heat produced by complete combustion. Operating a small engine at less than a 1.0 λ factor results in proportionately lower combustion gas temperatures and lower overall engine temperatures. However, this also produces a decrease in overall engine performance and variation in the exhaust gas emission levels. Engineers consider overall engine emissions, performance, and durability when selecting the optimal λ factor.

Engine Emissions

An engine operating at a 1.0 λ factor exhausts 12% H_2O in the form of water vapor or steam and 14% CO_2. This process is expressed in a chemical equation as $C_8H_{18} + O_2 \rightarrow CO_2 + H_2O$. However, combustion of gasoline in a typical engine involves many more chemicals and elements. Gasoline additives are required to enhance fuel quality and performance. See Figure 5-4.

GASOLINE ADDITIVES	
Additive	Function
Anti-icers	Prevent fuel from freezing in lines
Anti-oxidants	Reduce gum formation in stored gasoline
Corrosion inhibitors	Minimize corrosion in fuel system
Detergents	Reduce/remove fuel system deposits
Fluidizer oils	Control intake valve deposits
Lead replacement additives	Minimize exhaust valve seat wear
Metal deactivators	Minimize effects of metals present in gasoline

Figure 5-4. Gasoline additives are required to enhance fuel quality and performance in different applications and environments.

Atmospheric air normally contains approximately 20.95% O_2, 78.08% nitrogen (N_2), and additional elements. All of these elements become involved during combustion. Hydrocarbon molecules are broken down into H and C atoms and combine with the atmospheric O_2. Exhaust gas content is primarily determined by the amount of atmospheric O_2 used in the combustion process. An engine operated at a .6 factor (less than the stoichiometric ratio) causes an O_2 deficiency which is most responsible for the production of carbon monoxide (CO) and an increase in HC.

As the λ factor reaches or exceeds 1.0, N_2 in atmospheric air becomes a factor as the leaner air-fuel ratio creates additional heat in the combustion chamber. With additional heat, the N_2 combines with available O_2 to form oxides of nitrogen (NOx). A λ factor too low can cause excess hydrocarbons and result in a deficiency of atmospheric O_2 and the production of CO.

Carbon Monoxide. *Carbon monoxide (CO)* is a toxic (poisonous) gas produced by incomplete combustion of gasoline or other HC-based fuels. Carbon monoxide is a product of incomplete combustion and is formed when an HC molecule is broken during combustion with a deficiency of available O_2. The limited amount of O_2 available allows only one O atom to bond to a C atom, creating CO.

An engine operated with an air-fuel ratio rich in fuel (air deficient) with < .9 λ factor has a positive correlation between O_2 deficiency and CO produced. The richer the air-fuel ratio, the more CO is created. When an engine is operated at 1.0 λ factor, CO emission is very low and independent of the air-fuel ratio. The CO produced is affected by the distribution efficiency of the charge in the combustion chamber. This includes crevices and dead areas that do not allow the flame front to spread efficiently throughout the combustion chamber. See Figure 5-5. OHV engines, by design, have fewer dead areas which increases overall engine combustion efficiency.

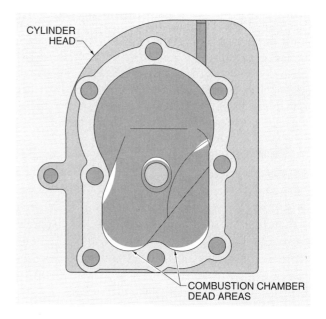

Figure 5-5. Combustion chamber dead areas that do not allow the flame front to spread through the combustion chamber are indicated by lower temperatures.

Hydrocarbon Emissions. Hydrocarbon emissions are commonly caused by incomplete combustion. Like CO, an engine operated with an air-fuel ratio rich in fuel (air deficient) with < .9 λ factor has a positive correlation between O_2 deficiency and HC produced. Other factors that increase HC emissions independent of the CO emission include the combustion chamber design and/or the presence of lubricating oil in the combustion chamber.

A common misconception is that HC emissions are caused by unburned, partially burned, or raw fuel. However, HC emissions consist primarily of methane, ethane, acetylene, and other hydrocarbons not readily oxidized at normal engine operating temperatures. High HC production is a product of an air deficiency that results in lower combustion chamber temperatures. At lower combustion chamber temperatures, it is more difficult to oxidize HC in the fuel. The H and C atoms remain bonded and are expelled through the exhaust.

As combustion temperature increases in response to a leaner (more air added) air-fuel mixture, the H and C atoms of the HC molecule eventually oxidize proportionately. Some environmental studies indicate that HC emission is an important consideration related to the presumed effects on the ozone layer in the lower portion of the atmosphere.

Oxides of Nitrogen. *Oxides of nitrogen (NOx)* is a term assigned to several different chemical compounds consisting of nitrogen (N) and oxygen (O). The "x" in NOx indicates an unknown atomic count due to the unstable nature of the compound. NOx emissions are quantified in parts per million (ppm). The three most common NOx compounds are nitric oxide (NO), nitrogen dioxide (NO_2), and dinitrogen monoxide (N_2O).

Nitric oxide (NO) is an oxide of nitrogen that is created in small amounts in nature and is somewhat toxic and colorless. This compound is much less toxic than other oxides of nitrogen. One source of NO occurs naturally when lightning is released in the atmosphere. Intense heat (1800°F) from a lightning bolt causes a chemical reaction in which a single N atom in the atmosphere bonds with a single O atom to create NO. This chemical reaction also occurs during combustion. The temperature of the initial flame front growing in the combustion chamber may reach 3000°F. This temperature can cause a normally benign N_2 atom to bond with an O atom during the short, instantaneous increase in flame front temperature.

Nitrogen dioxide (NO_2) is an oxide of nitrogen that is created in the combustion chamber during the instantaneous increase in an advancing flame front that is toxic and reddish-brown in color. Most of the same chemical reactions that produce NO also produce NO_2. Nitric oxide (NO) is produced at higher temperatures in the combustion chamber, with NO_2 produced at slightly lower temperatures. When NO_2 is discharged into the atmosphere, ultraviolet radiation and heat from the sun provide sufficient energy to separate the N_2 and the O_2 molecules and cause them to recombine into other forms of NOx.

Dinitrogen monoxide (N_2O) is an oxide of nitrogen that is commonly known as laughing gas. Although the quantity of N_2O produced in small engines is small within the proper environment, it is a part of the NOx family.

A Briggs & Stratton Model ZZ engine was used to power a portable generator for a field communications center during World War II.

Oxide of nitrogen emissions, based on a .9 λ factor, have a negative correlation with CO and HC. At a .6 λ to .7 λ factor (rich air-fuel ratio), the amount of NOx generated at lower combustion temperatures is statistically insignificant. Oxides of nitrogen only begin to form with an increase in combustion temperature.

The CO and HC emissions decrease as the λ factor moves closer to > .8 and the ratio of O_2 to HC increases. See Figure 5-6. More of the available HC molecules become involved in the combustion process. The more HC and O_2 that reacts, the cleaner and hotter the exhaust gases become. At the 1.0 λ factor, heat generated during the process causes more N_2 molecules to readily combine with available O_2 atoms to form NOx.

Octane

Octane is the ability of a fuel sample to resist engine knock and/or ping. All fuels are tested in a standard test engine and rated for engine knock in comparison to other fuel samples. The combustion process in an engine must be precisely controlled or inefficient combustion occurs. Knocking and/or pinging sounds during engine operation indicate inefficient combustion. These sounds are caused by intense vibration from multiple flame fronts traveling across the piston and detonating small portions of fuel before the engine reaches its designed point of combustion. Specifications and guidelines detailing performance characteristics of gasoline are based on ANSI/ASTM D4814-94D, *Specification for Automotive Spark-Ignition Engine Fuel.*

The octane requirement for an engine is primarily based on the compression ratio. As the compression ratio increases, higher octane fuels are required. High compression ratios cause the charge to become hotter. This temperature increase requires a more stable fuel to reduce the chance of autoignition. *Autoignition* is the spontaneous combustion of the charge commonly caused by low octane fuel or excessive compression ratio.

Antiknock Index. Octane quality (rating) is indicated by the antiknock index. The *antiknock index (AKI)* is the numerical value assigned to gasoline that indicates the ability to eliminate knocking and/or pinging in an operating engine. The more stable the charge during combustion, the higher the AKI. All gasoline is tested and assigned an AKI number using the research octane number and the market octane number. The *research octane number (RON)* is the octane number that affects engine knock at low to medium speed. The *market octane number (MON)* is the octane number that affects engine knock at high speed and performance in severe operating conditions and under load.

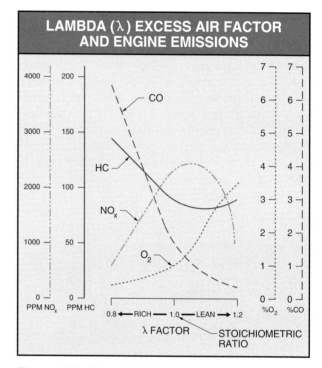

Figure 5-6. At a 1.0 λ factor, NOx and O_2 have a negative correlation to CO and HC emissions.

The AKI number is found by applying the formula:

$$AKI = \frac{R + M}{2}$$

where

AKI = antiknock index

R = research octane number

M = market octane number

For example, what is the AKI for gasoline with an RON of 92 and an MON of 82?

$$AKI = \frac{R + M}{2}$$

$$AKI = \frac{92 + 82}{2}$$

$$AKI = \frac{174}{2}$$

$$AKI = \mathbf{87}$$

The AKI is commonly posted on the gasoline pump. The AKI provides an accurate indication of the resistance to engine knocking over the entire range of engine speed, in severe operating conditions, and under load. Engines are tested by the manufacturer to determine the AKI required for proper engine operation. For example, a minimum AKI of 77 is recommended for all Briggs & Stratton L-head engines. A minimum AKI of 85 is recommended for all Briggs & Stratton overhead valve (OHV) engines.

The AKI can be raised or lowered by further refining of the gasoline or by chemical additives. Further refining removes more undesirable chemical compounds for increased stabilization when the fuel is compressed and burned. Chemical additives such as tetraethyl lead (TEL), which is used to produce leaded fuel, also greatly improved octane ratings. Leaded fuel became popular after World War I, but has gradually been phased out in compliance with stricter emission standards.

Tetraethyl Lead. The addition of tetraethyl lead (TEL) improves the stability of gasoline by causing the flame front to remain intact as it crosses the piston head. Tetraethyl lead dramatically reduces the incidence of autoignition of the compressed charge. Tetraethyl lead was added to gasoline in a ratio of approximately one part TEL to 1300 parts gasoline (1300 : 1). This very effective additive was joined by several chemical scavengers to facilitate the removal of most lead compounds formed during combustion. These additional additives include chlorine (Cl) and bromine (Br). These agents combine with the trace lead left in the combustion chamber and form gaseous lead compounds at the prevailing temperatures present during combustion.

One new lead compound produced in the reaction was lead oxide (PbO_2), which provided an advantage in engine operation. Although this compound was almost completely removed by the scavenger elements, the remaining amount attached itself to the valve train of the engine and created a chemical barrier on valve train parts. Lead oxide is somewhat caustic, but it provides some protection from more active compounds created in the combustion chamber. This allowed the use of less expensive valve train components because of the protective coating formed by residual lead oxide.

Tetraethyl lead was never intended to be an additional lubricant for the valve train of the engine, and the phasing out of TEL in gasoline had little effect on the small engine industry. The industry had already been using the durable, wear-resistant valve train components required for the inherently high combustion chamber and engine operating temperatures. Tetraethyl lead is no longer used in domestic gasoline, and other gasoline additives have been developed to control octane ratings.

Volatility

Volatility is the propensity of a liquid to become a vapor. All liquids have a certain measurable propensity to become a vapor based on the liquid and ambient temperature. The ultimate measure of volatility of a liquid occurs when the transition to a vapor occurs at the boundary of the liquid and the atmosphere or boiling point. All liquids boil at different temperatures.

Vaporization is the process in which a liquid is sufficiently heated to change states of matter from liquid to a vapor. Gasoline is very volatile and is blended to take advantage of its propensity to vaporize based on the environment in which it is used. Gasoline is blended to provide optimum performance and efficiency. If the ambient temperature is high, the gasoline must be blended to prevent it from vaporizing too easily. If the ambient temperature is low, the gasoline must be blended to make it vaporize more easily.

Liquid gasoline is not flammable. Only the vapor emitted from gasoline burns. The combination of gasoline vapor and atmospheric oxygen provides a medium for combustion in an internal combustion engine. Gasoline volatility affects startability and operating performance based on providing sufficient vapor at the correct ambient temperature for the combustion chamber. Industry standards for rating volatility of any given fuel use the distillation test and the Reid vapor pressure test.

Distillation Test. A *distillation test* is a test for determining the composition and volatility characteristics of the components of a given fuel sample. A distillation test is performed by boiling a fuel sample and measuring the rate of vaporization. Boiling points of various gasoline blends are recorded from initial boiling to complete vaporization of the entire sample. Time and temperature results are plotted on a graph.

The initial boiling point of gasoline (accounting for seasonal ambient temperature variations) ranges from 85°F – 105°F. The temperature at which the first 10% of the fuel sample evaporates indicates how well this particular blend of fuel reacts when starting in different temperature spectrums. The first 10% of vaporization indicates the rate of vaporization for the fuel sample.

Ransomes America Corporation
The engine used to power this sod cutter is rated at 5.5 HP and has a manual choke.

Reid Vapor Pressure Test. The *Reid vapor pressure (RVP) test* is a test used to determine the pressure produced from the vaporization process. In the test, a glass vessel (bomb) is cooled to 32°F. Fuel is poured into the vessel at a temperature of 32°F – 40°F. The glass vessel containing the fuel is immersed in 100°F water. The vessel is removed every 2 min and shaken to promote vaporization. A measurement is made of the accumulated pressure after shaking of the fuel produces no additional pressure increases.

The RVP test provides a base number identifying the tendency of the fuel to vaporize and the pressures produced. Gasoline blended for summer use has an RVP of approximately 8 psi at 100°F, and approximately 12 psi or more at 100°F for winter use, according to the expected temperatures in that geographical region. Gasoline with an RVP in excess of 12 psi is used in the severe weather of the northern climates.

Ambient temperature is the main factor in selecting the proper fuel volatility. See Figure 5-7. For example, using gasoline blended for summer in a cold environment results in hard starting. This occurs because the gasoline has been blended to vaporize at higher temperatures, and at lower temperatures less vapor is available per unit of gasoline. In addition, cold air is more dense and contains more oxygen per unit of air, causing a leaner air-fuel mixture.

GASOLINE VOLATILITY	
Low Volatility	**High Volatility**
Poor cold weather operation	Poor hot weather operation
Spark plug deposit buildup	Vapor lock
Combustion chamber deposit buildup	Poor fuel economy
Poor cold starting	Excessive fuel evaporation

Figure 5-7. Gasoline volatility is selected for optimum performance in specific weather conditions.

Although the engine may have ample liquid gasoline in the combustion chamber, there is a minimal amount of vapor available in cold winter temperatures. The engine is more difficult to start because of the lack of vapor and lean air-fuel ratio, despite the presence of the liquid gasoline in the combustion chamber. Using gasoline blended for winter use in the summer can result in the fuel vaporizing too rapidly. This can form undesirable vapor bubbles in the fuel system, resulting in vapor lock.

Vapor Lock. *Vapor lock* is the stoppage of fuel flow caused by internal pressure of a fuel vapor bubble that equals or exceeds the ambient fuel pressure. All gasoline contains vapor bubbles. Vapor bubbles are caused by agitation, heat, and fuel volatility. In normal conditions, the fuel system of an engine passes vapor bubbles without a problem. However, an increase in agitation, heat, and/or fuel volatility can cause restriction by vapor bubbles at critical points in the fuel system. See Figure 5-8.

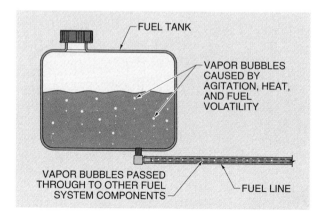

Figure 5-8. Vapor bubbles caused by agitation, heat, and fuel volatility can restrict fuel flow at critical points in the fuel system (vapor lock).

Restriction of fuel flow occurs as a result of pressure differences in the system. Vapor bubbles have a measurable internal pressure that determines the size and rate of expansion. When the internal pressure of the vapor bubble equals or exceeds the pressure of the fuel, fuel flow stops. The most common areas of vapor bubble fuel restriction are the small orifices of the fuel system such as the pilot jet orifice. As fuel flow is slowly restricted by growing vapor bubbles, engine temperature increases from the leaner air-fuel mixture. Fuel temperature also increases with engine temperature, adding to the problem until vapor lock stops the engine.

Reformulated (Oxygenated) Gasoline

Reformulated (oxygenated) gasoline (RFG) is gasoline that contains chemical additives to increase the amount of oxygen present in the gasoline blend. In 1990, President Bush signed into law the Clean Air Act of 1990. This act mandated that gasoline used in the nonattainment zones must contain an additional 2.7% of oxygen by 1995. The classification of nonattainment zone was based on the CO levels in a geographic region during the winter. Reformulated (oxygenated) gasoline was introduced to reduce CO emissions of all gasoline-powered internal combustion engines. Additional oxygen in the gasoline increased combustion efficiency by achieving combustion efficiency closer to the stoichiometric ratio. Oxygen was added by blending oxygenates into the gasoline. The most common of these fuel oxygenates is alcohol.

Alcohol. *Alcohol* is a fuel, or fuel additive, used to enhance the octane rating of gasoline. In the 1970s, ethyl (grain) alcohol was used as an alternative fuel and renewable energy source. The addition of alcohol to enhance octane ratings proved to be an inexpensive method of increasing overall performance of the gasoline. Alcohol mixes easily with gasoline without the use of an additional blending agent.

Alcohol burns with less Btu than gasoline. A 1.0 λ factor for alcohol is 9 : 1, and it has a richer air-fuel ratio than 14.7 : 1 for gasoline. The addition of alcohol provides more oxygen, resulting in a leaner overall air-fuel ratio. This produces increased combustion chamber temperature, combustion gas temperature, and overall engine operation temperature. The oxygen content of alcohol affects the oxygen present in the air-fuel mixture.

The alcohol additives commonly used in gasoline are ethanol and methanol. *Ethanol* is an alcohol additive that is distilled from fermented grain and used in gasoline as an octane enhancer. Ethanol is produced from common grain crops such as corn through a process similar to the manufacture of beverage alcohol. Ethanol used in gasoline is denatured (changed) to make it unfit for drinking. Ethanol can be used in limited quantities in small engines. For example, Briggs & Stratton engines can operate satisfactorily using a gasoline blend including up to 10% by volume of ethanol.

Methanol is an alcohol additive that is distilled from methane gas and used in gasoline. Methanol, or wood alcohol, contains a very high concentration of oxygen and is an excellent solvent. Methanol was used in combination with other chemical compounds as an octane enhancer in the 1980s. The use of a methanol-blended gasoline in a Briggs & Stratton engine may result in deterioration of fuel system components such as rubber fuel lines and carburetor components. In addition, methanol may dissolve the protective sealing material applied to the carburetor at the factory. Methanol contains up to 50% more oxygen than ethanol, which can cause extreme overheating. Briggs & Stratton does not recommend the use of any gasoline blend containing methanol.

Alcohol has a high affinity for water and absorbs water from any substance that it contacts, including air, rubber, or other material in the fuel system. For example, a water bubble found in a fuel tank is not always caused by a defective fuel can. A water bubble can be formed from alcohol in the fuel combining with humidity in the air.

Most alcohols act like solvents and may attack some materials used in the fuel system of the engine. Some alloys used in carburetors are resistant to oxidation in the presence of gasoline. However, oxidation in the presence of alcohol additives may dissolve portions of these alloys and accumulate in the fuel bowl, lines, or fuel filter. Accumulated dissolved material can cause air-fuel mixture malfunction and the plugging of small orifices in the carburetor.

Alcohol blended in gasoline can cause problems in two-stroke cycle engines. Alcohol can remove the lubricating qualities from the gasoline-oil mixture, causing lubrication-related failures. A significant number of two-stroke cycle engine failures are caused by overheating due to lean air-fuel ratios (from alcohol additives) and the lack of lubrication.

Methyl Tertiary Butyl Ether. *Methyl tertiary butyl ether (MTBE)* is a nonalcohol oxygenate fuel additive derived from a chemical reaction of methanol and isobutylene. This reaction does not contain methanol, and it eliminates all of the unfavorable characteristics associated with methanol. The use of MTBE has increased since 1991. When blended with gasoline at 15% by volume, MTBE provides the mandated 2.7% increase in oxygen. The octane rating of gasoline is also increased with MTBE. Gasoline blends containing MTBE or other similar ethers have a distinctive odor. Gasoline blends containing up to 15% MTBE provide satisfactory performance. Although there is insufficient information regarding threats to public health from MBTE use, studies are currently under way to assess possible negative environmental effects and health risks. MBTE is no longer used in the United States.

Liquefied Petroleum Gas (LPG)

Liquefied petroleum gas (LPG) is a gaseous fuel that consists of propane, propylene, butane, and butylene in various mixtures. In general, for LPG fuels in the United States, the LPG mixture consists primarily of propane. LPG is produced as a by-product of natural gas processing and petroleum refining. LPG has been used as a transportation fuel around the world for more than 60 years. Most of the LPG used in the United States is produced domestically.

LPG can be used as a gaseous fuel alternative to gasoline or diesel fuel, and is well suited to motorized vehicles because it can be stored in a liquid state and used in a gaseous state. This allows for efficiency in storage density, which permits acceptable refueling intervals. A common application of LPG-fueled engines is in storage facilities or terminals for forklifts and other material handling equipment because of its reduced emissions. The use of LPG provides cleaner and quieter combustion, which also results in a longer service life compared with gasoline-fueled engines. A gasoline-fueled engine must be retrofitted with LPG storage and fuel delivery hardware.

Unlike gasoline-fueled engines, LPG-fueled engines do not have air-fuel ratios affected by altitude change. LPG-fueled engines use a mixing valve along with a regulator that is controlled by atmospheric pressure. Gasoline-fueled engines use a carburetor in which the main jet is not sensitive enough to compensate for a change in atmospheric pressure. LPG requires higher ignition temperatures, and burns slower than gasoline. This requires special sparkplugs and ignition timing adjustments for maximum combustion efficiency. Additionally, fuel accessibility, cold weather performance, and retrofitting costs must be factored in when considering LPG as a fuel.

AIR PRESSURE DYNAMICS

The atmosphere of Earth is similar to the water in the ocean in that it exerts pressure. All life on the surface of Earth exists at the bottom of this ocean of air. The human body has adapted to the constant pressure of the atmosphere. This constant invisible pressure is rarely felt because the pressure inside the human body is equal to the atmospheric pressure outside. With equal pressures inside and outside the body, there is no difference in pressure sensed.

Atmospheric pressure is the result of the actual weight of the air surrounding the earth. A single air column with a cross-sectional area of 1 sq in. extended into the atmosphere results in a weight of 14.7 lb at sea level. This column produces a pressure of 14.7 psi at the end of the column closest to Earth. See Figure 5-9.

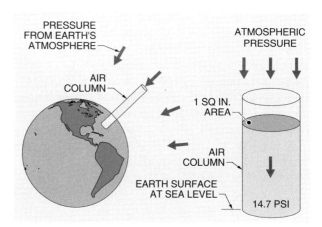

Figure 5-9. Atmospheric pressure is the result of the actual weight of the air surrounding the Earth.

Weather and physical characteristics create low- or high-pressure areas in the atmosphere. These pressure changes cause changing weather patterns and wind. The average value of 14.7 psi changes at different elevations in proportion to the density and temperature of the air. As the elevation increases, atmospheric pressure decreases.

The fuel system in an internal combustion engine utilizes many principles of air pressure dynamics. Air as a gas possesses certain physical properties similar to those of liquids. The main differences between liquid and gaseous states of matter are the distance between the molecules comprising the matter and the incompressibility of the liquid. A gas contains many molecules that are far apart, reducing the cohesive force between the molecules. The cohesive force attempts to keep all molecules in close proximity to each other. An increase or decrease in cohesive force can be achieved by changing the state of the material. A solid material has the greatest cohesive force with molecules in closest physical proximity to each other. This force helps retain the shape of a solid material.

A liquid material has less cohesive force between its molecules. This allows the liquid to assume the shape of a container. Cohesive force of a liquid can be measured by the surface tension of the liquid. A gas has very little cohesive force between its molecules. This allows a gas to assume the shape of a container and increase and decrease in pressure to fill a given (sealed) container.

The dynamics of air are similar to the dynamics of any moving liquid. Because both liquids and gases flow, they can both be defined as fluids. All fluids respond in the same way when a pressure differential exists. Fluids flow from high-pressure areas to low-pressure areas, depending on the pressure differential. See Figure 5-10. When pressure is applied to a fluid, the pressure is the same value throughout the entire liquid regardless of where the measurement is taken. This applies to all fluids as long as there is no motion. Once a fluid is in motion, pressures and pressure measurement rules change.

CARBURETOR OPERATION PRINCIPLES

A *carburetor* is an engine component that provides the required air-fuel mixture to the combustion chamber based on engine operating speed and load. The carburetor is one of the most misunderstood, misdiagnosed, and repaired components of an internal combustion engine. Although carburetors vary in design, all carburetors function using the same basic physics of air flow dynamics and pressure differentials.

The simplest carburetor is made from a tube attached to the intake port of the engine. The tube is used to direct air flow into the combustion chamber as the piston moves down the cylinder bore with the intake valve open. Piston movement creates a low pressure or vacuum in the combustion chamber. Following a basic rule of air flow dynamics, the air moves from the area of higher pressure outside the engine to the area of lower pressure inside the combustion chamber.

Fuel vapor is supplied to the combustion chamber by attaching a fuel supply (fuel bowl) to the tube. The connection between the fuel bowl and the tube is made through an emulsion tube. See Figure 5-11. An *emulsion tube* is a small, hollow, cylindrical component placed in the carburetor with one opening submerged in the fuel bowl and the other opening projecting through the inner wall of the tube. The liquid fuel is forced from the fuel bowl into the air stream of the tube from pressure differences created by applying Bernoulli's principle.

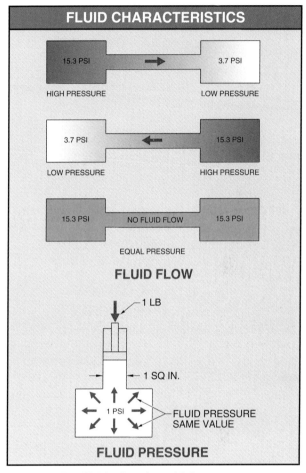

Figure 5-10. Fluids flow from high-pressure areas to low-pressure areas and exert pressure of the same value throughout a system.

Bernoulli's Principle

Bernoulli's principle is a principle in which air flowing through a narrowed portion of a tube increases in velocity and decreases in pressure. Bernoulli's principle is named after Daniel Bernoulli, a Swiss physicist in the 1700s. Bernoulli's principle is based on the equation of continuity. The equation of continuity states that if the cross-sectional area of a hollow tube is multiplied by the velocity of a fluid flowing through the tube, the product is equal at any given point along the tube. See Figure 5-12.

108 SMALL ENGINES

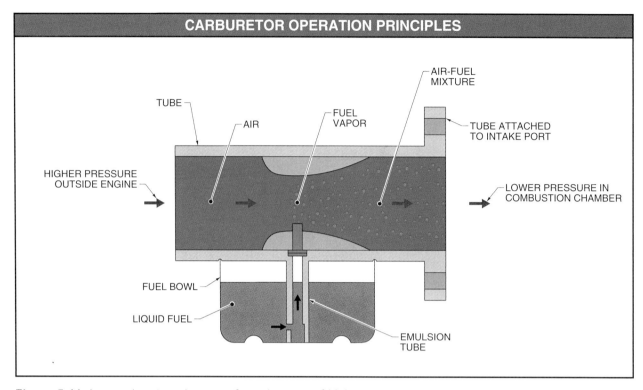

Figure 5-11. In a carburetor, air moves from the area of higher pressure outside the engine to the area of lower pressure inside the combustion chamber.

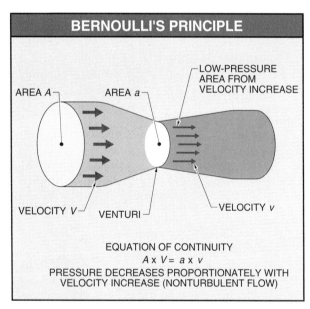

Figure 5-12. Bernoulli's principle is based on the equation of continuity, which states that if the cross-sectional area of a hollow tube is multiplied by the velocity of a fluid flowing through the tube, the product is equal at any given point along the tube.

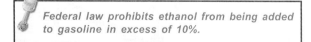

Federal law prohibits ethanol from being added to gasoline in excess of 10%.

In Bernoulli's principle, no more energy can be obtained than is in the system. However, pressure changes in the system can be used as energy to produce work. Applying Bernoulli's principle, air velocity must increase to maintain the same volume of air. The pressure of the air must decrease proportionately as air velocity increases. Pressure is the only quantity that is changeable in the equation. Energy used to increase the air velocity comes from a decrease in pressure at the venturi. Bernoulli's principle is used on an airplane wing. The airplane can fly because of the pressure difference created over and under the wing. Lift is created by lower atmospheric pressure above the wing and higher atmospheric pressure below the wing. See Figure 5-13.

The shape of the wing disrupts the normal air flow and makes the air on the top of the wing travel farther. The greater distance requires the air to travel faster to meet the air at the end of the wing. The faster the air travels, the greater the pressure decrease and pressure difference from air below the wing. The area of high pressure produced under the wing has enough force to lift the airplane into the air. In a carburetor, air flowing into the combustion chamber must flow through the venturi. A *venturi* is a narrowed portion of a tube. The venturi is shaped and functions in the same way as two airplane wings. Air velocity must increase when passing through the restricted orifice of the venturi to maintain the same volume of air passed through the carburetor to the combustion chamber. The low-pressure area created is less than atmospheric pressure, and fuel is forced up the emulsion tube to enter the air stream.

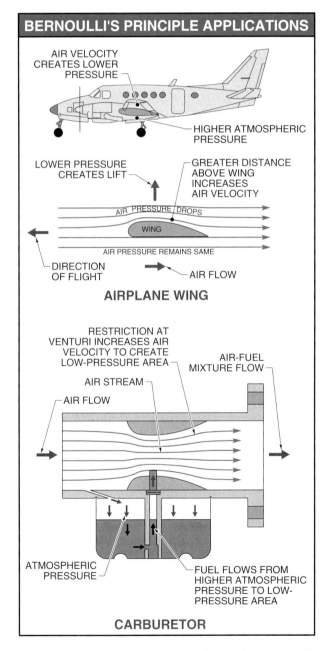

Figure 5-13. Bernoulli's principle is used to create lift from pressure difference to fly an airplane and to supply fuel into the air stream of the carburetor.

Carburetor Operation

A carburetor operates by utilizing pressure differences created by features designed into the carburetor. Air is introduced into the throat of the carburetor. The *throat* is the main passage in the carburetor which directs air from the atmosphere and air-fuel mixture to the combustion chamber. As air moves through the throat and is restricted by the venturi, a low-pressure area is created at the point of greatest restriction. One opening of the emulsion tube is located directly at the low-pressure area. As moving air passes through, ambient atmospheric pressure is exerted on the fuel in the fuel bowl, causing it to rise in the emulsion tube. Fuel is discharged from the emulsion tube and enters the flow of air through the carburetor. As air velocity increases, the pressure difference increases, and more fuel is forced up the emulsion tube. This process continues until there is no pressure differential between the fuel in the fuel bowl and the venturi. See Figure 5-14.

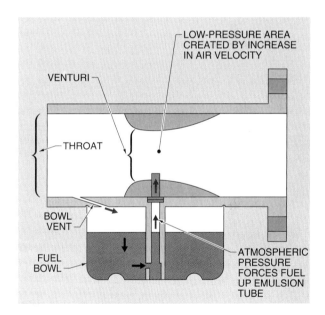

Figure 5-14. Continuous atmospheric pressure is applied on the fuel in the fuel bowl through the bowl vent.

A carburetor must provide continuous atmospheric pressure on the fuel in the fuel bowl using a bowl vent. A *bowl vent* is a passage drilled into the carburetor connecting the fuel bowl to the atmosphere. The bowl vent allows ambient atmospheric pressure to be continually applied on the fuel in the fuel bowl. Without a bowl vent, fuel would not be forced out from the fuel bowl into the air stream, regardless of the low-pressure area created at the venturi.

Bernoulli's principle and moving vehicles

When a large truck passes an automobile on the highway, the automobile may experience a violent push toward the passing truck. The pushing force is attributable to Bernoulli's principle, based on pressure differences of air flow between the vehicles. As the truck passes the automobile, a large volume of air is forced into a small space between the vehicles and increases in velocity. As the air increases in velocity, the pressure drops, creating a low-pressure area between the vehicles. Higher pressure on the opposite side of the vehicles overcomes the low-pressure area, creating a force which attempts to push the vehicles together.

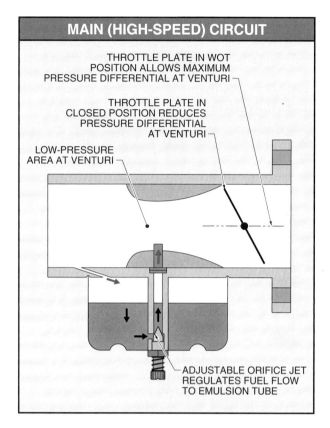

Figure 5-15. The throttle plate position controls the pressure differential at the venturi to limit the amount of fuel entering the engine.

Speed Control. A simple carburetor allows an engine to operate at one speed but with no control of the amount of fuel delivered to the combustion chamber. Engine speed, air and fuel volume, and horsepower are regulated by the position of a throttle plate. A *throttle plate* is a disk that pivots on a movable shaft, regulating air and fuel flow in a carburetor. The throttle plate shaft rotates inside throttle plate shaft bushings or carburetor body to change the throttle plate position. The throttle plate is located in the throat of the carburetor on the engine side of the venturi.

When closed, the throttle plate limits the amount of air entering the carburetor. This also limits the amount of fuel entering the engine by reducing the air flow, which reduces the pressure differential at the venturi. The pressure differential between the atmosphere and the end of the emulsion tube decreases proportionately with a reduction in air flow. See Figure 5-15.

With the throttle plate in the wide open throttle (WOT) position, the engine receives the maximum volume of air, resulting in a maximum pressure differential at the venturi. As the throttle plate is slowly closed, the engine speed decreases. This engine speed decrease is a result of the decrease in air flow and the resulting decrease in pressure differential forcing fuel up the emulsion tube into the air flow.

In addition to a throttle plate, fuel control is enhanced with a jet. A *jet* is a fuel-limiting device that regulates fuel flow to the emulsion tube. The jet is installed in or near the base of the emulsion tube of a carburetor and is submerged in the fuel retained by the fuel bowl. Prior to 1990, most Briggs & Stratton engines used an adjustable jet, which allowed a small engine service technician to adjust fuel flow to the base of the emulsion tube. Most Briggs & Stratton carburetors now use a fixed orifice jet. The *main (high-speed) circuit* is the path from the fuel bowl to the emulsion tube created by the fixed orifice jet in the carburetor. Engine performance is improved with better control of the fuel entering the air stream.

Engine speed at idle is controlled by the idle circuit. The *idle circuit* is the path from the fuel supply to a small hole in the throat on the engine side of the throttle plate that provides the fuel required at idle speed. Without an idle circuit, the farther the throttle plate is closed, the slower the engine operates. As the throttle plate moves closer to closed position, the engine eventually stalls from the lack of fuel. This reduces the pressure differential at the venturi and stops fuel flow up the emulsion tube. See Figure 5-16.

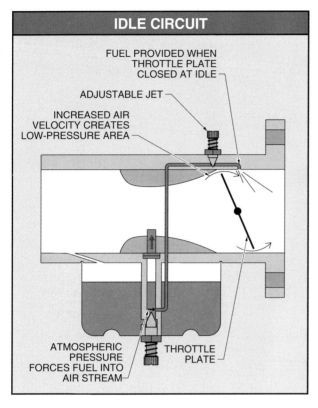

Figure 5-16. The idle circuit provides fuel required at idle speed to prevent stalling from decreased air flow from a closed throttle plate.

In the idle circuit, the throttle plate closes and air continues to pass through to the combustion chamber. The small opening between the throttle plate and the carburetor body causes the air to accelerate as it passes through. A low-pressure area is created at the throttle plate edge and the inner surface of the emulsion tube.

A machined passage in the carburetor body connects the fuel supply to a small hole in the carburetor on the engine side of the throttle plate. Atmospheric pressure on fuel in the fuel bowl forces fuel up through the small hole in the carburetor body to enter the air stream. This provides the small amount of fuel required at idle speed. An adjustable jet can be added for fine adjustment of the idle circuit. The final position of the throttle plate at idle is determined by the true idle setting. *True idle* is the carburetor setting when the throttle plate linkage is resting against the idle speed adjusting screw after idle air-fuel mixture adjustment.

Cold Engine Starting. Heat from an operating engine provides additional vaporization of the charge for proper engine operation. During cold engine starting, the cylinder walls do not provide the heat necessary to assist in the further vaporization of the liquid fuel droplets in the combustion chamber. An air-fuel ratio rich in fuel is only needed during cold starting, not for normal operation. Additional fuel is provided to the combustion chamber during cold starting by a choke or primer system.

A *choke plate* is a flat plate placed in the carburetor body between the throttle plate and air intake that restricts air flow to help start a cold engine. The choke plate restricts air flow into the engine during the intake stroke. A strong low-pressure area is produced throughout the entire carburetor as the piston moves during the intake stroke. The atmospheric pressure applied through the bowl vent forces a large volume of fuel up the emulsion tube and into the carburetor to aid in cold starting. See Figure 5-17.

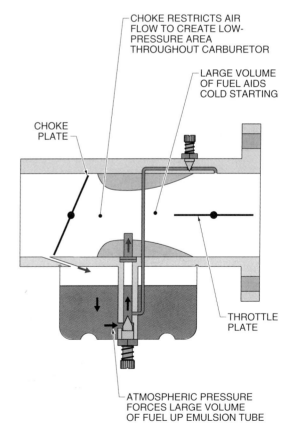

Figure 5-17. As the piston moves during the intake stroke with the choke plate in the closed position, a strong low-pressure area is created throughout the entire carburetor.

A *primer system* is a rubber bulb that is depressed to force a metered amount of fuel into the venturi to help start a cold engine. The fuel is then introduced into the combustion chamber within the air stream once the engine begins to operate. A primer system eliminates the need for a choke.

Primer systems on Briggs & Stratton engines consist of a dry bulb primer system or a wet bulb primer system. A *dry bulb primer system* is a primer system consisting of a rubber bulb filled with air connected to the fuel bowl by a passageway. This system extends the bowl vent from the carburetor to the bulb, with a single hole used to allow atmospheric pressure into the fuel bowl. Some dry bulb primer systems use drilled jets to allow a specific amount of air to pass through. The primer bulb is depressed by placing a finger over the hole in the bulb and collapsing the bulb into the bulb retainer. This action applies a force greater than atmospheric pressure to the fuel in the fuel bowl. A metered amount of fuel is forced up the emulsion tube and into the venturi.

A *wet bulb primer system* is a primer system consisting of a rubber bulb filled with fuel connected to the fuel bowl by a passageway. This system works in a similar way as a dry bulb primer system. However, instead of air, the rubber bulb is full of fuel at all times. When the bulb is depressed, the contents of the bulb are forced directly into the venturi. As the bulb returns to its original shape, fuel is drawn from the fuel reservoir through a one-way check valve to replenish the fuel in the bulb. See Figure 5-18.

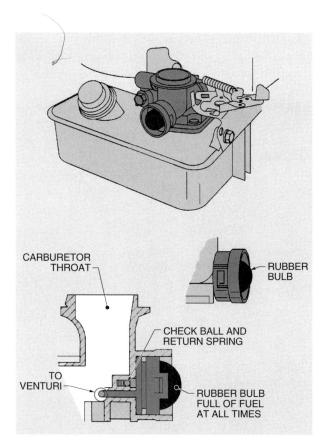

Figure 5-18. The primer bulb system forces a metered amount of fuel up the emulsion tube and into the venturi to help during cold starting.

Air Bleeds

An *air bleed* is a passage in the carburetor that directs air and atmospheric pressure into the main and idle circuits to facilitate the mixture of air and fuel. The *emulsion tube well* is the cavity surrounding the emulsion tube. The main circuit air bleed to the emulsion tube increases fuel economy and helps decrease exhaust emissions. Additional air and atmospheric pressure in the emulsion tube area allows fuel moving up the emulsion tube to begin mixing with air. An air bleed also acts a vent for the emulsion tube well to allow a continuous flow of fuel up the emulsion tube and into the air stream. See Figure 5-19. Most carburetors now have an air bleed for both the main and idle circuits. This allows a more efficient use of the available fuel in the carburetor.

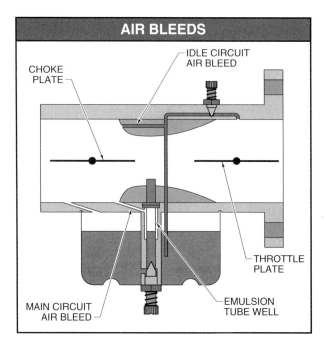

Figure 5-19. Air bleeds provide air and atmospheric pressure into the main and idle circuits to enhance fuel economy and reduce exhaust emissions.

When the engine is accelerated from idle to top no-load speed, there is a short delay in fuel transfer from the idle circuit to the main circuit. To improve acceleration in the engine and to better control fuel delivery at partial load, some transitional holes are added to the idle circuit to allow fuel to be delivered as the throttle plate begins to open. As each hole in succession is uncovered by the moving throttle plate, more fuel is delivered to the air stream, which improves acceleration as well as fuel delivery during light to moderate load.

 Every inch of fuel lift distance requires 1 sec of engine cranking to move the fuel through an empty fuel line and fuel pump into the carburetor.

Emulsion Tube and Jets

The emulsion tube functions with a fixed orifice jet or an adjustable orifice jet to meter, mix, and deliver the proper amount of air and fuel to the engine. The emulsion tube contains a series of holes along its length. These holes allow air to pass through the emulsion tube during engine operation. The fixed orifice jet plays an important role in the delivery of air and fuel by limiting the amount of fuel that is allowed to refill the emulsion tube when the engine is operating. The fuel level in the emulsion tube varies based on throttle plate position or load.

When an engine is operating at top no-load speed, the fuel level in the emulsion tube is the same as the level in the fuel bowl. Fuel allowing the engine to continue to operate is fed through the idle circuit via one or more transitional holes. See Figure 5-20. The engine consumes very little fuel and the fixed orifice main jet allows enough fuel to enter the idle circuit via the emulsion tube to maintain an equal level with the fuel in the fuel bowl.

Under a light load such as a turning mower deck blade, an engine must produce more torque and therefore consumes more fuel. As the mower deck is engaged or another light load is applied, the throttle plate opens beyond the transitional holes and begins to draw fuel from the emulsion tube. The engine now consumes more fuel per unit of time than the fixed orifice main jet allows to pass through it. The fuel level in the emulsion tube drops to a slightly lower level than the level in the fuel bowl. The fixed orifice main jet limits the amount of fuel that can enter the emulsion tube to refill it. When the engine uses more fuel than the fixed orifice main jet allows back into the emulsion tube, fuel level in the emulsion tube decreases.

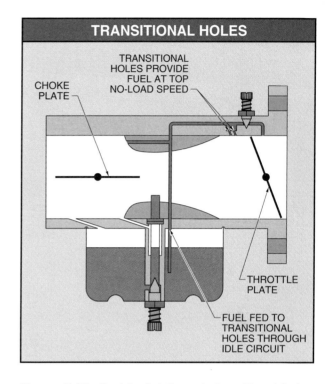

Figure 5-20. Fuel is fed through transitional holes when an engine is operating at top no-load speed.

The level in the emulsion tube drops farther when a moderate load is applied to the engine. There is a significant difference between the fuel level in the fuel bowl and the fuel level in the emulsion tube. The limiting factor is the fixed orifice main jet, which allows a specific amount of fuel to pass per unit of time compared to the amount of fuel the engine is consuming. The main circuit air bleed comes into effect at this point. As the fuel level in the emulsion tube decreases, holes in the emulsion tube are progressively exposed to the air entering through the main air bleed. This allows more pathways for air and fuel to begin to mix to improve the vaporization process for greater efficiency. See Figure 5-21.

Under maximum load, the throttle plate is at WOT and the engine is producing maximum torque. The engine requires all of the fuel that can be supplied through the fixed main jet orifice. The fuel level in the emulsion tube drops until there is no measurable fuel level in the emulsion tube. The fuel then flows directly through the fixed orifice jet into the emulsion tube.

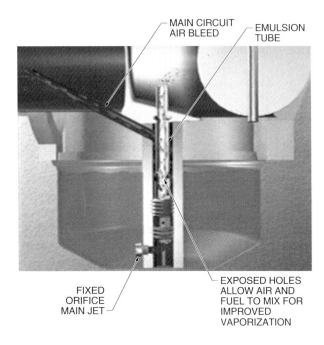

Figure 5-21. Holes in the emulsion tube are progressively exposed to air as the fuel level in the emulsion tube decreases to improve the vaporization process.

Pilot Jet. A *pilot jet* is a carburetor component that contains a fixed orifice jet that meters and controls fuel flow to the idle circuit of the carburetor. The pilot jet is a common component on recent carburetors that include an idle circuit. The pilot jet contains a small fixed orifice at one end which restricts the fuel flow from the fuel bowl by causing fuel to pass through the orifice to reach the idle circuit and transitional hole reservoir. The *transitional hole reservoir* is a cavity that supplies fuel to the idle mixture screw and orifice and the transitional holes. The transitional hole reservoir found in some carburetors is covered with a Welch plug. A *Welch plug* is a hemispherically curved metal cover that expands and seals to the shape of a cavity when impacted. A hole is incorporated in the body of the pilot jet that aligns with the idle air bleed passage.

When the engine is at idle and is consuming very little fuel, the pilot jet orifice has little effect. When the engine is under a light to moderately light load, the pilot jet orifice limits the amount of fuel that reaches the transitional holes in the carburetor body. In addition to limiting the fuel flow, the pilot jet also provides the connection for the volume of air passing through the idle air bleed orifice and may provide a place for air and fuel to mix before delivery to transitional holes. The pilot jet functions like a small emulsion tube. See Figure 5-22. When the engine is under moderate load, the fuel level in the pilot jet well decreases, allowing the maximum amount of air into the circuit to mix with the fuel.

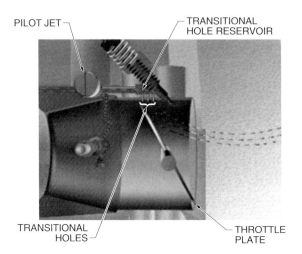

Figure 5-22. A pilot jet meters and controls fuel flow to the idle circuit of the carburetor.

Fuel Bowl Level Regulation

The addition of a needle, seat, and float allows the carburetor to regulate the amount of fuel that is stored in the fuel bowl. A *needle and seat* are components used together that provide a tapered seal to regulate the flow of fuel into the carburetor. A *float* is a carburetor component that floats at a specific level to regulate the opening and closing of the needle and seat. The needle is attached to the float, which rises and falls as the fuel level in the fuel bowl is consumed or replenished. The position of the float regulates the flow of fuel into the fuel bowl for consistent delivery to the engine.

Fuel Bowl Vent Design

Fuel bowl vents allow atmospheric pressure to be continuously applied to fuel in the fuel bowl. Fuel bowl vent designs commonly used on Briggs & Stratton engines are external and internal. An *internally vented carburetor* is a carburetor that has the fuel bowl vent located between the air filter and the venturi of the carburetor. The vent area and the air going to the vent are kept clean by the air filter. This system also provides a compensating function for the air-fuel mixture fed to the engine. See Figure 5-23.

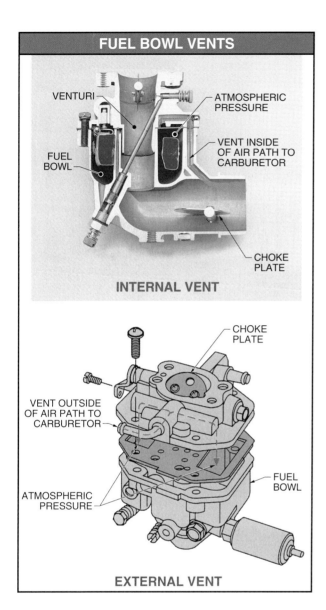

Figure 5-23. An internal vent carburetor has the fuel bowl vent located between the air filter and the venturi of the carburetor. An external vent carburetor has the fuel bowl vent located outside the air path of the carburetor.

The compensating function of an internal vent carburetor occurs as a result of pressure differences created. The air filter prevents dirt and dust from entering the engine, but in the process restricts air flow. This restriction causes a slight pressure variation between the atmospheric pressure and the pressure on the carburetor side of the air filter. The pressure in the carburetor can drop to several tenths of a psi lower than atmospheric pressure. This small variation in pressure can be significant when adjusting the carburetor.

Air flow restriction in the air filter increases with the amount of dirt and dust trapped. When an air cleaner filter becomes filled to 50% of its capacity with dirt, pressure on the fuel in the fuel bowl drops proportionately. This drop in pressure may not always be apparent. For example, a pressure of 12.0 psi on the fuel is still enough pressure to move or lift the fuel into the air stream. Although the pressure differential is reduced significantly between the atmosphere pressure (14.7 psi at sea level) and the venturi in this example, the pressure differential is still enough to move some fuel. Due to the limited amount of air passing through the venturi and the decrease in atmospheric pressure fuel in the fuel bowl, the amount of fuel delivered to the venturi decreases approximately proportional to the amount of obstruction in the air filter element. There is a proportional relationship between the pressure difference and the amount of lift the pressure difference can provide. With a decrease in pressure difference and resulting available work, the amount of fuel lifted into the air stream decreases.

The volume of fuel introduced into the air stream is decreased and the air-fuel ratio becomes leaner as the air filter becomes dirtier. However, the leaner air-fuel ratio is compensated by the reduction of air flow caused by air flow restriction by the dirty air filter. The air-fuel ratio remains relatively constant as the effect of air flow decreases and pressure decreases on fuel in the fuel bowl proportionately compensate for each other.

Murray Inc.

This edger/trimmer is powered by an engine rated at 3.5 HP at 3600 rpm equipped with a manual choke.

116 SMALL ENGINES

An engine with an internal vent carburetor continues to operate with no symptoms of trouble such as dark smoke or poor fuel economy. Power loss can become evident when the air filter becomes almost completely clogged, depending on the power requirements of the application. Power requirements close to the maximum power output of the engine show an air filter problem more quickly. Maximum performance of an internal vent carburetor requires proper air filter maintenance. A new air filter should always be installed before adjusting any Briggs & Stratton carburetor.

An *external vent carburetor* is a carburetor that has the fuel bowl vent located outside the air path of the carburetor. This provides an unobstructed passage to the fuel bowl. The passage commonly consists of a rubber elbow attached to a metal nipple pressed into the side of the carburetor body. External vent carburetors produce the common symptom of dark smoke when the air filter becomes clogged. Dark smoke is produced because air flow is restricted by the air filter but the pressure remains at ambient atmospheric pressure on the fuel in the fuel bowl. This causes a richer air-fuel ratio and excess fuel is not completely burned, causing dark smoke.

Altitude Compensation

As altitude increases, available atmospheric pressure decreases, primarily due to the change in air density. *Air density* is the mass of air per unit volume. Air density decreases exponentially with increases in altitude. For example, 1 cu yd of air at sea level has an approximate mass of 2.75 lb. At 32,736′ above sea level, the same 1 cu yd of air has a mass of .88 lb. With this decrease in mass at the same volume, the air density decreases.

Air density determines the quantity of available elements in the air. With less air density, less oxygen is available to be mixed with fuel and delivered to the combustion chamber. This results in a richer air-fuel ratio in engines operating at higher altitudes. Generally, at elevations above 5000′, a smaller diameter fixed orifice main jet is required to compensate for the change in air density. A smaller diameter fixed orifice main jet increases performance due to a more efficient air-fuel ratio. However, the change in air density still results in reduced horsepower. Engine horsepower decreases by 3.5% for each 1000′ above sea level.

Some carburetors have a fixed orifice main jet that is not removable for high altitude compensation. The main air bleed orifice cup can be replaced or removed to allow more air into the emulsion tube area (if equipped). See Figure 5-24. The manufacturer's repair manual should be consulted before removing any air bleed orifice cups.

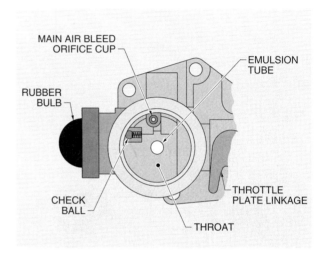

Figure 5-24. Some carburetors have a main air bleed orifice cup that can be replaced or removed to allow more air into the emulsion tube area for high altitude compensation.

Atmospheric pressure and air density

Some small engine service technicians believe that the lower atmospheric pressure at higher elevations compensates for reduced air density. Following this logic, the carburetor should consume less fuel based on the lower pressure differential present in the carburetor and there is no need for altitude compensation. However, this logic is incorrect, as the decrease in air density per 1000′ is more significant than the decrease in the atmospheric pressure per 1000′. Therefore, altitude compensation such as using a smaller diameter fixed orifice main jet is required for acceptable performance at higher elevations.

CARBURETOR DESIGN

All carburetors operate using the same basic principles. The positions of various components and parts may differ, but the basic principles remain constant. The size, orientation, and complexity of the carburetor varies based on engine design, application, and/or emission requirements. Common carburetor designs include updraft, downdraft, sidedraft, and multiple barrel. See Figure 5-25.

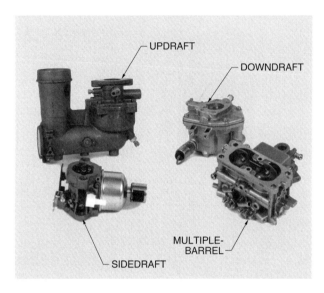

Figure 5-25. Carburetor size, orientation, and complexity varies based on engine design, application, and/or emission requirements.

An *updraft carburetor* is a carburetor that has the air intake opening below the fuel bowl. The updraft carburetor is a compact design that has been used for many years on engines and applications with limited room for air cleaner assemblies, fuel tanks, and PTO attachments. A *downdraft carburetor* is a carburetor that has the air intake opening above the fuel bowl. A downdraft carburetor is commonly used on horizontal shaft engines.

A *sidedraft carburetor* is a carburetor that has an air intake opening above the fuel bowl and parallel to a horizontal plane. A sidedraft carburetor is used on the Briggs & Stratton Model 250000 Series engine, which is commonly known as the crossover engine. This engine has a sidedraft carburetor located on the side opposite the intake valve to allow a large engine to fit in a small engine compartment.

A *multiple-barrel carburetor* is a carburetor that contains more than one venturi. It may have a common or shared intake opening. The multiple-barrel carburetor allows greater calibration accuracy and consistency. An example of two carburetors which represent distinctively different features commonly used are the Vacu-Jet carburetor and the Pulsa-Jet carburetor. Carburetor designs vary based on engine design and power requirements of the application. Different carburetor features are required to meet design and power requirements.

Vacu-Jet Carburetor

Although the Briggs & Stratton Vacu-Jet carburetor is no longer installed on current production engines, it is still common in the field. The Vacu-Jet carburetor has some unique features and adjustment procedures. For example, the Vacu-Jet carburetor uses the fuel tank as a very large fuel bowl. A single pick-up tube extends down into the tank and acts as the connection between the carburetor and the fuel supply. See Figure 5-26.

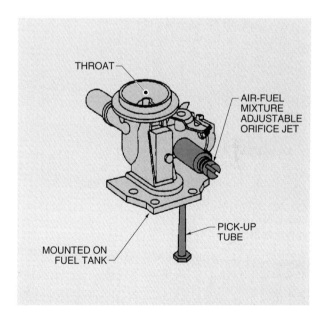

Figure 5-26. The Vacu-Jet carburetor uses the fuel tank as a very large fuel bowl with a single pick-up tube to the fuel supply.

Vacu-Jet carburetors do not contain a float, emulsion tube, transitional holes, needle and seat, or pilot jet. The air-fuel mixture is supplied through two holes in the side of the carburetor throat. Instead of an emulsion tube, fuel is drawn directly up the pick-up tube into a cavity cast in the side of the carburetor. An adjustment needle and a jet are located between the cavity and the fuel delivery holes in the reservoir. Fuel is delivered to the reservoir and forced through the space between the needle and the jet opening.

The needle and fixed orifice combination is an adjustable orifice jet used to meter the flow of fuel delivered to the combustion chamber. An *adjustable orifice jet* is an assembly used to regulate passage of a fluid through an opening. It consists of a needle, fixed orifice, sealing packing, and packing nut. The term adjustable orifice jet replaces the older term needle valve, which was commonly used in the field. Once fuel moves through the adjustable orifice jet, it enters a cavity, and is delivered to the engine through one of the two metering holes based on the throttle plate position and load. This fuel feed system is called suction feed. See Figure 5-27.

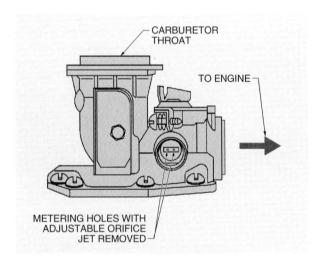

Figure 5-27. In a Vacu-Jet carburetor, fuel is delivered from the adjustable orifice jet through one of two metering holes.

At idle, a small hole closest to the air intake feeds fuel when the throttle plate is in alignment. Under load, the throttle plate moves closer to WOT and uncovers a second larger hole to supply additional fuel required. The small hole bleeds some air into the reservoir to provide limited vaporization of the fuel before it leaves the reservoir through the larger hole. A fuel pump is not required with a Vacu-Jet carburetor. Fuel is lifted from the fuel tank and delivered to the engine by the pressure differentials between the atmosphere, the combustion chamber, and the venturi (in later versions).

The fuel level in the tank affects the air-fuel ratio because the engine has a limited amount of low pressure or vacuum with which to lift the fuel from the tank. The value of the low-pressure area at the venturi is based on engine speed. See Figure 5-28. The only variable is the vertical distance the fuel must be moved to reach the air stream. The farther the fuel must be moved, the more work required.

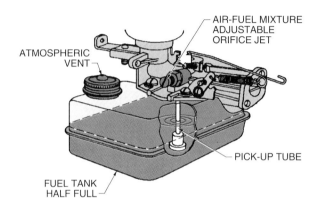

Figure 5-28. The Vacu-Jet carburetor should be adjusted with the fuel tank half-full to compensate for high and low fuel level conditions in the fuel tank.

If the fuel tank is completely full, the pick-up tube in the carburetor is also full to that level. The distance the fuel must move to reach the air stream in the carburetor is very small. As the fuel is consumed by the engine, the fuel level in the fuel tank decreases. For example, if the fuel tank is half-full, the distance the fuel must be moved increases from 1″ to 4″. To move the same amount of fuel a larger distance requires additional work. The available work is fixed, thus the air-fuel ratio changes with the level in the fuel tank. With a nearly empty fuel tank, a lean air-fuel ratio is created. With a full fuel tank, a rich air-fuel ratio is created. The Vacu-Jet carburetor should be adjusted with the fuel tank half-full to obtain the best compromise of air-fuel ratios at high and low fuel level conditions in the fuel tank.

The Vacu-Jet carburetor contains a small check ball in the pick-up tube. A *check ball* is a component that functions as a one-way valve to allow fuel to flow in one direction only. As the engine turns over when starting, the check ball provides a positive seal to stop fuel from running back down the tube as the low-pressure area dissipates in the carburetor and combustion chamber. The high-pressure area pushes fuel up the pick-up tube in pulses corresponding to the piston movement. Without the check ball, fuel has difficulty reaching the carburetor and combustion chamber at less than optimum starting speeds.

Pulsa-Jet Carburetor

The Pulsa-Jet carburetor uses a pressure-actuated fuel pump to eliminate problems caused by the variable fuel lift distance in the Vacu-Jet carburetor. The Pulsa-Jet carburetor lifts fuel from the fuel tank to the fuel cup using a pump. The *fuel cup* is a reservoir located high inside the fuel tank. The fuel cup is a holding area for fuel to be delivered to the carburetor. See Figure 5-29. The fuel pump used with a Pulsa-Jet carburetor provides greater efficiency than the suction feed system used on a Vacu-Jet carburetor. The fuel pump operates by vacuum created by the intake stroke of the piston.

The diaphragm flexing motion increases and decreases volumes within the chambers of the fuel pump. A *diaphragm* is a rubber membrane that separates chambers and flexes when a pressure differential occurs. In the process, the diaphragm compresses the spring during the intake stroke of the piston. The vacuum created by diaphragm motion causes the flapper valve on the inlet side of the pump to lift off its seat. This allows fuel to flow up the suction pipe into the cavity in the pump cover. When the vacuum decreases after the intake stroke, spring tension pushes the diaphragm out, causing gasoline to be pressurized. Spring action tension also causes the outlet flapper valve to lift off its seat, allowing the fuel to flow from the cavity to the fuel cup in the fuel tank.

Vacu-Jet/Pulsa-Jet Carburetor Automatic Choke System. Later model vertical shaft, tank-mounted Vacu-Jet and Pulsa-Jet carburetors feature an automatic choke system. This system shares a common diaphragm with the fuel pump (on Pulsa-Jet carburetors only), but is independent of the fuel delivery system. The automatic choke system uses a manifold vacuum to operate. Between the carburetor and the fuel tank is a diaphragm of fuel-resistant rubber that includes the linkage and choke spring. See Figure 5-30. The linkage is connected to the choke plate near the air inlet of the carburetor. When the diaphragm is in place, the choke spring applies a constant force to the diaphragm and the linkage rotates the choke plate into the closed position.

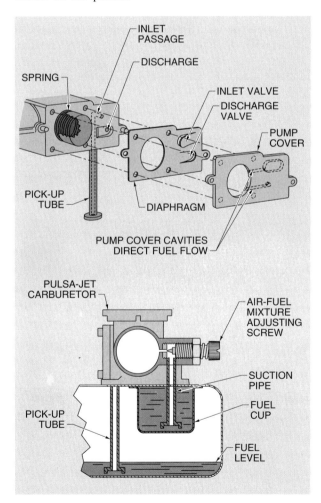

Figure 5-29. In a Pulsa-Jet carburetor, a pump lifts fuel in the fuel tank to the fuel cup, which serves as a holding area for fuel delivered to the carburetor.

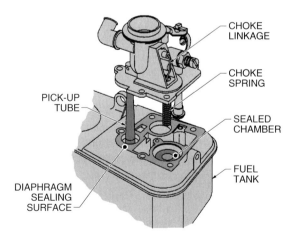

Figure 5-30. The choke spring applies a constant force to the diaphragm and linkage, rotating the choke plate into the closed position.

During the starting process, the moving piston produces a vacuum (manifold vacuum) within the carburetor. This vacuum is directed below the diaphragm into a sealed chamber through a drilled passage beginning near the base of the throttle shaft in the throat of the carburetor. Cleanliness and integrity of the seal between the diaphragm and the sealed chamber are required for proper operation. Some carburetors have a small, stamped groove as the passage from the carburetor to the vacuum chamber. The groove acts as a timing device to control the rate at which the vacuum signal enters the sealed chamber.

Manifold vacuum builds during the starting cycle, affecting the sealed chamber. The carburetor side of the diaphragm is exposed to the ambient atmospheric pressure. Atmospheric pressure compresses the choke spring and actuates the choke linkage when the low-pressure area in the sealed chamber is sufficiently low. The linkage arm moves the choke plate toward the open position. The final position of the choke plate is determined by the choke spring rate, free length, and the vacuum present in the low-pressure area of the sealed chamber. The manifold vacuum increases and the choke plate remains in the open position when the engine begins operating.

This choke system also can alter the air-fuel ratio during engine operation. For example, when an engine encounters a load such as high grass, the engine responds with a decrease in rpm. The decrease in rpm reduces the vacuum in the sealed chamber, allowing the spring force to partially overcome the atmospheric pressure. The force from the compressed spring results in a partial closing of the choke plate to decrease the air delivered to the engine. The richer air-fuel ratio helps overcome the increase in load. When the engine returns to normal rpm or the load is removed, the choke plate returns to open position.

Some Vacu-Jet/Pulsa-Jet Automatic Choke carburetors are equipped with a small bimetal choke spring. The bimetal choke spring is a coiled spring consisting of two dissimilar metals. Each metal has a different coefficient of thermal expansion, causing the choke spring to expand or contract with temperature change. The bimetal choke spring contracts to assist in closing the choke plate with a low ambient temperature. During a high ambient or engine temperature, the bimetal choke spring expands to assist in the opening of the choke plate to reduce restart problems caused by excessive fuel. See Figure 5-31.

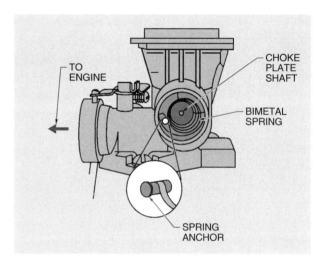

Figure 5-31. The bimetal choke spring contracts to assist in closing the choke plate at low ambient temperatures and expands to assist in opening the choke plate at high ambient temperatures.

Mi-T-M Corporation
This pressure washer discharges 3.6 gallons per minute (gpm) at 2500 psi.

More hydrocarbon emissions are discharged into the atmosphere per year from outdoor grilling of food than the combined exhaust emissions from lawn and garden equipment.

CARBURETOR SERVICE PROCEDURES

Cleanliness of the work area, tools, and hands are commonly overlooked when servicing a carburetor. The work space and tools should be clean and free of any dirt, oil, or other contaminants. Clean hands reduce the possibility of foreign matter introduced during service causing a carburetor to leak or malfunction. See Figure 5-32. The external casting of the carburetor should be thoroughly cleaned using a petroleum distillate. After disassembly, each orifice should be cleaned and inspected for debris or contaminants using a combination of a commercial aerosol carburetor cleaner and compressed air.

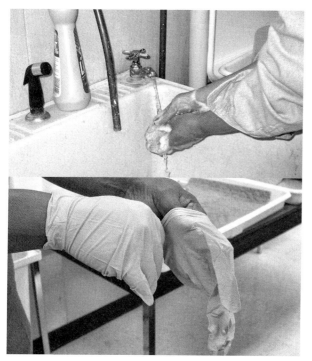

Figure 5-32. Clean hands reduce the possibility of contaminants being introduced during carburetor service. Protective gloves are worn to protect hands from carburetor cleaner.

Warning: To prevent injury, proper ventilation and protective equipment are required when using aerosol cleaners and compressed air.

Welding torch tip cleaners, wire tags, pipe cleaners, or other mechanical devices can cause damage to orifices and should not be used to clean carburetors. This damage may be missed by the naked eye and can cause future problems. Carburetor components should be disassembled and reassembled with care to prevent damage. For example, carburetor manufacturers have determined that even slight overtightening of an emulsion tube or fixed orifice jets can alter the operation of a carburetor. Soaking of carburetor components in commercial carburetor cleaners should be limited. For example, a Briggs & Stratton carburetor should never be soaked for more than 30 min. All rubber seals and components must be removed before soaking.

Parts Removal

Removal of all or some carburetor parts during service is based on the required service procedures and economics. Some small engine service technicians remove throttle plates, throttle shafts, and Welch plugs for carburetor cleaning. The cost of disassembly, cleaning, and reassembly must be compared with the cost of a new carburetor. Labor rates and replacement part costs may exceed the cost of the installation of a new carburetor. Each case must be considered individually to provide the best service solution for the customer.

Service time required on carburetors also varies depending on the way the carburetor is manufactured. For example, carburetors and components may be fabricated and assembled by the manufacturer using specialized equipment and techniques. Some throttle plates and shafts are machined with a laser for accuracy in alignment with transitional holes in the carburetor casting. Removal and reassembly of the throttle plate and shaft may require more time than other carburetor components and must be considered in the total cost of service.

Foreign Matter

Foreign matter in a carburetor can be generally categorized by color and relative size of the particulate. See Figure 5-33. Foreign matter may be found in the fuel bowl, behind inlet seats or Welch plugs during service. Determining the identity and source of the foreign matter can expedite service procedures. Common foreign matter includes blue/green sediment, silver/aluminum particulates, black spheres, and water.

Blue/green sediment present in the fuel bowl most commonly occurs in zinc body carburetors. The blue/green material is primarily copper, which is added to the zinc to allow easier machining of parts during manufacture. Blue/green sediment indicates that a chemical (usually alcohol) has breached the chemical seal applied at the factory and has extracted some of the copper in the casting alloy. The carburetor must be replaced if blue/green sediment is present.

Figure 5-33. Determining the identity and source of carburetor foreign matter can expedite service procedures.

JLG Industries, Inc.

This articulating boom lift can lift up to 500 lb to a height of 45′ with power supplied by a 21 HP engine.

Silver/aluminum colored particulates present in the fuel bowl appear as a scaling or aluminum rust. This material occurs in high ambient temperatures with some alcohol fuel blends or the extended presence of water. Proper cleaning with approved materials and procedures usually remedies the problem.

Small black spheres present in the fuel bowl are actually tiny pieces of rubber from fuel lines. These may have been introduced by the improper service technique of plugging a fuel line with a threaded bolt. Small pieces of rubber fuel line are shredded by the threads of the bolt and settle in the fuel bowl. Small black spheres are also caused by repeated removal and replacement of the fuel filter over an extended period of time. Small black spheres can cause a variety of leakage and performance problems and must be removed with approved cleaning materials.

Water present in the fuel bowl is most commonly the result of using gasoline blended with alcohol, such as ethanol. Water can cause rusting of the fuel bowl, throttle shaft, choke shaft, and/or other steel carburetor components. Water can be removed by using compressed air or aerosol cleaners. However, removal of residual rust on components can be difficult and tedious. In some cases, the component can be removed from the body of the carburetor and cleaned using fine emery paper or similar abrasive material. The carburetor should be replaced if the rust is present throughout the carburetor.

Fuel Filter

A *fuel filter* is a fuel system component that removes foreign particles by straining fuel from the fuel tank. Fuel filters commonly used for small engines include the mesh screen fuel filter and the paper element fuel filter. A *mesh screen fuel filter* consists of a single plastic screen that strains out particles in the fuel tank or suspended in the fuel. The size of the openings in the mesh screen is based on the number of holes per square inch. For example, the pick-up tube on Vacu-Jet and Pulsa-Jet carburetors has a filter mesh screen rated at 200 or 200 openings per square inch. A filter mesh screen rated at 100 has 100 openings per square inch. The higher the mesh screen number, the smaller the openings. A 100 filter mesh screen allows larger-sized particles to pass, compared to the smaller openings in a 200 filter mesh screen.

Filtering capacity of fuel filters may also be expressed in microns. A *micron* (μ) is a unit of area measurement equal to one thousandth of a millimeter (.001 mm). The number of microns indicates the actual size of the opening in the filter medium. The smaller the micron number, the finer the filtering capability of the fuel filter.

Briggs & Stratton mesh screen fuel filters are classified by color code as white or red and are commonly used on gravity feed fuel systems. See Figure 5-34. A *gravity feed fuel system* is a fuel system that uses the location of the tank to provide head pressure to force fuel to flow to the fuel reservoir or fuel bowl of the carburetor. *Head pressure* is the force derived from the mass of a contained liquid such as fuel stored in a fuel tank. Head pressure is solely from the volume of the fuel that is above the inlet fitting of the carburetor. A gravity feed fuel system does not supply pressure from a fuel pump to deliver fuel. A white mesh screen fuel filter is rated at 75μ. A red mesh screen fuel filter is rated at 150μ.

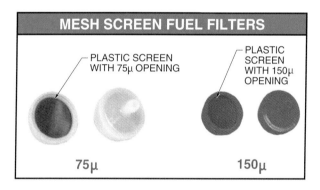

Figure 5-34. Mesh screen fuel filters used on gravity feed fuel systems are color coded to indicate filtering capacity in microns (μ).

A *pleated paper fuel filter* is a paper filter element that consists of multiple folds or pleats to strain out particles suspended in the fuel. This filter has a clear plastic casing and is rated at 60μ. A gravity feed fuel system using a pleated paper fuel filter must have at least a 1″ vertical height difference between the fuel tank outlet fitting and the carburetor fuel inlet fitting. See Figure 5-35. The 1″ vertical height difference provides a sufficient amount of pressure for maintaining required fuel flow.

In some cases, the fuel filter is used as a vapor trap to prevent vapor bubbles from entering the carburetor. As fuel flows through the fuel system, a gasoline vapor bubble forms in the fuel filter. The size of the vapor bubble is dependent on the ambient temperature, vibration of the application, and volatility of the fuel. The fuel filter may appear to have little or no fuel in it at certain times because of the vapor bubble size and pressure. If the vapor bubble is large, the internal pressure of the bubble may reduce the fuel flow volume in proportion to the bubble size. Although the fuel filter on an operating engine appears empty, fuel still flows through the lowest part of the filter.

Figure 5-35. A pleated paper fuel filter has a paper filter element and is rated at 60μ.

Fuel Pump

A *fuel pump* is an engine component that pressurizes the fuel system to advance fuel from the fuel tank to the carburetor. Fuel pumps are most commonly used on engine applications that do not have sufficient head pressure due to fuel tank location to allow consistent performance of the engine. Most Briggs & Stratton fuel pumps use pressure differentials produced in the crankcase for operation.

Crankcase pressure for the fuel pump can be accessed by a fitting on the dipstick tube, a hollow bolt or fitting in the side of the crankcase, or a fitting on the crankcase breather assembly. A rubber hose connects the crankcase to the fuel pump for sending the vacuum pulse. The fuel pump is designed with multiple chambers and diaphragms. When the piston moves towards TDC, a low pressure (vacuum) is created in the crankcase. The low pressure is applied on one side of the fuel pump diaphragm. It flexes and draws fuel into one primary chamber of the fuel pump. See Figure 5-36.

In an engine operating for 1000 hours at 3600 rpm, the piston completes 432,000,000 strokes.

124 SMALL ENGINES

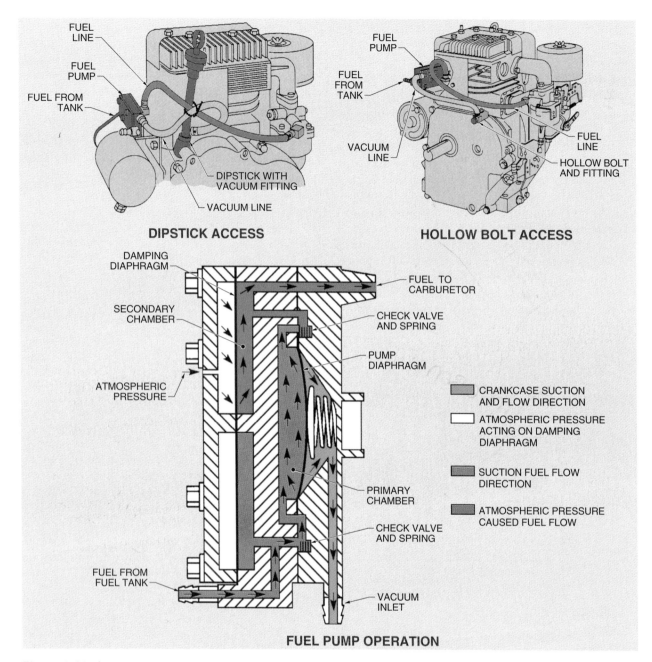

Figure 5-36. Crankcase pressure is used to apply pressure on one side of the fuel pump diaphragm.

Two check valves (flapper valves) allow fuel into the primary chamber and seal the chamber. A *check valve* is a valve that allows the flow of material in one direction. The check valve prevents the reversal of fuel flow when the vacuum ceases and high crankcase pressure occurs as the piston moves towards BDC. High crankcase pressure flexes the fuel pump diaphragm in the opposite direction. Force from the diaphragm, with assistance from an internal spring, forces fuel into the secondary chamber. The secondary chamber is separated by the damping diaphragm.

A *damping diaphragm* is the diaphragm in a fuel pump that flexes from pressurized fuel to increase volume in the fuel section of the secondary chamber. At the same time, fuel is pumped through the secondary chamber to the carburetor. When low crankcase pressure occurs as the piston moves towards TDC, atmospheric pressure and negative pressure in the primary chamber forces the damping diaphragm in the opposite direction. This discharges remaining fuel from the secondary chamber and provides fuel flow when there is low pressure in the crankcase.

GOVERNOR SYSTEM

CHAPTER 6

A governor system maintains a desired engine speed for maximum safety and performance. Governor systems used on small air-cooled engines are the pneumatic governor system, mechanical governor system, and electronic governor system. Troubleshooting governor systems requires verification of proper operation of governor system and carburetion system components.

GOVERNOR SYSTEM OPERATION PRINCIPLES

A *governor system* is a system that maintains a desired engine speed regardless of the load applied to the engine. Most small air-cooled engines have a governor system to control engine speed and torque for maximum performance and safety. Excessive engine speed can result in severe damage and possible injury. Maximum engine performance is obtained through consistent control of engine speed and throttle plate position for the applicable torque required to overcome any given load.

A governor system operates in the same way as the cruise control system on an automobile. With the cruise control activated, an automobile maintains a preset speed while traveling down the highway. The speed of the automobile is a function of engine speed (rpm) and applied load. An *applied load* is a resistive force opposing engine forces. See Figure 6-1.

For example, the cruise control set at 65 mph functions as a governor to maintain this speed until the system is deactivated. The cruise control responds to changing loads, such as traveling up the steep grade of a mountain. As the engine speed begins to decrease in response to the load, the throttle plate is opened to allow more air-fuel mixture into the combustion chamber. The engine is then able to produce the additional torque required to overcome the applied load to sustain the set speed.

 Snow throwers require a governor system that compensates for load variations from snow drifts or wet snow.

125

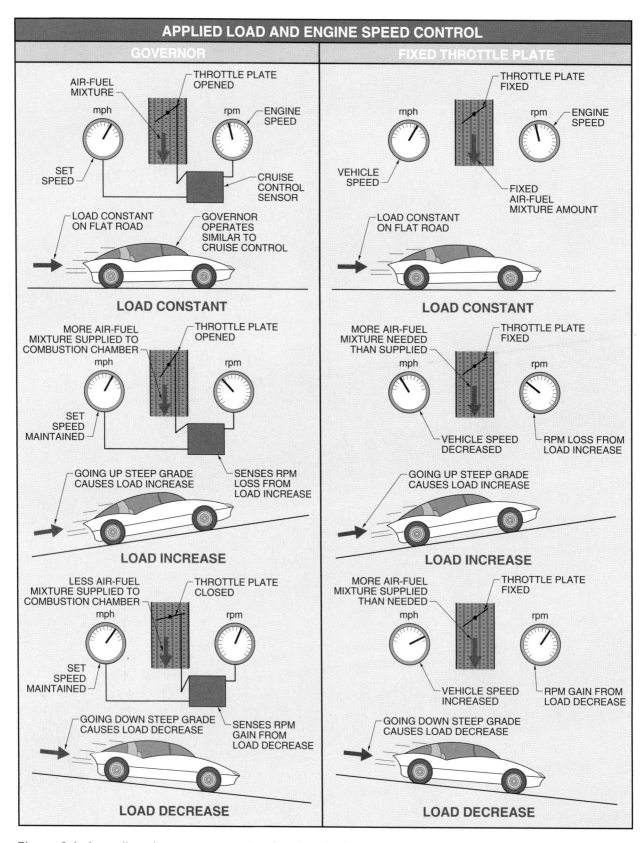

Figure 6-1. A small engine governor system functions in the same way as cruise control on an automobile to maintain engine speed regardless of the applied load.

As the automobile continues up the grade, the cruise control increases the throttle plate opening to compensate for the increased applied load. If the available engine torque is greater than the applied load of the grade, the automobile maintains the set speed. If the available engine torque is less than the applied load of the grade, the engine slows and the automobile loses speed.

After reaching the top of the mountain and traveling down the steep grade, the applied load decreases and the throttle plate position is changed to an almost closed position. As the automobile continues downward, the throttle plate remains in the almost closed position until the cruise control senses a decrease in engine rpm caused by an increase in load, such as another hill. Like cruise control on an automobile, governor systems on small engines use a movable throttle plate to vary the volume of air-fuel mixture supplied to the combustion chamber in response to an increase or decrease in applied load. In contrast, a small engine with a fixed throttle plate is subject to undesirable engine speed variation.

For example, if an automobile with a fixed throttle plate position has an applied load increase, the engine responds by decreasing engine speed. The air-fuel mixture previously used to maintain engine speed is now used to produce torque. This reduces the energy available to maintain engine speed. The fixed throttle position also limits the amount of air-fuel mixture that can enter the combustion chamber. With an applied load decrease, the same amount of air-fuel mixture supplied exceeds the torque required, and engine speed increases. With a fixed throttle plate, engine speed increases or decreases are approximately proportional to the applied load.

Evolution of Governor Systems

Early internal combustion engine governor systems used only the applied load for speed and torque control. This system included a primitive carburetor without a throttle plate or a carburetor with a fixed throttle plate position preset by a cable or fastener. See Figure 6-2.

In this early governor system, the engine is set to operate at a constant speed and is producing speed and torque to do work. Under no-load conditions at this speed, energy produced by combustion of the air-fuel mixture produces enough torque to overcome the internal dynamic friction of the engine. Any excess energy from the combustion of the air-fuel mixture results in an increase in engine speed. If the applied load is small, the engine produces just enough torque to overcome the load. For this reason, the throttle plate of the engine is set for a slightly open position to allow a small amount of air into the carburetor. See Figure 6-3.

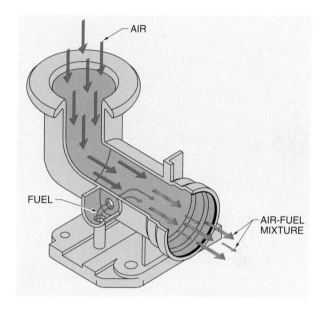

Figure 6-2. Early governor systems included a carburetor without a throttle plate and used applied load for speed and torque control.

With a small amount of air allowed to pass through the carburetor, a proportionally small amount of fuel is drawn from the carburetor bowl into the venturi. The throttle plate position of an engine operating at 3600 rpm under no-load conditions is almost closed, rather than wide open as most people expect.

Under load, most engines experience a decrease in speed whether they have a fixed or a movable throttle plate. This decrease in speed is caused by the physical dynamics of the engine. This changes when an engine is operating at top no-load speed. *Top no-load speed* is the top speed setting an engine achieves without any parasitic load from equipment components. A *parasitic load* is any load applied to an engine that is over and above the frictional load of an engine, such as a lawn mower blade.

 A throttle plate half open at maximum load equals approximately 75% load.

128 SMALL ENGINES

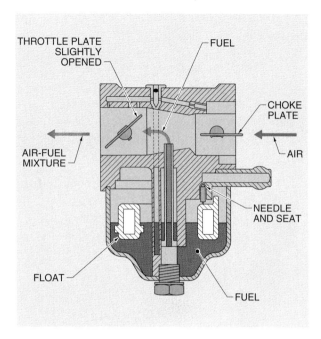

Figure 6-3. A fixed throttle plate carburetor is set to allow a small amount of air into the carburetor.

An engine operating at top no-load speed has a short but finite amount of time to fill the combustion chamber, ignite the air-fuel mixture, and evacuate the exhaust gases from the combustion chamber. The faster an engine operates, the shorter the time available to complete these events. Under significant load, engine speed must decrease to allow enough time to ingest the air-fuel mixture required to overcome the load. Early carburetors without a throttle plate or with a fixed throttle plate position controlled engine speed and torque, but did not compensate for variable loads. This severely limited engine performance under variable loads.

Governor System Components

Governor system components commonly consist of a speed-sensing device and a governor spring. A *speed-sensing device* is a governor system component attached through linkage to the throttle plate of the carburetor to sense changes in engine speed. The speed-sensing device monitors engine rpm and functions to maintain the slowest possible engine speed. The *governor spring* is a governor system component that pulls the throttle plate toward the wide open throttle (WOT) plate position.

When the governor spring is installed and in the extended position, it works against the speed-sensing device, slowing engine speed. Both governor system components and forces in the engine function together to obtain equilibrium, or governed speed. The *governed speed* of an engine is the speed obtained at the balance point between the forces of the speed-sensing device and the governor spring. See Figure 6-4.

The speed-sensing device monitors any increase or decrease in engine speed. This provides a measurable and predictable force solely based on engine speed. Force on the speed-sensing device initiates attempts to completely close the throttle plate, resulting in an engine speed decrease to idle. Without a counteracting force, the speed-sensing device affects the position of the linkage that actuates the throttle plate toward closed position.

The governor spring also initiates a throttle plate position adjustment. It is designed to provide a force sufficient to open the throttle plate to WOT position. The result of a WOT position of the throttle plate is increased horsepower (a function of speed) or increased torque if a sufficient load was applied to the engine. The governor spring provides a measurable and predictable force, opposing the force of the speed-sensing device. The force of the governor spring provides the adjustability and movement of the throttle plate.

The equilibrium obtained between the speed-sensing device and governor spring results in the desired engine speed and torque output. A properly-operating governor system responds to engine rpm deviations from top no-load speed by opening or closing the throttle plate as required to maintain the desired engine speed. Governor systems are designed to respond to specific application loads. Examples of specific application loads include equipment used to pump water, generate electricity, or cut grass. Governor systems used on small air-cooled engines are the pneumatic governor system, mechanical governor system, and electronic governor system.

Pneumatic Governor Systems

A *pneumatic governor system* is a system which uses force from moving air (pneumatic force) produced by rotating flywheel fins to sense engine speed. The pneumatic force actuates a linkage and spring connected to the throttle plate to maintain a desired engine speed. The pneumatic governor system is a simple and dependable governor system.

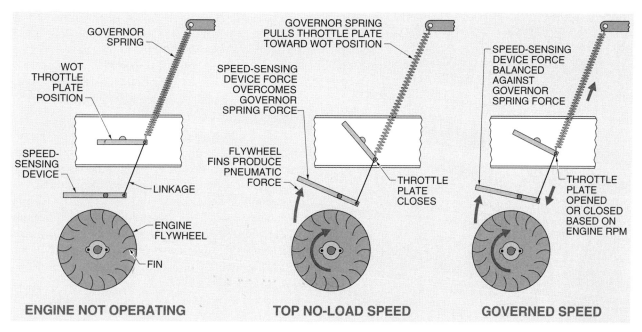

Figure 6-4. Forces from the speed-sensing device and governor spring are balanced to control throttle plate position in a governor system.

When the engine is started, the flywheel begins to rotate with increasing speed. See Figure 6-5. Fins on the flywheel are designed to move and direct ambient air under the blower housing. As flywheel speed increases, air speed and resulting pneumatic force increase proportionately. Pneumatic force is directed toward the governor blade. A *governor blade* is a movable metal or plastic blade which deflects air from flywheel fins to act as the speed-sensing device of a pneumatic governor system. The governor blade is attached to the same throttle control linkage as the governor spring.

As engine speed increases, pneumatic force on the governor blade increases proportionately to actuate the throttle control linkage. When force from the governor blade on the throttle control linkage exceeds the force applied by the governor spring, engine speed decreases. Forces produced by the governor blade and the governor spring are directly opposite each other. Based on the throttle control position, governed speed may or may not be the top no-load speed or the governed speed of the engine.

If engine speed decreases from an increased load, the governor spring force becomes greater than the pneumatic force on the governor blade and the throttle plate is opened proportionately. The opening of the throttle plate allows more air-fuel mixture into the combustion chamber to increase the torque output required to overcome the increased load.

John Deere Worldwide Commercial & Consumer Equipment Division

This walk-behind lawn mower is powered by a 5 HP engine with a cast aluminum alloy engine block and primer system for cold starting.

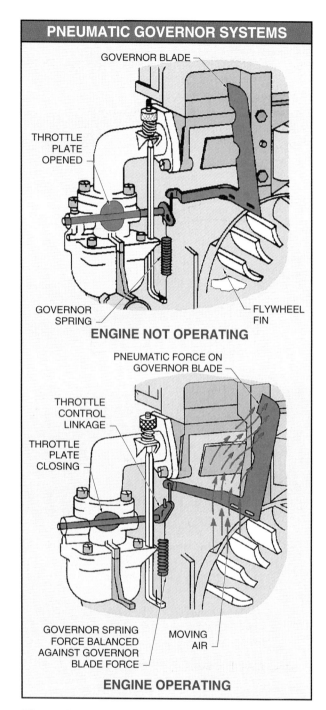

Figure 6-5. A pneumatic governor system senses engine speed by the amount of pneumatic force directed on the governor blade.

using a spring rate specific to the engine speed range required for the application load requirements. *Spring rate* is the force necessary to stretch the spring one unit of length in in., mm, or other units from its free length. *Free length* is the overall dimension of the spring when unloaded. Spring rates are expressed as lb/in., oz/in., gm/mm, or gm/in.

A governor blade is manufactured using various materials in different configurations. Governor blades may be mounted in vertical or horizontal positions. See Figure 6-6. For example, a horizontal shaft engine commonly uses a sheet metal vertical governor blade and pivot. This style of pneumatic governor system is referred to as a vertical (hang-down) air vane governor. Governor blade material, mass, or angle of air deflection affects overall governor performance on the engine. Any change in these variables can cause engine speed control problems.

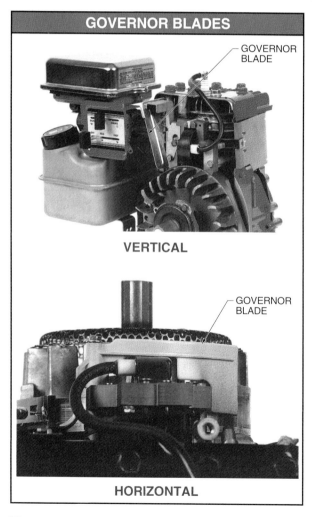

Figure 6-6. Governor blade size and shape is based on engine design and application requirements.

With the engine OFF, the throttle control can be set to stretch the governor spring to hold the throttle plate in WOT position. The governor spring is the only force on the throttle plate when the engine is OFF. Force applied by the governor spring is calibrated

Pneumatic Governor Geometry. Pneumatic force from rotating flywheel fins on a governor blade varies depending on engine speed, governor blade position, and governor blade configuration. For example, if there is a pneumatic force of 1 lb on the governor blade, the 1 lb force is multiplied by the difference between the distances from the fulcrum (pivot point) of the governor blade. The pivot point is located in such a way as to make a long lever arm of the governor blade and a short lever arm of the connection to the throttle plate linkage of the carburetor. The pivot point position is located to obtain the proper mechanical advantage.

The pivot point position commonly results in a mechanical advantage of approximately 4 : 1 to compensate for the forces produced by the governor spring and other opposing forces connected to the throttle plate linkage. See Figure 6-7. The 4 : 1 ratio equates to the distance from the pivot point to the end of the long lever arm compared to the distance from the pivot point to the end of the short lever arm.

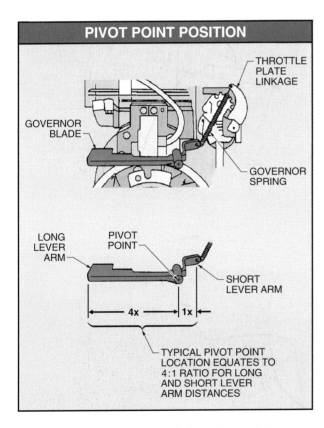

Figure 6-7. The location of the pivot point on the governor arm helps produce the mechanical advantage required to overcome opposing forces in the governor system.

Other engine operation forces can affect the ability to close the throttle plate. Air drawn in through the carburetor creates a variable force dependent on throttle plate position. With the throttle plate in the nearly closed top no-load speed position, the force on the throttle plate from air passing through the carburetor is relatively great. As the throttle plate opens to allow more air into the carburetor, forces on the plate decrease significantly and assist in affecting the throttle plate linkage position. See Figure 6-8.

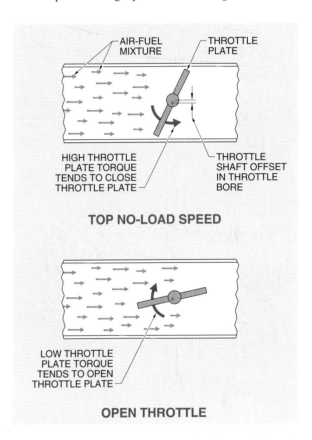

Figure 6-8. Torque produced by the force of the air-fuel mixture impacting the throttle plate varies with throttle plate position.

Most throttle shafts are slightly offset in the bore of the carburetor by .011″ – .018″. As air entering the carburetor contacts the throttle plate, force is applied to the surface of the plate. When the throttle plate is in the top no-load position, the offset or longer side of the throttle plate creates greater torque, which tends to close the throttle plate. When the throttle plate is moved toward WOT position, torque is reduced, which tends to open the throttle plate. The mechanical advantage from the long lever arm of the governor blade compensates for these variable forces. See Figure 6-9.

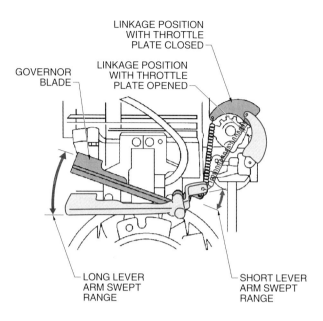

Figure 6-9. The pivot point position of the governor blade produces a smaller swept range on the short lever arm of the governor blade.

The increase in mechanical advantage of this pivot point position results in a loss of motion. For example, the full range of motion of the governor blade produces a much smaller swept range of motion of the short lever arm connected to the throttle plate linkage. The shorter the lever arm, the less sensitivity produced by the governor system. A large movement of the governor blade is required to produce a perceptible motion of the short lever arm.

Some pneumatic governor systems use a curved governor blade to enhance the performance of the system. The curved blade shape provides a greater force differential based on governor blade position. A *force differential* is the measured difference in forces acting on a single object. The curved governor blade amplifies the force differential by the curve on the leading edge of the governor blade. When pneumatic force decreases, the curved governor blade moves closer to the flywheel from the existing governor spring pressure at the top no-load speed setting. The leading edge of the governor blade then moves in a swept arc that allows the leading edge to be closer to the flywheel than the leading edge of a flat governor blade. With the governor blade closer, more air is allowed to escape past the blade, which reduces the pneumatic force.

When a curved governor blade is at top no-load speed position, it provides the governed speed and stability of a flat governor blade. See Figure 6-10. When the engine has an applied load, force from the governor spring exceeds the pneumatic force on the governor blade. Engine speed decreases and the curved governor blade reacts by moving closer to the flywheel.

As the leading edge of the curved governor blade moves inward, the pneumatic force contact area is reduced. Redirection of the pneumatic force causes a more rapid inward motion toward the center of the flywheel. This results in a faster response of the throttle plate to a more open position. When the curved governor blade is moved away from the center of the flywheel, the blade is exposed to the normal, predictable forces of the air.

The force differential is increased as the governor blade moves closer to the flywheel and the force is decreased by redirection of the air flow. By reducing the force at the governor blade in this manner, the curved governor blade moves close to the flywheel more quickly than a flat governor blade. The curved governor blade design allows the engine to respond more quickly to engine speed decreases from loads.

 Pumps commonly use an adjustable fixed speed control to maintain consistent rpm except for normal speed change due to load.

Mechanical Governor Systems

A *mechanical governor system* is a governor system that uses a gear assembly that meshes with the camshaft or other engine components to sense and maintain the desired engine speed. Engine speed is sensed by a gear-driven component that meshes and rotates at approximately crankshaft speed. See Figure 6-11. Mechanical governor systems do not utilize air from flywheel fins and are the most common governor systems used on small air-cooled engines built in the last decade.

The governor gear assembly consists of a governor gear, an independently hinged set of flyweights, governor cup, and a governor gear shaft. The rotation of the governor gear causes the hinged flyweights to move outward. As the governor gear spins faster, perpendicular inertia is increased and directed through a lever on the bottom of the flyweights. Inertia causes the flyweights to apply pressure on the governor cup. See Figure 6-12.

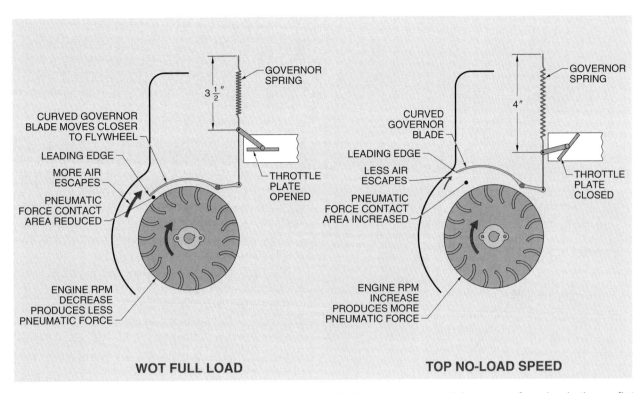

Figure 6-10. The curved governor blade responds more quickly to engine speed decreases from loads than a flat governor blade.

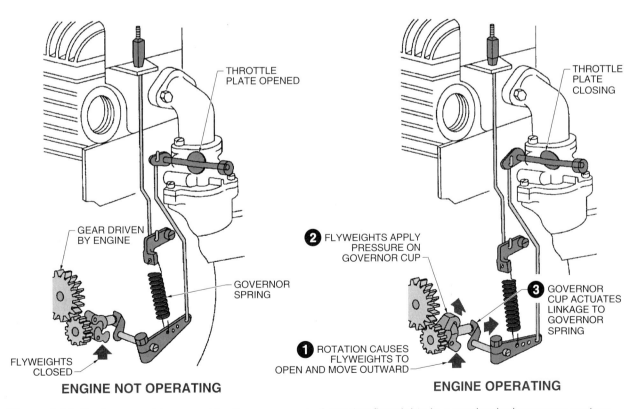

Figure 6-11. Engine speed is sensed by the movement of rotating flyweights in a mechanical governor system.

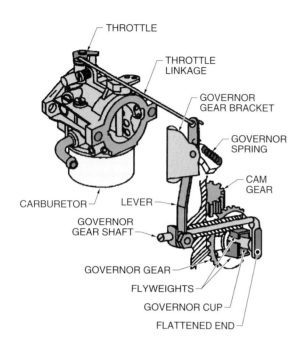

Figure 6-12. The governor cup applies pressure on the governor gear shaft to actuate the throttle linkage.

The governor gear shaft is anchored to the governor gear bracket, crankcase cover, or sump of the engine. The governor cup slides up against and applies force to a flattened end of a governor shaft that extends out of the crankcase. A lever is attached to the governor gear shaft. Various styles of levers and linkage assemblies connect the governor gear shaft to the throttle shaft and plate.

The governor spring and linkage in a mechanical governor system is essentially the same as the pneumatic governor system. In stretched position, the governor spring applies force to open the throttle plate to WOT position. When the engine reaches the preset top no-load speed, the forces of the governor spring and mechanical governor gear flyweights equalize. If the load on the engine is suddenly removed, the engine would begin to overspeed because of the excess amount of air-fuel mixture entering the combustion chamber.

The increased engine speed resulting from the sudden removal of the load causes the flyweights to exert a greater force at the governor cup and against the governor shaft. The increase in force exceeds the governor spring force and the throttle plate begins to close. The closing of the throttle plate limits air-fuel mixture flow into the combustion chamber, reducing engine speed.

When engine speed decreases due to increased load, the flyweights apply less force on the governor cup and governor shaft. This reduction in force allows the governor spring to slightly open the throttle plate to allow more air-fuel mixture into the combustion chamber. The engine then produces more torque to overcome the load.

Mechanical Governor Geometry. On a standard governor lever arm with multiple linear holes, the governor spring can be moved farther from the pivot point of the governor lever arm to decrease sensitivity and/or increase speed. See Figure 6-13. Before changing the governor spring position, the governor design should be analyzed and the repair manual should be referenced to determine the consequences of the change.

When the governor spring is moved away from the pivot point on the governor lever arm, changes occur in the length of the governor spring and in the mechanical advantage of the governor lever arm. Stretching the governor spring to the new position increases the applied force of the spring. This increases the top no-load speed of the engine caused by the increased force required to reach equilibrium with force from the governor flyweights.

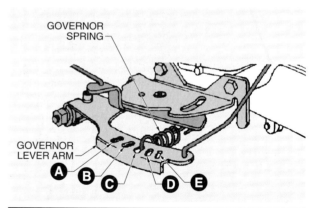

GOVERNOR SPRING SPEED ADJUSTMENT			
Position	rpm Range*	Position	rpm Range*
A	2700-3100	D	3700-3800
B	3200-3300	E	3900-4000
C	3400-3600		

* fine adjustment between ranges made with tab-bending tool

Figure 6-13. Sensitivity and speed can be adjusted by moving the governor spring position.

Mechanical advantage of the governor lever arm is increased by the greater distance of the governor spring from the pivot point. This causes an increase in the amount of force differential (the same force over a greater distance) at the governor spring to cause an equal amount of governor lever arm rotation. Force reduction at the flyweights caused by a load-induced reduction in rpm must be greater to create an equal amount of governor lever arm rotation. This results in a greater rpm loss before the governor responds. For proper speed control, the top no-load speed must be readjusted whenever there is a change in governor spring position.

Some governor systems have governor lever arms with non-linear hole patterns. See Figure 6-14. For example, a wedge-shaped governor lever arm has a protrusion toward the governor spring and anchor. In the wedge, holes are drilled in a pattern which provides a closer relationship between the holes furthest from the spring anchor. This design reduces the spring length differential between holes compared to the standard linear holes and governor lever arm design. As the spring is moved up the governor lever arm, less rpm increase is gained per hole.

The geometry of the connections between the governor lever arm and governor spring can also affect proper governor operation. For example, on some larger engines, rotating the governor spring anchor tab toward the governor shaft produces the effect of reduced spring rate. The more severe the angle produced, the greater the reduction in spring rate, providing the engine rpm remains constant. See Figure 6-15.

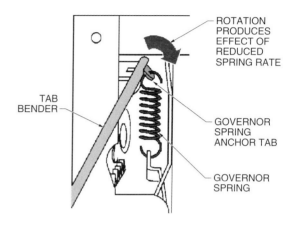

Figure 6-15. Bending the governor spring anchor tab can increase governor system response by producing the effect of a reduced spring rate.

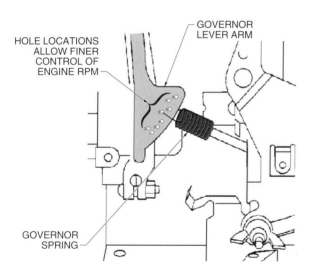

Figure 6-14. A non-linear hole pattern on a governor lever arm reduces the spring length differential between holes to allow finer control of engine rpm.

The same principles apply with regards to the spring position as on a linear hole lever arm. Sensitivity is gained or lost by the mechanical advantage provided by the change in lever arm length. The only difference between the two styles of governor lever arm is that the non-linear holes provide a more uniform distance between the hole and the spring anchor.

Ariens Company

This tiller is powered by a 5 HP engine that features a horizontal shaft, mechanical governor, and manual friction speed control of the carburetor.

The actual spring rate does not change as a result of bending the governor spring anchor tab. The new angle formed between the governor lever arm and the anchor tab causes a decrease in the effective governor lever arm length. This produces a reduced force differential between the flyweights and the governor spring to increase the speed of governor system response. The result is a lesser decrease in rpm from top no-load speed to a new applied load.

Long Governor Lever Arms. On some engines designed specifically for generator operation, a longer than normal governor lever arm is incorporated into the existing governor system. The long governor lever arm allows a larger swept area at the throttle plate linkage connection to the carburetor. See Figure 6-16. A small amount of rotation at the governor shaft results in greater swept arc at the carburetor linkage point. Conversely, any moderate motion of the throttle plate caused by the force of the air-fuel mixture passing through the carburetor does not influence the governor shaft. Thus, a long governor lever arm provides less sensitivity to throttle plate forces.

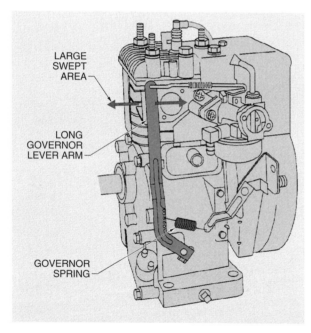

Figure 6-16. The large swept area produced by a long governor lever arm increases engine rpm control accuracy in generator applications.

Washing machines were put on stilts to fit the engine under the tub until 1931, when a low-profile Briggs & Stratton L-head, ½ HP Model Y engine was introduced to fit under washing machine tubs.

Electronic Governor Systems

An *electronic governor system* is a governor system that uses a limited angle torque (LAT) motor in place of the governor spring and speed-sensing device used in a mechanical governor system. A *limited angle torque (LAT)* motor is a direct current (DC) motor used to control governor system components in an electronic governor system. A LAT motor responds to load changes using changes in voltage applied to the inductive field coil to quickly open or close the throttle plate. A LAT motor contains an armature, an inductive field coil, and governor return springs. See Figure 6-17. An *armature* is a rotating part of an electric motor consisting of a segmented iron core or permanent magnet mounted on the motor shaft. An *inductive field coil* is a coil of wire attached to a segmented iron core that produces a magnetic field when current is passed through it. The segmented iron core efficiently directs the magnetic field around the armature.

In a top no-load position, armature position is maintained by the magnetic field produced by voltage applied to the inductive field coil. The polarity of the magnetic field in the inductive field coil is controlled by the direction of the windings and current flow. On a LAT motor with a segmented iron core armature, the magnetic field produced in the inductive field coil attracts the magnetic field on the armature shaft to rotate the armature and move the throttle plate. On a LAT motor with a permanent magnet armature, a magnetic field having the same polarity is created. The two magnetic fields repel each other when in close proximity, causing the armature shaft to rotate toward a position of dissimilar polarity with the inductive field coil. An increase in voltage to the inductive field coil increases the magnetic field strength exponentially.

During normal operation, voltage applied to the LAT motor regulates throttle plate movement. When the engine is shut down or during an electrical failure, the throttle plate position is controlled by governor return springs. A *governor return spring* is an electronic governor system component that applies force to close the throttle plate. This is exactly the opposite of the normal operation of a governor spring in other governor systems. Governor return springs provide a fail-safe function to prevent overspeeding during an electrical failure by returning the throttle plate to idle position. All electronic governor systems must have a fail-safe return to idle mechanism.

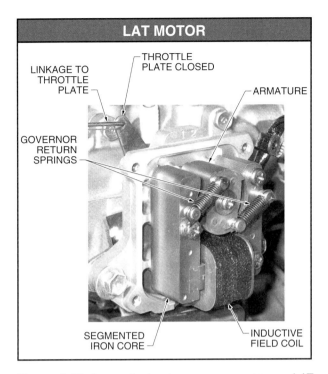

Figure 6-17. In an electronic governor system, a LAT motor actuates throttle plate linkage position by current sent from the controller.

An electronic governor system utilizes an electronic controller to count the number of ignition armature impulses per second. Ignition armature impulses vary with engine speed. An increase in load causes engine speed and the resulting ignition armature impulses per second to decrease. Based on the number of ignition armature impulses, a computer program provides an increase or decrease in voltage to the LAT motor to close or open the throttle plate. An electronic controller on this governor system incorporates a dither effect to increase governor efficiency. A *dither effect* is an effect based on the theory that less time is required to accelerate mass that is already in motion. The dither effect keeps the throttle shaft in constant but imperceptible motion. The constant motion allows the governor system to react quickly to changes in load conditions.

The electronic governor system is a relatively new governor system for small engines and provides greater sensitivity and control of engine speed than a mechanical governor. However, unexpected or unaccounted friction from binding in the governor and throttle linkage can cause problems. An electronic governor system responds differently to engine load and speed if any binding occurs. To prevent binding, the carburetor may be modified during manufacture to use ball bearings to support and reduce friction on the throttle shaft. This assures the freedom of movement required to maintain precise rpm control for increased engine performance and efficiency.

GOVERNOR DROOP

Governor droop is the amount of rpm decrease between the top no-load governed speed and the rpm where power is delivered. For example, when a snow thrower is moved from a clean sidewalk to a large snow drift, there is a decrease in engine speed. This decrease is governor droop, and it occurs before the engine has compensated for the increased load. Governor droop should not be confused with the rpm difference between top no-load and maximum or peak torque. A *governor droop curve* is a combination of rpm decrease and maximum brake horsepower (BHP) curves. See Figure 6-18.

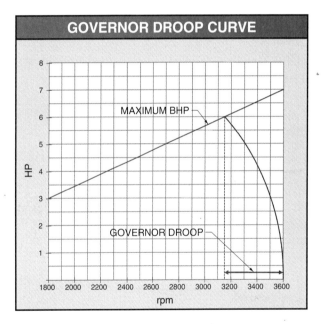

Figure 6-18. A governor droop curve indicates the maximum BHP a governed engine can deliver.

Governor systems limit the amount of horsepower available by controlling engine speed (rpm). The maximum horsepower a governed engine can deliver occurs at the intersection of the rpm droop curve and the maximum BHP curve. The rpm that produces maximum horsepower is always less than the top no-load rpm.

A governor droop curve provides information regarding specific engine speed requirements and the need to set the top no-load speed of an engine above the rpm that delivers the power required by the load. A generator, or any application that requires operation at a specific engine speed, requires the proper governor setting for satisfactory operation of the equipment. The governor droop of an engine is not fixed and can be changed after the engine is designed and produced.

When generating electricity, engine rpm must remain close to constant or damage may occur to electronic devices. The governor system could be altered to provide more sensitivity. However, this would render the engine governor system and the engine unusable for anything other than a generator application.

A governor system designed for very precise speed control may react to any and all load changes by hunting or surging at speeds other than top no-load speed. *Hunting* is the undesirable quick changing of engine rpm when set at a desired speed. *Surging* is the undesirable slow changing of engine rpm in a cyclical pattern when set at a desired speed. Hunting and surging are similar malfunctions and are commonly used together when describing undesirable engine rpm variation.

A governor system set for very precise control on a rotary lawn mower would cause hunting and surging with the slightest load change. Minor load changes from grass length or thickness would produce excessive rpm variation. When used to drive a generator, the same governor system provides desired close control of engine rpm. Governor design and sensitivity must address the control of loads and rpm according to application requirements. Manufacturer's specifications should be consulted before making any governor system adjustments.

Governor Droop Control

Governor droop control on a standard production engine requiring precise speed is accomplished by making the governor system think that the engine is operating at a high rpm. This is done by stretching the governor spring by adjusting the tab setting to change the force applied to the governor lever arm. The angle of the governor spring is also changed to be more acute. This provides a more sensitive governor system to help control the inherent governor droop in the engine. See Figure 6-19.

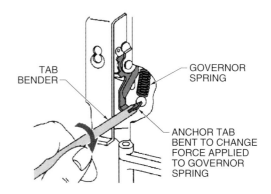

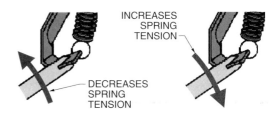

Figure 6-19. Changing the tab setting to stretch the governor spring provides greater sensitivity to help control governor droop.

Speed control based on load

If a water pump is attached to an engine that has no governor system, the pumping action and mass of water can be used to control engine speed. When the pump is primed and the engine is started, the work required to lift the water into the pump and discharge the water is the applied load. Engine speed is controlled by the specific volume of water pumped per minute.

Engine speed is also changed by the introduction of an additional load or the absence of a load. For example, if the pump is placed on a platform 10′ above the ground, the additional work required to lift the water to a higher location results in a decrease in engine speed and decrease of water discharged.

Without a governor system, the engine cannot adjust for changing loads. If all of the water has been pumped, the engine will experience a sudden decrease in load. With the load removed, and the throttle plate in a fixed position, combustion of the same amount of air-fuel mixture results in increased engine speed.

When the governor spring anchor assembly is rotated for constant speed applications, the spring angle is changed dramatically, which results in an effective decrease in spring rate. The decrease is a result of the deviation from perpendicular spring position to an acute spring position. This decreases the force necessary to stretch the spring.

Engineers can calculate the exact governor droop for any given engine. Excess engine speed above the top no-load speed to produce the required governor droop is calculated and incorporated into the governor design. This provides the precise control necessary for applications such as a generator engine to maintain the correct rpm and produce the proper cycles of electricity.

GOVERNOR SENSITIVITY

Governor sensitivity is the result of the interrelationship of the geometry of the governor lever arm, governor spring rate, and the force generated by the flyweights. Each of these components is designed for a specific application. For example, on an engine designed primarily for generator use, the governor spring rate is less than a standard engine application. The governor lever arm may be longer to provide a larger linear travel for the same arc, and the governor spring may be located closer to the pivot point of the governor shaft. The mass of the flyweights may also be altered to provide the combination of forces required by the application.

Governor Springs

Governor spring properties which affect governor sensitivity are spring rate and free length. Spring rate is most commonly expressed as a weight in lb/in. or oz/in. Governor system component operation requires proper governor spring rate and free length. Spring rate can be used to determine force exerted regardless of size or material. For example, a spring rated at 3 lb/in. with a free length of 1″ exerts a force of 3 lb when stretched to 2″. The same spring exerts a force of 4.5 lb if stretched to 2.5″, and exerts a force of 6 lb if stretched to 3″.

The relationship between the force applied and the linear distance a spring is stretched is proportional. Force required to stretch or compress a spring a certain distance can be calculated using Hooke's Law.

Hooke's Law. *Hooke's Law* is a law that states that the amount of stretch or compression (change in spring length) is directly proportional to the applied force ($F \sim \Delta x$). For example, two times the applied stretching force applied proportionally results in two times the linear distance. Any spring maintains this relationship until the spring has exceeded its elastic limit. *Elastic limit* is the last point at which a material can be deformed and still return to its original physical dimensions. Once permanently deformed, the predictable relationship of the spring force and length is irreparable. Therefore, a stretched governor spring that has been shortened does not function according to original specifications. See Figure 6-20. Hooke's Law is applied using the formula:

$F = k(x_1 - x_0)$

where

F = force (in lb)

k = spring rate (in lb/in.)

x_1 = new spring length (in in.)

x_0 = original spring length (in in.)

For example, a spring has a free length of 2″ and spring rate of 2 lb/in. If a 2 lb stretching force is applied, the spring stretches 1″ to a new spring length of 3″. What force must be applied to the free length to stretch the spring to a new spring length of 4″?

$F = k(x_1 - x_0)$

$F = 2 \times (4 - 2)$

$F = 2 \times 2$

$F = $ **4 lb**

If a force of 2 lb equals 1″ of stretch to a new spring length of 3″, then a force of 4 lb equals 2″ of stretch to a new spring length of 4″.

Spring Rate Change. A common assumption is that installing a governor spring with a greater spring rate opens the throttle plate faster to reduce the governor droop of the engine. This assumes the force of the governor closes the throttle plate and the governor spring opens it. However, a change to a greater spring rate does not improve sensitivity and increases the governor droop of the engine.

Spring rate change and its effect on sensitivity is best illustrated by a pneumatic governor system. See Figure 6-21. When a governor blade is at top no-load speed position, there is a constant force applied. An increased load on the engine causes decreased engine speed, change of air velocity, and a change in force applied on the governor blade.

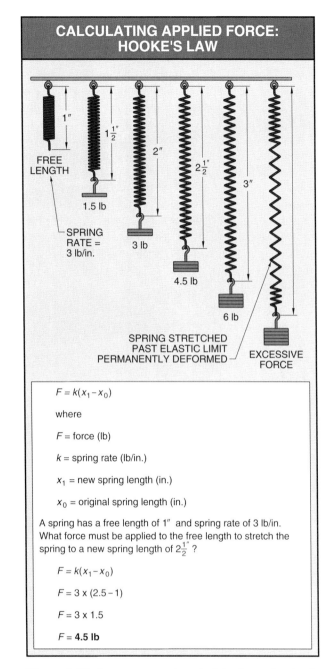

Figure 6-20. A governor spring stretched past its elastic limit is permanently deformed and must be replaced.

For example, consider a governor blade that has only two positions: top no-load speed and WOT full-load speed. The force F_2 at top no-load speed in this example is 5 lb regardless of spring rate. Maximum governor spring length (maximum spring force) is 4″ at top no-load. In WOT full-load speed position, the governor spring is stretched to $3\frac{1}{2}″$. Free length of the governor spring is 3″. In the equation, the force F_1 is higher with the lower spring rate at the WOT position. Consequently, WOT occurs because of greater pneumatic force from higher air velocity on the governor blade. Higher air velocity at the governor blade results in higher rpm at full load.

The lower spring rate requires a 1 lb decrease in force to move the throttle plate from top no-load position to WOT full-load position. A smaller rpm decrease would provide the 1 lb force change. The equation shows that a 4 lb/in. governor spring requires a 2 lb change in force ($F_2 - F_1 = 2$ lb) to move the governor blade to WOT full-load position. The smaller the decrease in rpm, the smaller the rpm droop.

In most applications, a governor system with high sensitivity is not required. In some cases, such as a rotary lawn mower, a larger governor droop is acceptable to ensure a stable governing system. With a rotary lawn mower, engine speed to produce power is a much lower priority when compared to a generator engine.

Lever principle used in governor lever arms and spring rate

A father and son are playing on a seesaw. The seesaw consists of a single wooden board with a fulcrum (pivot point) at the center. The board balances in perfect equilibrium with no persons sitting on either end of the board.

The father weighs 180 lb. The son weighs 90 lb. When the father and son sit on each end of the board, the son does not have enough mass to overcome the mass of the father and is raised above the pivot point. To compensate for the difference in mass, the fulcrum or pivot point is moved closer to the father to increase the distance from the fulcrum to the son. This provides the son with a longer portion of the board.

The increased distance from the fulcrum results in enough mechanical advantage for the 90 lb son to raise the 180 lb father above the fulcrum. However, the son must travel further to raise the father. The father moves in a smaller short swept area while the son moves through a greater swept area. The work required to move the father or son is the same, with the force redirected.

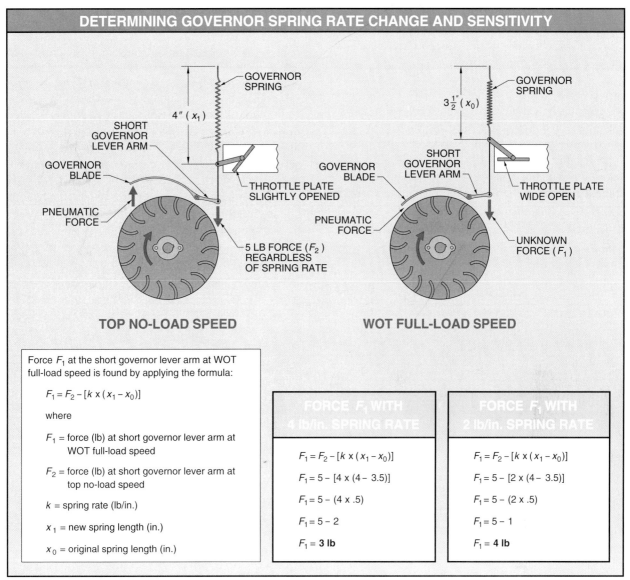

Figure 6-21. Governor system sensitivity and governor droop can be controlled by using a governor spring with a different spring rate.

Governed Idle

Governed idle is a governor system function that allows an engine to accept light to moderate loads at idle speed without stalling the engine. For example, a lawn tractor with the throttle control set at true idle is backed out of a garage. If the engine is not capable of producing sufficient torque to move the lawn tractor, the load produced from moving the tractor causes the engine to lose rpm and possibly stall. This is caused by the inability of the engine to produce sufficient torque at the slow idle speed rpm, as well as the inability of the governor to respond to the light to moderate load.

The lack of governor response is due to the fact that at true idle speed the main governor spring has no effect on the governor system. The main governor spring at true idle speed is at its free length position and applies no appreciable tension. Most governor springs in true idle speed position also have a small amount of clearance between the governor spring anchor and the governor linkage. Consequently, when the engine slows due to the increased load, the governor spring is unable to quickly respond by opening the throttle plate.

To solve this problem, a governed idle spring is added to the existing governor system to provide a spring-initiated idle speed. A *governed idle spring* is a governor system component that has a low spring rate that offsets some of the force from the speed-sensing device to improve governor sensitivity at low rpm. See Figure 6-22. In addition, the governed idle spring removes any lost motion caused by clearance between the governor spring and the governor linkage. With the engine operating at idle speed, the governed idle spring is set to produce a light force opposing the speed-sensing device. This allows the governor system to be responsive to light to moderate loads regardless of the main governor spring position.

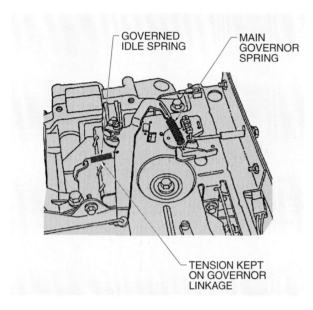

Figure 6-22. A governed idle spring removes lost motion between the governor spring and governor linkage to improve governor sensitivity at low rpm.

 A common misconception is that the primary purpose of a governor system is to prevent engine overspeeding.

Whenever a governed idle governor system is used, two rules must be applied:

- The idle speed screw on the carburetor must be set at less than governed idle speed. This is accomplished by holding the throttle control linkage against the idle speed screw when adjusting.

- The governed idle speed setting must be performed before the top no-load speed is set. When a governor system contains more than one spring, the spring forces are added together.

Hysteresis

Hysteresis is the undesirable motion of governor system components caused by engine vibration and governor system friction characteristics. Hysteresis can result in erratic governor system response. When an engine produces vibration, waves carrying the vibration energy are transmitted away from and sometimes through the source of vibration. These waves have a certain frequency pattern or signature. When an engine is operating, the components farthest from the source of the vibration wave are the most affected. Some vibration is common with applications driven by small engines.

There are instances where vibration waves may encounter other vibration waves caused by a natural vibration, a machine part, or equipment connected to the engine. For example, a loose mounting bolt on an engine begins to rotate as the vibration waves pass through it. If the natural vibration frequency of the mounting bolt is the same frequency as the vibration waves passing through it, the mounting bolt eventually unscrews itself from the engine.

If the vibration waves are of the same frequency, the object goes into a state of resonance. *Resonance* is the state of the vibration wave frequency being equal to the natural vibration wave frequency of the component. This state is amplified by the addition of energy from the same frequency waves to produce a much stronger single pulse. This pulse can cause hysteresis of the governor spring. Hysteresis is caused as the governor spring oscillates radially and axially to create a phantom spring force increase or decrease based on extension from the free length. See Figure 6-23. The changing spring force causes a proportional amount of governor system instability.

If the frequency of the two vibration waves are exactly 180° out of phase from each other, they cancel each other out, resulting in little, if any, vibration. This is the desired effect of using a crankshaft counterweight to balance an engine. Crankshaft counterweights produce a vibration wave which counteracts vibration waves created by other moving parts in the operating engine.

 In 1982, the Consumer Product Safety Act required that lawn mowers be equipped with a safety system that automatically stops cutting blade rotation when the handle of a walk-behind mower is released.

Governor System **143**

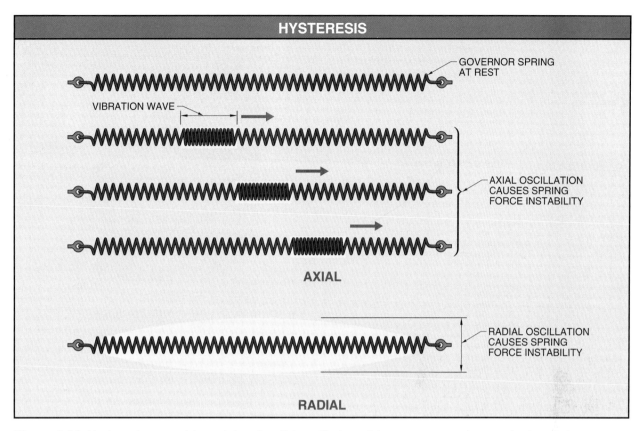

Figure 6-23. Hysteresis caused by axial and radial oscillation of the governor spring results in erratic governor system performance.

SERVICE PROCEDURES

Service procedures for governor systems are usually performed after verifying proper carburetion system operation. Governor system replacement parts are chosen by engine type number. Top no-load speed specifications are provided by the manufacturer. After new governor system parts are installed, top no-load speed must be checked with a tachometer. A static governor (engine not operating) adjustment eliminates possible causes of governor system problems to reduce the amount of service required.

Mechanical Governor Adjustment

All Briggs & Stratton mechanical governor systems have a static governor adjustment that must be performed before any troubleshooting. See Figure 6-24.

Mechanical governor systems are typically adjusted using the procedure:

Billy Goat Industries, Inc.

This truck loader vacuums turf debris using a 16 HP engine featuring overhead valves, cast iron cylinder sleeves, and a mechanical governor.

144 SMALL ENGINES

1. Loosen the governor arm clamping nut at the base of the external governor lever arm.
2. Grasp the governor linkage attached to the carburetor throttle plate and push the throttle plate into the idle position.
3. Move the linkage from idle to WOT position and note the direction of movement of the external governor arm (clockwise – counterclockwise).
4. Place linkage in WOT position and hold.
5. Using a screwdriver or appropriate tool, turn the governor shaft in the direction that the arm moved when the linkage was actuated from idle to WOT.
6. Hold the linkage and governor shaft and torque the governor arm crimping bolt to the proper specification.

This adjustment procedure places all governor linkage and components in a known position. With the rotation of the governor shaft, the governor cup is pushed back on the governor gear shaft toward the governor gear. This collapses the flyweights into the maximum throttle plate opening position. The throttle plate in the carburetor is then placed in the WOT position and the bolt tightened accordingly. The motion of the governor arm from idle position to WOT position identifies the direction the governor shaft must turn for the governor system to respond to a full-load condition.

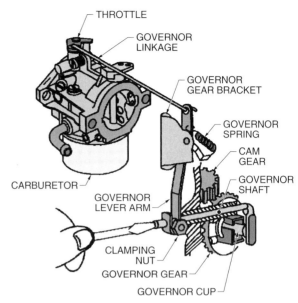

STATIC GOVERNOR ADJUSTMENT

Figure 6-24. A static governor adjustment is required before any mechanical governor system troubleshooting is performed.

In 1908, Stephen F. Briggs and Harold M. Stratton began an informal partnership to manufacture and market a six-cylinder, two-cycle automobile engine.

Ransomes America Corporation
The engine used to power the Cushman® Bellhop personnel mover is rated at 14 HP or 10.4 kW (10,400 W).

In the 1920s, the Red-E Engine Company purchased engines from Briggs & Stratton for use on their reel-type mower and other garden and agricultural products.

ELECTRICAL SYSTEM

CHAPTER 7

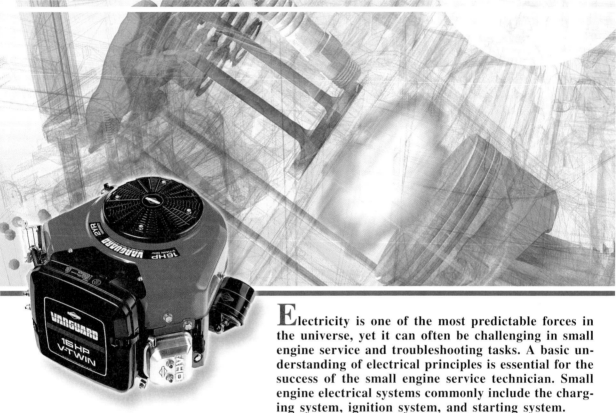

Electricity is one of the most predictable forces in the universe, yet it can often be challenging in small engine service and troubleshooting tasks. A basic understanding of electrical principles is essential for the success of the small engine service technician. Small engine electrical systems commonly include the charging system, ignition system, and starting system.

ELECTRICAL PRINCIPLES

Electricity is a predictable force, yet it can often be challenging in service and troubleshooting tasks. Electricity is different from other small engine systems where a mechanical action is observable, because it cannot be seen. In addition, the possibility of electrical shock is a concern when working with any electrical system.

With the exception of a 120 V starter motor used on a few small engines, most small engine electrical systems are not capable of inflicting a severe electrical shock. However, caution should always be exercised when working on any electrical system to prevent any possible injury. Basic electrical principles are used in all components of an electrical system. A thorough understanding of basic electrical principles and characteristics of specific engine electrical systems reduces the possibility of injury.

Electricity is energy created by the flow of electrons in a conductor. A *conductor* is a material that allows the free flow of electrons. The flow of electrons occurs at the atomic level. Wires and other conductors that conduct electricity are made of atoms. An *atom* is a small unit of a material that consists of protons, electrons, and neutrons. *Protons* are the parts of the atom that have a positive electrical charge. *Electrons* are the parts of the atom that have a negative electrical charge. *Neutrons* are the parts of the atom that are neutral, and have no electrical charge. The *nucleus* is the center of the atom, which consists of protons and neutrons.

145

An atom resembles a solar system. Electrons revolve around the nucleus like planets around the sun. For example, a copper atom contains 29 protons and 35 neutrons in its nucleus, and 29 electrons in four orbits. The attraction of positively-charged protons and negatively-charged electrons keeps the atom together. As in a magnet where opposite poles attract and similar poles repel, the negative electrical charge of the electron attracts the positive charge of the proton. In a copper atom, there is only one free electron in the outer orbit. See Figure 7-1.

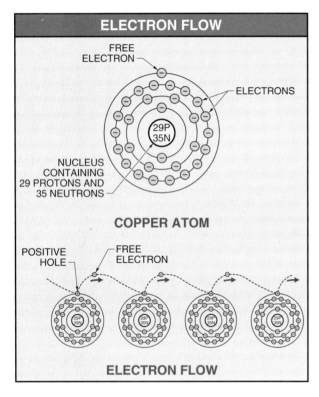

Figure 7-1. Electron flow in a copper conductor occurs as free electrons jump from the outer orbit of one atom to another.

A *free electron* is an electron that is capable of jumping in or out of the outer orbit. When an electron jumps from the outer orbit, it leaves a positive hole that attracts a nearby negative electron. For example, in a cable to a de-energized starter, free electrons jump around at random. When the starter is energized, free electrons jump from negative to positive.

Briggs & Stratton maintains over 4000 active service part numbers for original replacement parts.

The distance from the electron to the nucleus of the atom varies according to the number of electrons in the atom. The further away from the nucleus, the less attractive force is exerted by the nucleus on the electron. If one or more electrons are located in the outermost orbit of larger atoms, less energy is required to dislodge them from the orbit.

Voltage

In order to cause a flow of electrons in a conductor, there must be electrical pressure. *Voltage* is the amount of electrical pressure in a circuit. A *circuit* is a complete path that controls the rate and direction of electron flow on which voltage is applied. Circuits commonly consist of a voltage source, a pathway for electrons to flow, and a load or loads. A *voltage source* is a battery or some other voltage-producing device. A *pathway* is a conductor (commonly copper wire), which connects different parts of the circuit. A *load* is a device that uses electricity, such as the starter motor, lights, or other application accessories.

Voltage is measured in volts. A *volt (V)* is the unit of measure for electrical pressure difference between two points in a conductor or device. Voltage causes electrons to move in a circuit. The voltage required to make electrons flow in a circuit is similar to the water pressure required to make water flow in a pipe. Electrons flow only when an imbalance of potential exists between two areas. Electron flow is similar to the flow of fluids from an area of high pressure to an area of low pressure. Although it is common to have voltage without electron flow, it is not possible to have electron flow without voltage.

For example, a pressure gauge on an air compressor reads a specific amount of pressure regardless of whether the pressurized air is being used by a pneumatic tool. The pressure gauge reading is the difference between pressure from the compressor tank and the atmospheric pressure in the shop. When applied to electricity, the same is true for voltage. Voltage in a circuit is the measurement of the difference between the voltage source and any other point in a circuit. See Figure 7-2.

Although a difference in voltage (pressure) causes the flow of electrons through the circuit, voltage never actually moves. For example, in an air compressor, the difference causes the pressurized air to flow from a high pressure area to a low pressure area. The air moves, but the pressure remains. Likewise, voltage does not flow through a circuit. Voltage causes electron flow through the circuit.

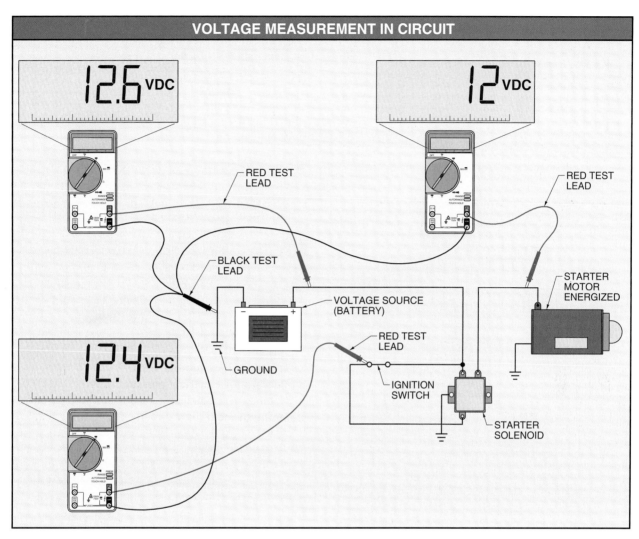

Figure 7-2. Voltage measured in a circuit is the measurement of the difference between the voltage source and any other point in a circuit.

Voltage is produced any time there is an excess of electrons at one terminal of a voltage source and a deficiency of electrons at the other terminal. The greater the difference in electrons between the two terminals, the higher the voltage. The amount of voltage required in a circuit depends on the application. The voltage source for most small engine applications is a 12 V battery. The battery supplies the pressure to move electrons through a conductor as long as there is a difference in voltage in the circuit between the voltage source and the component.

 Electrical pressure in a circuit (volts) is named after Alessandro Volta, who built the first modern electric battery in 1800.

MTD Products Inc.
The engine used to power this lawn tractor features a dual circuit alternator.

Current

Current is the flow of electrons moving past a point in a circuit. Current was abbreviated with the letter I after intensity by early scientists. Current may be alternating or direct. *Alternating current (AC)* is the flow of electrons that reverses direction at regular intervals. *Direct current (DC)* is the flow of electrons in one direction only. See Figure 7-3. For electricity to flow, there must be a difference in electrical potential. *Polarity* is the state of an object as negative or positive. AC reverses its negative and positive potentials, changes polarity, and alternates. DC does not change its polarity except in some specialized circuits.

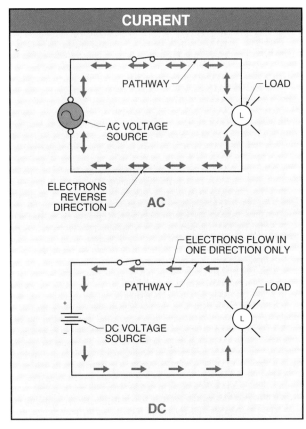

Figure 7-3. Current is the flow of electrons moving past a point in a circuit and can be AC or DC.

Current is measured in amperes. *Amperes (A)* or amps is the measurement of the number of electrons flowing through a conductor per unit of time. The flow of electrons measured in amps is similar to the flow of water through a pipe measured in gallons per minute (gpm).

Early scientists working with electricity believed that current flow occurred from the positive side to the negative side of a battery. It has since been proven that current flows from the negative side to the positive side of a battery. However, it is common in the outdoor power equipment industry to describe current flow using the traditional positive-to-negative current flow convention.

Current flow and heat

A wrench that accidentally contacts the positive side of the battery on one end and ground on the other end becomes hot quickly. Heat is produced as the wrench provides an unimpeded path for battery current at 12 V. The high volume of current causes the conductor (wrench) to generate heat.

Resistance

Resistance (R) is the opposition to the flow of electrons. Conductors, components, and devices in a circuit all have resistance. Resistance is measured in ohms (Ω). Resistance is similar to the blockage of fluid flow through a pipe from a reduced diameter or kinks. Current flow in an electrical circuit always results in heat. The amount of current flowing through a circuit is dependent on the applied voltage in the circuit and total resistance of the pathway and components included in the circuit. Most electrical components in a circuit provide a measurable amount of resistance.

Current flows easily through conductors having little resistance. Conductors contain free electrons to allow current flow. Metal is generally a good conductor. The best metal conductors are gold and silver. However, the high cost of these metals prohibits their use except in highly specialized equipment. The most common conductor used on small engines is copper wire.

Wire must safely carry the current that is flowing through the circuit. The amount of current a wire can safely carry is determined by wire size. Wire is sized by using the American Wire Gauge (AWG) numbering system. The smaller the AWG number, the larger the diameter of the wire. For example, a No. 12 wire is 80.8 mils (thousandths of an inch), and No. 22 wire is 25.35 mils. Generally, the larger the diameter of the wire, the less resistance and the more current it can safely carry. See Figure 7-4. Conductors are commonly covered with a material that serves as an insulator. An *insulator* is a material through which current cannot flow easily. Insulators provide great resistance and are commonly made from rubber, plastic, and paper. For example, wire is commonly coated with plastic.

WIRE SIZE AND RESISTANCE

AWG NUMBER	DIAMETER*	Ω/1000' AT 68°F
12	80.8	1.6
14	64.1	2.5
16	50.8	4.0
18	40.3	6.4
20	32	10.2
22	25.35	16.2

* in mils

Figure 7-4. Wire is sized using the American Wire Gauge (AWG) numbering system.

Electron movement in a conductor

Free electrons move randomly rather than in a straight line through a conductor. When under the influence of a magnetic field, electron movement is controlled by the magnetic field. The actual movement is through a series of collisions, each transferring motion and direction to the next. The speed of a colliding electron is near the speed of light, but the actual measured speed in a copper wire is approximately 3'/hr. For example, if the movement of a single electron could be tracked, it would require approximately 4 hrs to travel from a battery terminal in an automobile wiring harness to the opposite battery terminal.

Excessive current in a wire can result in overheating. Overheating can cause the melting of insulation on the wire. The loss of insulation can lead to bare wires contacting another conductor and causing a short circuit. A *short circuit* is an undesirable complete circuit path that bypasses the intended path and has very little resistance.

The patent for the Gas Engine Igniter was granted in 1910. It was the first successful product manufactured by Briggs & Stratton.

AC Voltage

AC voltage is the most common type of voltage used to produce work. AC voltage is produced by a generator. A *generator* is an electrical device that produces an AC sine wave as a wire coil is rotated in a magnetic field or as magnets are rotated inside a wire coil. An *AC sine wave* is a symmetrical waveform that contains 360°. The wave reaches its peak positive value at 90°, returns to zero at 180°, increases to its peak negative value at 270°, and returns to zero at 360°. See Figure 7-5.

A *cycle* is one complete wave of alternating voltage that contains 360°. A *pulse* is half of a cycle. An AC sine wave has one positive pulse (B+) and one negative pulse (B−) per cycle. The designation of B+ or B− is used to indicate the current flow direction. The letter B stands for bias (direction) of current flow. The B can also stand for battery, which was the only source of electricity available to the early scientists. *Frequency* is the number of complete electrical cycles per second (cps). *Hertz (Hz)* is the international unit of frequency equal to one cycle per second. For example, frequency of 5 cycles in a second is equal to 5 Hz. The United States and many other countries use AC voltage having a frequency of 60 Hz. Other countries in the world use AC voltage having a cycle frequency of 50 Hz for residential power.

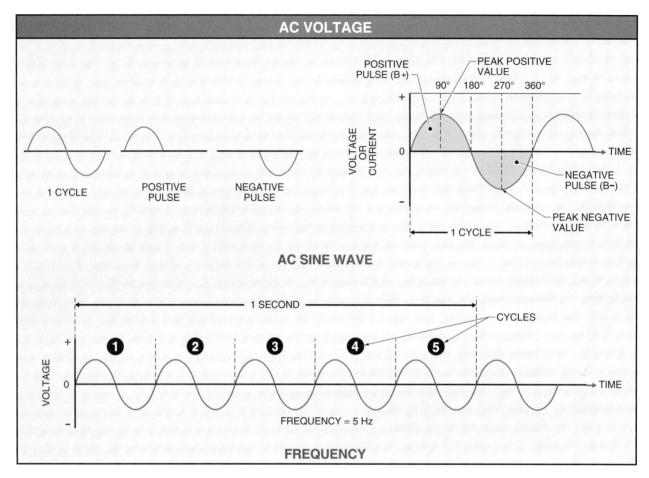

Figure 7-5. An AC sine wave is a symmetrical waveform that contains 360°. Current flow direction alternates in each cycle.

DC Voltage

All DC voltage sources have a positive and a negative terminal. Current in a DC circuit flows in one direction only. See Figure 7-6. Positive and negative terminals establish polarity in a circuit. All points in a DC circuit have polarity. A common source of DC voltage is a battery. In addition to obtaining DC directly from batteries, DC is also obtained by passing AC through a rectifier. A *rectifier* is an electrical component that converts AC to DC by allowing the current to flow in only one direction. DC voltage obtained from a rectified AC voltage supply varies from almost-pure DC voltage to half-wave DC voltage.

Wacker Corporation
The ignition system is shut down if a low oil condition occurs on this walk-behind trowel.

 The international unit of frequency (Hertz) is named after Heinrich Hertz, who demonstrated the existence of electromagnetic waves (cycles) in 1887.

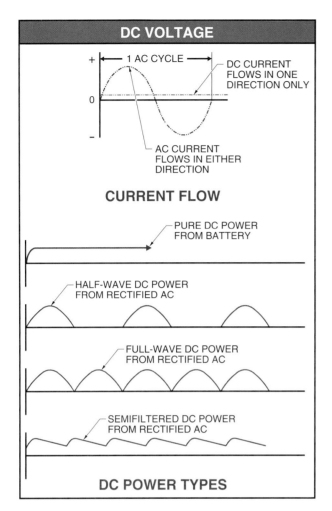

Figure 7-6. Current in a DC circuit flows in one direction only. All points in a DC circuit have polarity.

Series Circuits

A *series circuit* is a circuit that has two or more components connected so that there is only one path for current flow. Total circuit current flows through each component in a series circuit. Switches are often connected in series to control the entire circuit and to develop circuit logic. A *switch* is any component that is designed to start, stop, or redirect the flow of current in an electrical circuit. When connected in series, all switches in a circuit must be closed before any current flows. Opening one or more of the switches stops all current flow. See Figure 7-7.

Switches can be manual, mechanical, or automatic. A *manual switch* is a switch that is operated by a person. For example, ignition switches and light switches are manual switches. A *mechanical switch* is a switch that is operated by the movement of an object. For example, a safety switch installed under the seat of a lawn tractor is a mechanical switch. The switch is closed when the operator is seated, and open when the operator is off the seat. The switch can be used to prevent engagement of the blade without the operator seated for safety.

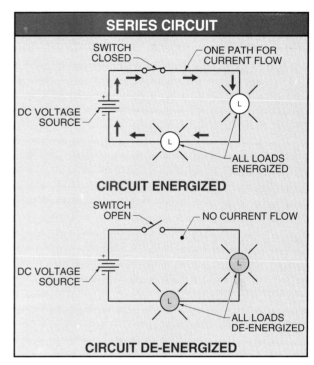

Figure 7-7. There is only one path for current flow in a series circuit.

An *automatic switch* is a switch that stops the flow of current any time current limits are reached. These switches include pressure, temperature, level, and flow switches. For example, on some small engine low oil pressure warning systems, an automatic switch senses low oil pressure and de-energizes the ignition circuit to shut down the engine.

Parallel Circuits

A *parallel circuit* is a circuit that has two or more paths (branches) for current flow. See Figure 7-8. Loads and current flow in different branches are isolated from each other. For example, on a lawn tractor electrical system, the headlights are wired in parallel to allow a headlight to continue working if there is a break in the wire or if the other headlight burns out. Switches are often connected in parallel to develop circuit logic. When connected in parallel, one or more switches must be closed before current flows. All closed switches must be opened to stop current flow.

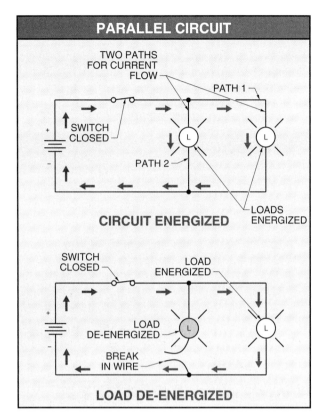

Figure 7-8. A parallel circuit is a circuit that has two or more paths (branches) for current flow.

MTD Products Inc.

A Magnetron® ignition system is used on all current production Briggs & Stratton single-cylinder and two-cylinder engines.

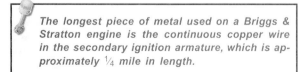

The longest piece of metal used on a Briggs & Stratton engine is the continuous copper wire in the secondary ignition armature, which is approximately ¼ mile in length.

Series/Parallel Circuits

A *series/parallel circuit* is a circuit that contains a combination of components connected in series and parallel. The component(s) function in series or parallel, depending on location in the circuit. See Figure 7-9. When switches are connected in series and parallel, all series switches and one or more parallel switches must be closed before current flows. Any one or more switches connected in series, or all switches connected in parallel, must be opened to stop current flow. For example, on a small engine electrical system, the ground wire terminal is routed through the ignition switch in series, but can also pass through an interlock system which, in certain conditions, provides an alternate path to ground.

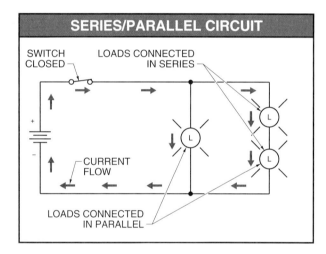

Figure 7-9. A series/parallel circuit contains a combination of components connected in series and parallel.

Limiting Current Flow

In a normal circuit, the amount of current flow is determined by the size of the load. However, if there is a problem such as a short circuit or overloaded circuit, higher than normal current flows, resulting in an overcurrent condition. An *overcurrent condition* is a condition that occurs when the amount of current flowing in a circuit exceeds the design limit of the circuit. An overcurrent condition is dangerous and can cause damage to the equipment. Fuses are wired in series and automatically stop the flow of current if a short circuit or overcurrent condition occurs.

A *fuse* is an overcurrent protection device with a thin metal strip that melts and opens the circuit when a short circuit or overcurrent condition occurs. The metal strip quickly melts and opens the circuit when the current flowing through this strip is greater than the rating of the fuse.

Electrical Prefixes

In electrical systems, numbers are commonly used when taking measurements, specifying equipment, and troubleshooting. Numbers used in electrical systems can range from very large numbers to very small numbers. In order to simplify larger and smaller numbers, metric prefixes are used. Prefixes can be used with any electrical quantity such as volts (1000 V = 1 kV), amps (1000 A = 1 kA), ohms (1000 Ω = 1 kΩ), watts (1000 W = 1 kW). For example, the rated output of a generator can be expressed as 4000 W or 4 kW. The most commonly used prefixes in the electrical and electronic fields are mega, represented by an uppercase letter M; kilo, represented by a lowercase letter k; milli, represented by a lowercase letter m; micro, represented by the Greek letter µ; and pico, represented by a lowercase letter p. See Appendix.

Converting Units

To convert a base unit number that does not include a prefix to a unit that includes a prefix, the decimal point in the base unit number is moved to the left or right and a prefix is added. For example, a small engine ignition system is capable of producing 15,000 V. To convert the base number to kilovolts (kV), the decimal point is moved three places to the left and the prefix kilo is added. In this example, 15,000 V is equal to 15 kV. See Appendix.

Ohm's Law

Electrical system values can be determined by using Ohm's law. *Ohm's law* is a law that states the relationship between voltage, current, and resistance in any circuit. Any value in this relationship can be found using Ohm's law. For example, in a circuit where voltage is constant, if resistance increases, current decreases. Likewise, if resistance decreases, current increases. See Appendix.

Drawings and Diagrams

Drawings and diagrams are commonly used when installing small engine electrical components and accessories and when troubleshooting. Drawings and diagrams show how individual parts of a component or engine system are interconnected. See Figure 7-10. Electrical devices and components can be shown on diagrams using electrical symbols. *Electrical symbols* are graphic illustrations used in electrical system diagrams to show the function of a device or component. See Appendix.

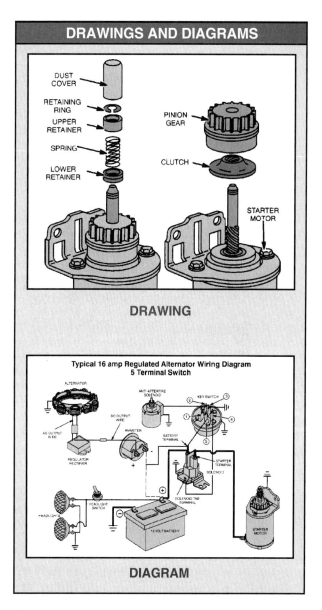

Figure 7-10. Drawings and diagrams are commonly used to show how individual parts of an engine component or system are interconnected.

Magnetism

Magnetism is an atomic level force derived from the atomic structure and motion of certain orbiting electrons in a substance. According to current theory, the orbiting of the electrons creates a very small but measurable magnetic field. A *magnetic field* is an area of magnetic force created and defined by lines of magnetic flux surrounding a material in three dimensions. *Magnetic flux* is the invisible lines of force in a magnetic field. A *magnet* is a material that attracts iron, cobalt, or nickel and produces a magnetic field. A magnet is comprised of an infinite number of molecular magnets arranged in an organized manner.

Magnetism at the atomic level

An electron that orbits and spins in one direction carries a specific magnetic field orientation. If another electron in the same orbit spins in the opposite direction, the two magnetic fields cancel each other. This is why most elements are not magnetic. However, some elements, such as iron, cobalt, or nickel, have some of the orbiting electrons spinning in the same direction. For example, in iron, there are four outer electrons spinning in the same direction that allow iron to be magnetic in certain situations. The combined magnetic field of the four outer electrons spinning in the same direction forms a magnetic domain. In nonmagnetic iron, magnetic domains are randomly aligned and have no specific orientation. A material only becomes magnetized when the magnetic domains align to form a strong magnetic field.

A magnet can be permanent or temporary. A permanent (hard) magnet is a magnet that retains its magnetism after a magnetizing force has been removed. See Figure 7-11. A temporary (soft) magnet is a magnet that can only become magnetic in the presence of an external magnetic field. When the magnetic field is removed, the magnetized material quickly returns to a nonmagnetic state. For example, an iron bar is a temporary magnet surrounded by a wound coil.

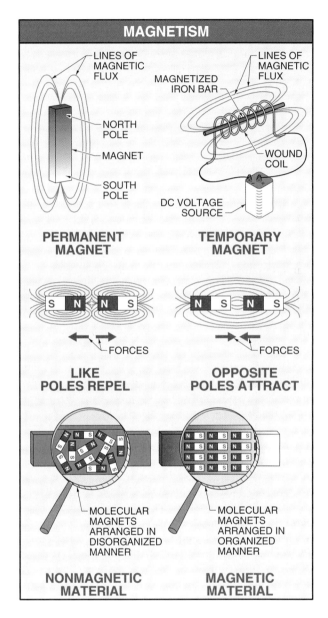

Figure 7-11. Permanent magnets retain magnetism after a magnetizing force has been removed. Temporary magnets only become magnetic in the presence of an external magnetic field.

All magnets have a north and south pole. The orientation of the poles allows lines of magnetic flux to emanate from one end and reenter the magnet at the opposite end, forming a magnetic circuit. The difference between a magnetic circuit and an electrical circuit is that an electrical circuit can be separated into positive and negative static charges. A magnet cannot exist without both north and south poles. Each piece of a broken magnet has a north and south pole orientation regardless of from where the piece came.

Like poles of two magnets repel each other. Opposite poles of two magnets attract each other. This is demonstrated by placing magnet poles next to each other. Like poles produce a resistive force and repel each other. Opposite poles produce an attractive force and attract each other.

CHARGING SYSTEM

The *charging system* is a system that replenishes the electrical power drawn from the battery during starting and accessory operation. An alternator, sometimes used in conjunction with a regulator/rectifier, maintains the battery at full charge and supplies electricity for accessories such as lights and electric lifts. Electricity produced by the chemical reaction in the battery is replaced using an alternator. Drawings and diagrams describing the charging system and components follow the traditional positive-to-negative current flow convention commonly used in the outdoor power equipment industry. Electrical components should always be installed according to engine manufacturer specifications.

Alternator

An *alternator* is a charging system device that produces AC voltage and amperage. Alternating current is commonly used to power lighting circuits in the outdoor power equipment industry. In most cases, AC output must be converted to DC output to power engine components such as electromagnetic clutches and starter motors, and to maintain the required charge in the battery. Small engine alternators use the induction principle to produce electrical power.

Induction. *Induction* is the production of voltage and current by the proximity and motion of a magnetic field or electric charge. When an electrical current passes through a conductor, it creates a magnetic field. When a magnetic field moves and/or varies in strength through a conductor, electrons are made to flow. The *induction principle* is a theory which states that with a conductor, any one of the following (current, a magnetic field, or motion) can be produced by the remaining two. See Figure 7-12. For example, current can be produced using a magnetic field and motion of a conductor. A magnetic field can be produced using the flow of current through a conductor. Motion of a conductor can be produced using a magnetic field and current.

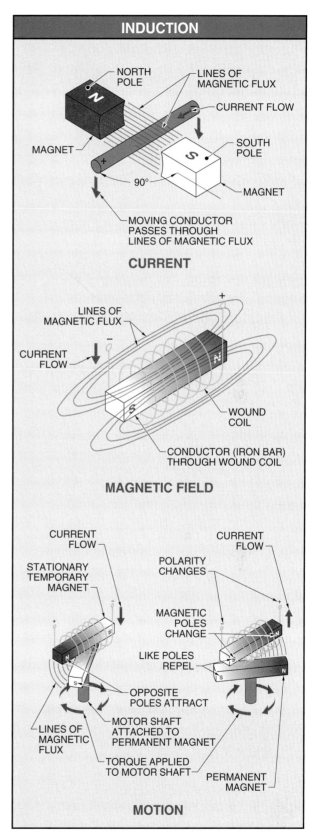

Figure 7-12. Induction can be used to produce current, a magnetic field, or motion.

Impact and loss of magnetic force

A common concern by small engine service technicians is sharp impact on magnetic material causing a possible loss of magnetic force. The magnetic material used in Briggs & Stratton products does not lose its magnetism from impact. However, a cracked or damaged magnet may cause a loss of integrity in the mount and become loose.

Current is induced in a conductor when a magnetic field is moving and/or interrupted. The magnet used to produce the magnetic field does not have to contact the conductor. Only the lines of magnetic flux must pass through it. Lines of magnetic flux encompass the magnet from the north pole to the south pole in a three-dimensional manner. When the lines of magnetic flux pass though a conductor, available free electrons are moved along the conductor in a complete circuit.

For maximum induced electron flow, the lines of magnetic flux must be perpendicular (90°) to the conductor. When these lines first influence the conductor, an induced voltage and current occurs. Without further motion, electron flow in the conductor ceases. Electron flow in the conductor continues with continued motion, varying of intensity, and/or interruption in the lines of magnetic flux.

Motion of the conductor through a stationary magnetic field or moving the magnetic field through a stationary conductor produces the same induced current. If the speed at which the lines of magnetic flux pass through the conductor is increased, the current induced increases proportionately. The polarity of the magnetic field determines the direction of current flow.

The size, content, and style of manufacture of a magnet dictates the force of the lines of magnetic flux. Generally, the larger the magnet, the greater the force. A strong magnetic field can be produced in a simple wire coil. A *coil* is a circular wound wire (winding) consisting of insulated conductors arranged to produce lines of magnetic flux. Coils are commonly used in small engine electrical systems. For example, coils in an ignition system are designed for a specific function. See Figure 7-13. The more current that flows through a coil, the larger and more powerful the magnetic field created. This increases the intensity and size of the lines of magnetic flux. Voltage and current produced by induction are determined by the length and diameter of the wire in a coil. A longer wire in a coil exposes more of the conductor to the moving magnetic field and pushes more electrons along the wire. Wire diameter affects the surface area of the wire exposed to the moving magnetic field and the amount of current that can be induced. A large-diameter wire induces more current than thin-diameter wire of the same length. The number of turns in the coil dictates the voltage produced and can be increased by using many turns of thin-diameter wire.

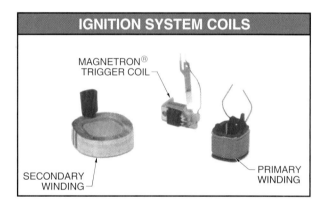

Figure 7-13. Coil design and size varies depending on the application.

Lamination Stack. A *lamination stack* is an electrical component that consists of thin iron layers used to focus and control the lines of magnetic flux. A lamination stack focuses the lines of magnetic flux perpendicular to wound wire on the lamination stack to achieve maximum current and voltage. See Figure 7-14. A lamination stack is used instead of a single piece of iron to help reduce undesirable eddy currents. An *eddy current* is undesirable current induced in the metal structure of an electrical device due to the rate of change in the induced magnetic field. Eddy currents reduce voltage output and generate excess heat. A lamination stack provides electrical insulation using small air gaps between laminations to control the formation of eddy currents. This reduces the temperature of the electrical component.

In addition to the alternator, induction is also used in other small engine electrical components such as a solenoid or starter motor. A *solenoid* is a device that converts electrical energy into linear motion. Components of a solenoid include a winding and an armature (plunger). Electricity applied to the winding of the solenoid produces a magnetic field. The magnetic field repels the plunger to produce linear motion. The linear motion of a solenoid is usually 3″ or less. Solenoids are commonly used to electrically control fuel flow and starter motors.

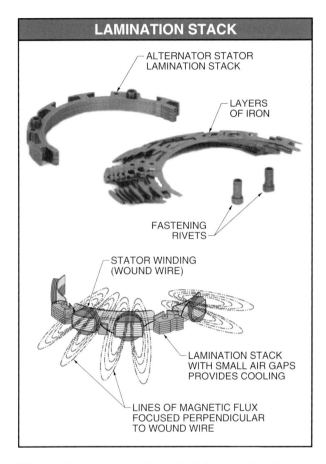

Figure 7-14. A lamination stack focuses the lines of magnetic flux perpendicular to wound wire on the lamination stack to achieve maximum current and voltage.

Alternator Operation. Flywheel magnets are commonly attached to the inside diameter of the flywheel with special epoxy cement. See Figure 7-15. The magnets are oriented so north and south poles alternate. This provides several strong magnetic fields under the flywheel with the stator. A *stator* is an electrical component that has a continuous copper wire (stator winding) wound on separate stubs exposing the wire to a magnetic field. On most stators, one end of the stator windings is attached to the stator lamination stack to provide a path for current flow through the engine block. The lamination stack is riveted together and provides the supporting structure for stator windings.

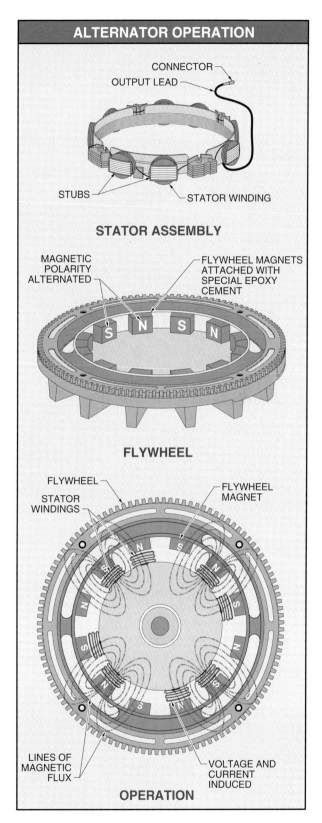

Figure 7-15. The alternator produces AC as the lines of magnetic flux from flywheel magnets pass through the stator windings.

As the flywheel rotates during engine operation, the lines of magnetic flux from the flywheel magnets pass through the stator windings. If the circuit is closed, current flows throughout the charging circuit. If the circuit is not closed, voltage is induced in the stator windings, but no current flows. Voltage is produced by the alternator whether the circuit is open or closed.

The flywheel magnets and stator windings produce current that flows in either direction (AC) based on the polarity of the specific magnet and stator winding. As the flywheel rotates, the flywheel magnets induce voltage and current in one direction, then in the opposite direction. The direction of the current flow is a function of the specific polarity of the magnet influencing the voltage induced in the stator winding. The lamination stack focuses the lines of magnetic flux and dissipates heat. In the process of producing current, heat is generated. Small air gaps between sections of steel in the lamination stack provide cooling by controlling eddy currents.

Diodes

A *diode* is an electrical semiconductor device that can be used to convert AC to DC. Diodes are made from a single crystal of silicon. During the manufacturing process, different elements are added to the silicon to make two areas containing either P-type material or N-type material. See Figure 7-16. *P-type material* is a portion of the silicon crystal that has an excess of protons and a deficiency of electrons. *N-type material* is a portion of the silicon crystal that has an excess of electrons and a deficiency of protons. P-type material is in need of electrons, and the N-type material is capable of giving up electrons. The *depletion region* is the region of a diode which separates P-type material and N-type material.

Diodes act as a one-way electrical check valve to allow current flow in one direction. If an external voltage is applied to the diode with the negative lead attached to the N-type material and the positive lead attached to the P-type material, the depletion region collapses and current flows. When the polarity is reversed, the depletion region grows and current flow stops.

When the polarity is reversed, the positive lead is attached to the N-type material and the negative lead is attached to the P-type material. Electron exchange occurs away from the depletion region from dissimilar polarity stopping current flow through the diode to act as a half-wave rectifier. A *half-wave rectifier* is an electronic device used in a charging system that converts AC to DC by blocking one-half of the AC sine wave to allow current to flow in only one direction.

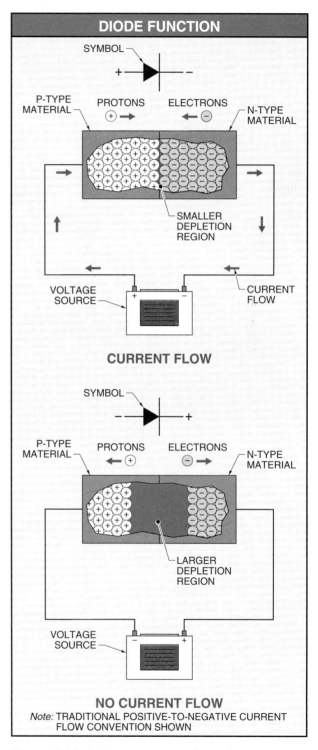

Figure 7-16. A diode converts AC to DC by acting as a one-way electrical check valve to allow current flow in one direction.

With the positive lead connected to N-type material (in series with the stator lead), the wave form is limited to the rise and fall of B+ voltage. See Figure 7-17. The diode blocks the B− portion of the sine wave to produce pulsating DC. *Pulsating DC* is DC voltage produced by rectifying (removing) one-half of an AC sine wave. The wave peaks provide a pulsed, but not smooth, transition from peak to peak. Pulsating DC, though different from smooth DC voltage provided by a battery, is still considered DC voltage.

Some alternators provide DC current through induction utilizing magnets on the outer diameter of the flywheel. The induction process works similar to the stator and magnets under the moving flywheel. This alternator has fewer magnets, and there are wider gaps between the magnets inducing current in the windings. The stator is mounted on the outside of the flywheel and the magnets only pass the stator windings once per revolution. This produces long periods of time between the intermittent voltage and current surges. Limited amounts of DC voltage and current are produced and a capacitor is required.

A *capacitor* is an electrical component that stores voltage. A capacitor increases the output of the stator windings and resists any change of voltage in a circuit by discharging or absorbing voltage whenever there is a variance. On applications requiring only a small volume of current, the capacitor is used in a circuit to absorb voltage during a pulse and to release voltage when there is no pulse.

Voltage Regulation

A *voltage regulation system* is a system that controls the amount of voltage required to charge the battery with a regulator/rectifier. A *regulator/rectifier* is an electrical component that contains one or more diodes and a zener diode. A *zener diode* is a semiconductor that senses voltage to measure the state of battery charge at the battery terminals. See Figure 7-18. A voltage regulation system decreases the current output of the alternator when the battery is charged and increases the current output of the system when the application uses battery power.

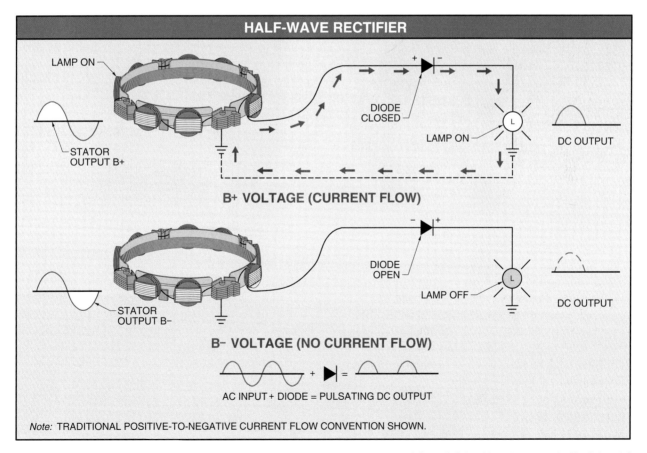

Figure 7-17. A half-wave rectifier is an electronic device that converts AC to DC by blocking one-half of the AC sine wave to allow current to flow in only one direction.

160 SMALL ENGINES

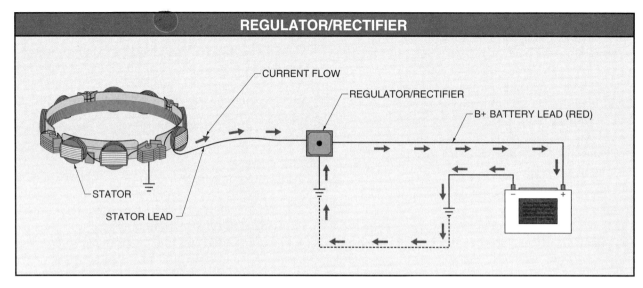

Figure 7-18. A regulator/rectifier is used in a 5 A – 9 A charging system to sense battery charge and voltage at the battery terminals.

Jacobsen Division of Textron Inc.

The 16 HP engine used to power this greens mower has a 29.3 cu in. displacement and a mechanical governor system.

If the battery is near or at full charge, the zener diode allows short pulses of current to flow to the battery. When the application is using battery power to drive a component such as an electromagnetic clutch, the zener diode senses the decrease in voltage at the battery. The zener diode begins emitting longer pulses of current to replace voltage used. When the battery is fully recharged, the zener diode returns to allowing short pulses of current.

All regulator/rectifiers must be firmly attached to the engine during testing to ensure a complete electrical circuit. An improperly grounded regulator/rectifier can result in inaccurate readings on the DMM or, in some cases, failure of the regulator/rectifier. For regulator/rectifier systems that produce 9 A or less, a DMM is connected in series. Regulator/rectifier systems that produce more than 10 A may require the use of a DC shunt to protect the DMM. See Figure 7-19.

All voltage-regulated systems need a battery with at least 5 V to operate properly. The zener diode is set to turn ON when it senses 5 V or more from the battery. If the battery does not have a minimum of 5 V, the output of the alternator system is 0 A.

The maximum rating of the alternator system is rarely reached during testing. If the alternator system is rated at 9 A, the maximum output of the system during field testing is likely to be less than 9 A. The reason the alternator system does not produce the maximum rating is based on the condition of the battery.

If battery terminal voltage is less than 5 V, a maximum output of the alternator is expected. However, when the alternator system is tested, the battery is partially charged so the regulator controls the current to bring the battery to full charge, not the maximum output of the alternator system. If the battery is nearly charged, the measured alternator output may be only 2 A – 3 A. If the battery is nearly dead, the output is commonly as high as 90% of the rated amperage.

 Briggs & Stratton has been granted nearly 700 patents since Stephen Briggs received a patent for the Gas Engine Igniter in 1910.

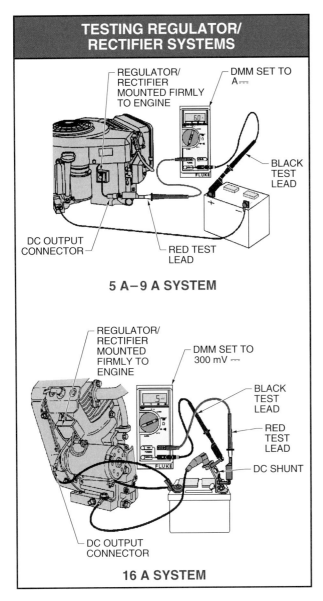

Figure 7-19. All regulator/rectifiers must be firmly attached to the engine during testing to prevent improper readings.

Full-Wave Rectification

Full-wave rectification is the process of rectifying AC and recovering the B– pulse of AC that the diode blocks. Full-wave rectification can be used to increase the charge going to the battery. Instead of blocking one-half of the current output of the AC alternator lead, a bridge rectifier is added to the circuit. A *bridge rectifier* is a device that uses four interconnected diodes to change one cycle of AC current into two DC pulses. This changes the current direction of the blocked pulse and redirects all current to move in the same direction. The regulator/rectifier uses a zener diode with a bridge rectifier to rectify AC and control the rate at which a charge is released. Each B+ pulse above the zero line is current flow in one direction. A B– pulse below the zero line is current flow in the opposite direction. See Figure 7-20.

The stator used with this style regulator/rectifier is unique to Briggs & Stratton alternator systems as both ends of the stator windings are routed from under the flywheel as two separate leads. The stator windings are insulated from the lamination stacks to prevent a complete circuit between the windings and the engine block. The common connection between the stator windings and the engine block is provided by the regulator/rectifier.

As the B+ portion of the stator output enters the regulator/rectifier, it encounters a bridge rectifier that allows current to pass through in one direction only. As current travels through the diodes, it passes through the battery, causing the necessary chemical reaction to provide energy for recharging. Once the B+ current has passed through the battery, it returns back to its source through the common connection on the regulator/rectifier housing. The charge reenters the regulator/rectifier through this connection and passes through two more diodes to complete the return to the stator windings.

The B– pulse passes through the regulator/rectifier in the same manner only in an opposite direction. The current enters the regulator/rectifier through the other stator lead, passes through two diodes, through the battery, and returns to the regulator/rectifier through the common connection provided by the regulator/rectifier housing. Once returning to the regulator/rectifier, the current passes through the other two diodes and returns to the opposite lead of the stator windings.

The result of redirecting the current is capturing almost all of the potential amperage in the system. If an unregulated alternator system is rated at 10 A, the alternator output is based on 5 A B+ and 5 A B– for a total of 10 A. With half-wave rectification, approximately one-half of the output of the alternator is lost. When full-wave rectification is incorporated in the alternator system, the B– current is redirected and converted into B+ current to capture almost all of the potential output of the system.

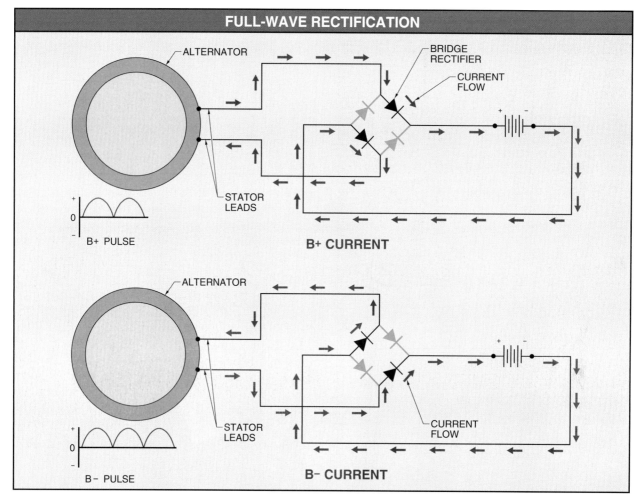

Figure 7-20. Full-wave rectification uses a bridge rectifier to change the current direction of the blocked pulse and redirect all current to move in the same direction.

Battery

A *battery* is an electrical energy storage device. Small engine electrical systems commonly include a lead-acid battery. A *lead-acid battery* is a battery that stores electrical energy using lead cell plates and sulfuric acid (H_2SO_4). *Electrolyte* is a mixture of water and sulfuric acid (H_2SO_4) used in a lead-acid battery. Battery condition is commonly checked using a hydrometer. See Figure 7-21. A *hydrometer* is an instrument used to measure the specific gravity of a liquid. *Specific gravity* is a comparison of the mass of a given sample volume compared to an equal volume of water. As a battery is discharged, electrolyte loses sulfuric acid and increases the percentage of water, resulting in a decrease of specific gravity.

A lead-acid battery produces voltage through chemical reactions initiated by current flow either into or out of the battery. Cell plates inside the battery are chemically treated with different materials. One plate is covered with spongy lead. The other plate is covered with lead dioxide. Spongy lead has a high degree of chemical reactivity when immersed in H_2SO_4. Sulfuric acid molecules break up into positively charged hydrogen ions and negatively charged sulfate ions. At the spongy lead plate, the sulfate ions react with the lead to form sulfate. The lead ions that leave the plate are positively charged, leaving the plate with a net negative charge.

 The word circuit comes from the Latin word cirularis, meaning circle.

Batteries are often rated by their reserve capacity and/or cold cranking amps. The *reserve capacity* of a battery is the amount of time a battery can produce 25 A at 80°F. *Cold cranking amps (CCA)* is the number of amps produced by the battery for 30 sec at 0°F while maintaining 1.2 V per cell. Corrosion of battery terminal connections can cause resistance, which reduces current flow. See Figure 7-22. Corrosion is best removed using a solution of water and baking soda.

Figure 7-21. Battery condition is commonly checked using a hydrometer.

Figure 7-22. Corrosion of battery terminal connections can cause resistance, which reduces current flow.

At the lead dioxide coated plate, another chemical reaction occurs simultaneously. This chemical reaction involves lead dioxide, sulfuric acid, and hydrogen ions. The products of this reaction are the lead sulfate remaining on the cell plate and water, which goes into solution with sulfuric acid. Electrons must be supplied by the lead dioxide plate to initiate the chemical reaction. The lead dioxide plate is left with a net positive charge.

Lead sulfate is created on both cell plates by the different processes and reactions in the battery. After a period of time, lead sulfate completely covers both of the plates. This causes the plates to become chemically similar, resulting in the battery being run down or completely discharged. To supply electrical energy, the battery must be recharged by adding electrons. Forcing electrons into the battery reverses the discharging process by returning the cell plates to their original condition. Electrons can then flow from negative to positive in a circuit because of having a negative charge and being attracted by the positive potential of the battery.

 An AC sine wave is created by the revolving motion inside an alternator, which produces a fluctuating voltage and current.

Battery Safety. Battery electrolyte is extremely corrosive and can cause severe burns to eyes and skin. Batteries must always be handled in an upright position. Electrolyte spilled on the skin must be immediately flushed off with cold water. Electrolyte splashed in the eyes must be immediately flushed out at an eye wash station or with cold water. A physician should be contacted immediately.

Warning: Always wear the required eye protection and protective clothing when servicing a battery.

Batteries must be serviced in a well-ventilated area. Batteries produce hydrogen gas that can cause an explosion if ignited by a spark or open flame. Always follow the manufacturer's recommended procedures for charging, installation, removal, and disposal.

IGNITION SYSTEM

The *ignition system* is a system that provides a high-voltage spark in the combustion chamber at the proper time. Ignition systems used in Briggs & Stratton engines are the breaker point ignition system and the Magnetron® ignition system. Although several variations of these systems exist, the function of each system is essentially the same. Single-cylinder engine ignition systems are less complex than multiple-cylinder engine ignition systems.

Breaker Point Ignition System

A *breaker point ignition system* is an ignition system that uses a mechanical switch to control timing of ignition. Breaker point ignition systems were common until the mid-1980s. The breaker point ignition system uses induction produced in an ignition coil to cause a spark to ignite the charge in the combustion chamber. All Briggs & Stratton flywheels contain magnetic inserts attached to their outer diameter. See Figure 7-23.

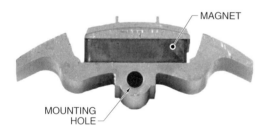

Figure 7-23. Magnetic inserts on the flywheel provide north and south polarity perpendicular to the crankshaft.

The magnetic orientation of the insert provides north and south polarity perpendicular to the crankshaft. The *ignition armature* is a component containing two or more coils which, when acted upon by a magnetic field, induce electrical energy. The ignition armature used on a breaker point ignition system contains two separate coils: the primary winding and the secondary winding.

Primary Winding. The *primary winding* is a coil that induces voltage in the secondary winding. One end of the primary winding is connected to a series of lamination stacks that bisect the ignition armature. The other end of the primary winding terminates at a connector or wire lead on the outside of the ignition armature body. A wire connects the primary winding connector to the breaker points.

Breaker points are an ignition system component that has two points (contact surfaces) that function as a mechanical switch. The contact surfaces are plated with nickel. When in contact with each other, the points act like a closed switch to allow electricity to flow to other parts of the circuit. See Figure 7-24. One point of the breaker points is held in a fixed position. The other point is retained by a spring and is mounted on a pivot. The points are held open by a point plunger. A *point plunger* is an ignition system component that holds the movable point contact surface away from the fixed contact surface to interrupt current flow to the primary winding. The point plunger is actuated by a flat spot machined into the crankshaft or camshaft. During most of the 360° rotation of the crankshaft, the points are held in open position. When the flat spot of the crankshaft is directly under the base of the point plunger, the point plunger retracts to allow current flow to the primary winding. Current flow through the primary winding creates a magnetic field which encompasses the primary and secondary windings.

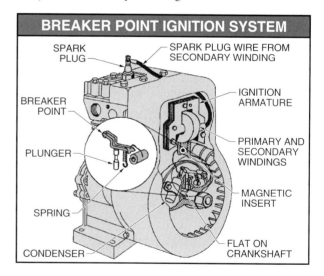

Figure 7-24. Breaker points function like a mechanical switch to control the flow of electricity to other parts of the ignition system circuit.

The theory of induction was first proposed by Michael Faraday in the early 1800s and became the basis for understanding the relationship between magnetism and the production of electricity.

Secondary Winding. The *secondary winding* is a coil in which high voltage is induced for use at the spark plug. The secondary winding has windings of copper wire the thickness of a human hair. The secondary winding contains over 1400′ of wire wrapped tightly around, but not in direct contact with, the primary winding. The ratio of turns of wire between the primary and secondary windings is approximately 1 : 60.

In a single-cylinder ignition armature, one end of the secondary winding is attached to the lamination stack of the ignition armature. The other end of the secondary winding is attached to the spark plug wire which is connected to the spark plug. The *spark plug* is a component that isolates the electricity induced in the secondary windings and directs a high voltage charge to the spark gap at the tip of the spark plug. See Figure 7-25. The spark plug shell secures the spark plug in the engine. The spark plug wire is connected to the spark plug terminal. The insulator prevents shorting between the center electrode and the grounding electrode. The *spark gap* is the distance from the center electrode to the ground electrode on the spark plug. The high-voltage spark jumps across the spark gap, igniting the air-fuel mixture in the combustion chamber.

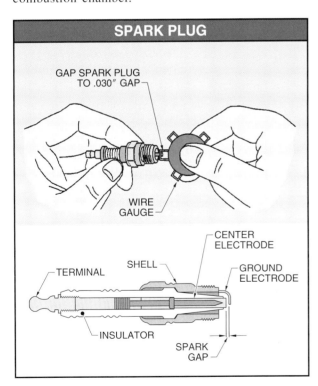

Figure 7-25. The spark plug directs a high voltage charge to the spark gap to ignite the charge in the combustion chamber.

The Start-Spark, developed by Briggs & Stratton in the 1930s, improved cold weather starting by increasing the heat and frequency of the spark.

Ignition Armature Induction. Any conductor that has current passing through produces self-inductance. *Self-inductance* is a magnetic field that is created around a conductor whenever current moves through the conductor. The size and intensity of a self-induced magnetic field is proportional to the amount of current passing through the conductor.

Breaker point ignition system operation begins when the flywheel magnets move toward the ignition armature. The ignition armature consists of a lamination stack, a primary winding, and a secondary winding. See Figure 7-26. Flywheel magnets produce a moving magnetic field that passes through the lamination stack. There is little, if any, magnetism flowing through the upper part of the lamination stack as the air gap provides high resistance. The approach of the magnetic field is timed with the flat spot machined in the crankshaft, causing the point plunger to fall, resulting in the closing of the breaker points.

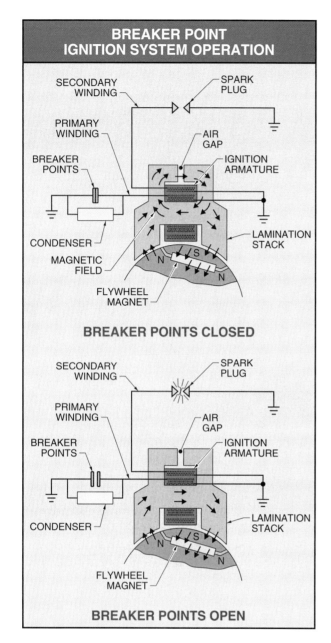

Figure 7-26. The opening of the breaker points causes the magnetic field in the primary winding to collapse to induce a high voltage in the secondary winding required for the spark plug.

The closed breaker points provide a path through the primary winding to ground completing a circuit and allowing the induction of current. As the magnet influences the primary winding, it induces voltage, resulting in current in the primary circuit. The primary circuit consists of the primary winding (one end attached to the lamination stack or engine ground), and the closed breaker points providing a complete circuit. The current induced in the primary winding flows from the primary winding through the breaker points and into the engine block, returning to its source.

As the first flywheel magnet moves closer to the ignition armature, the magnetic field from the flywheel magnets increases through the primary winding, causing an increased output of the inductive circuit. With this increase in current flow, the self-induced magnetic field of the primary winding reaches its maximum size and strength. Because the secondary winding is adjacent to the primary winding, the magnetic field in the primary winding induces a potential voltage in the secondary winding. There is no current flow in the secondary winding because there is no return path for the current to flow. A moving magnetic field can only induce current in the presence of a complete circuit.

As the alternating poles of the flywheel magnets continue to move past the ignition armature, the polarity changes. The polarity change influences the lamination stacks. Because of the self-inductance of the primary winding, a slight time lag allows the magnetic field induced by the polarity of the first flywheel magnet to remain. The breaker points open at a precise time. The opening of the breaker points severs the primary circuit, causing the magnetic field in the primary winding to collapse. The speed at which the primary winding magnetic field collapses is near the speed of light. The increase in speed induces a high voltage in the secondary winding. The voltage potential in the secondary winding increases to approximately 10,000 V – 15,000 V. This high voltage is sufficient to force a current flow between the spark gap of the spark plug to initiate combustion.

As the breaker points begin to open to sever the primary circuit, there is sufficient voltage and current flow in the primary winding circuit to jump the small gap between the breaker points. If this occurs, the magnetic field in the primary circuit does not abruptly collapse. To assist in the abrupt and accurate severing of the primary circuit, a condenser is connected to the primary winding lead.

A *condenser* is a capacitor used in an ignition system that stores voltage and resists any change in voltage. The condenser has the ability to store voltage similar to a battery. To ensure that the breaker points sever the circuit at the proper time, the condenser absorbs most of the energy induced in the primary windings. As soon as the breaker points have moved far enough apart to create additional resistance, the condenser absorbs any residual current. This reduces the possibility of a spark jumping the gap between breaker points. This increases the life of the breaker points dramatically. In addition, the condenser provides a more consistently-timed breaking of the primary circuit, allowing the magnetic field to collapse at the proper crankshaft position.

> ### Zener diode operation and pulse control
>
> *The operation of a zener diode can be compared to a microwave oven used to cook a frozen bagel. When the microwave oven is set to slowly defrost the bagel, microwaves are emitted in short pulses between long intervals. When the microwave oven is set to quickly cook the bagel, microwaves are emitted in long pulses between short intervals. A zener diode voltage regulator operates similar to a microwave oven by using variable pulse duration and intervals to regulate charging current to the battery.*

Magnetron® Ignition System

The *Magnetron® ignition system* is an ignition system that uses electronic components in place of breaker points and a condenser. The Magnetron® ignition system was introduced in the mid-1980s and provides greater efficiency and reliability than a breaker point ignition system. The Magnetron® ignition system works using the same basic ignition armature structure and design as a breaker point ignition system. The Magnetron® module replaces the function of the breaker points and the condenser. See Figure 7-27.

In ignition armature operation, current in the primary winding is built up in a few thousandths of a second. As the current flow builds, a magnetic field is created and expands around two windings overlapping each other. When the maximum current is reached, the magnetic field has surrounded the primary and secondary windings. To create the abrupt circuit interruption needed at the precise time of ignition, a Magnetron® module is incorporated into the ignition armature.

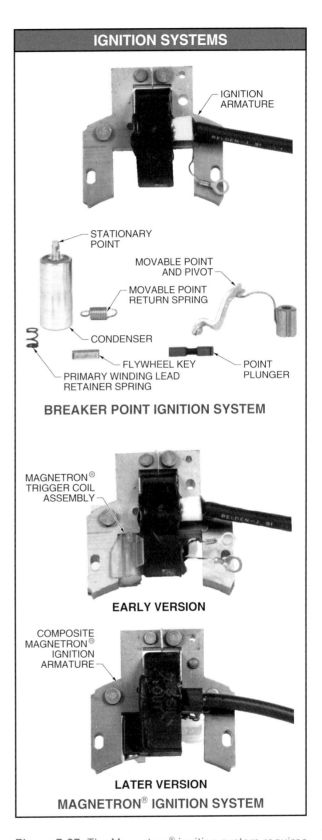

Figure 7-27. The Magnetron® ignition system requires fewer parts and less service than the breaker point ignition system.

Inside the Magnetron® module is a small coil of copper wire (trigger coil), a diode, and a transistor pair used for switching, to provide the means to collapse the magnetic field. The transistor pair is set in a normally open condition, preventing any current flow though the primary winding. The transistor pair performs the same function as the breaker points and condenser.

As the flywheel magnet insert approaches the Magnetron® trigger coil in the Magnetron® module, a small voltage is induced. See Figure 7-28. When voltage in the trigger coil reaches .65 V, the transistor pair turns ON or provides a closed circuit for the primary winding. This allows a 2 A to 3 A current to be induced in the primary winding and generate the magnetic field.

All flywheel magnets used with a Magnetron® ignition system have the same polarity sequence of north, south, north. The trigger coil and its circuits are polarity-sensitive. A *polarity-sensitive circuit* is a circuit that does not operate properly when exposed to the wrong polarity. When the flywheel turns past the optimum point of trigger coil electrical generation, flow in the trigger coil falls to zero. The polarity of the current begins to reverse itself due to the changing polarity of the flywheel magnets. The voltage drop in the trigger coil rapidly turns OFF the transistor, similar to throwing a switch. The lines of magnetic flux created by the primary windings collapse back to their source. The time it takes to collapse the field is 2 to 5 millionths of a second. This rapid decrease of the magnetic field coming from the primary windings induces a high voltage in the secondary winding. The 15,000 V to 20,000 V generated by the collapsing magnetic field causes a flow of current between the spark gap to ignite the charge in the combustion chamber.

The trigger coil allows the current to flow through the primary winding because the induced voltage of .65 V or greater in the trigger coil turns the transistor ON. The transistor stops all current flow when the voltage drops below .65 V. This is similar to the rise and fall of a voltage pulse on a sine wave.

The two types of Magnetron® ignition systems are the standard Magnetron® ignition system and the advancing-style Magnetron® ignition system. The standard Magnetron® ignition system starts with a set ignition timing and slowly retards spark as rpm increases. See Figure 7-29. This is caused by increased speed, creating more voltage in the trigger coil and a slower return to .65 V. The standard Magnetron® ignition system is identified by the small back gap in the lamination stacks at the top portion of the ignition module. This back gap contains a small piece of paper between the halves of the lamination stacks. The advancing-style Magnetron® ignition system is identified by a relatively wide back gap without any paper insert.

The advancing-style Magnetron® ignition system functions similar to the standard Magnetron® ignition system. The main difference is the addition of an SCR (silicon controlled rectifier). A *silicon controlled rectifier (SCR)* is a semiconductor that is normally an open circuit until voltage is applied, which switches it to the conducting state in one direction. In the advancing-style Magnetron® ignition system, the transistor pair is set in normally closed position.

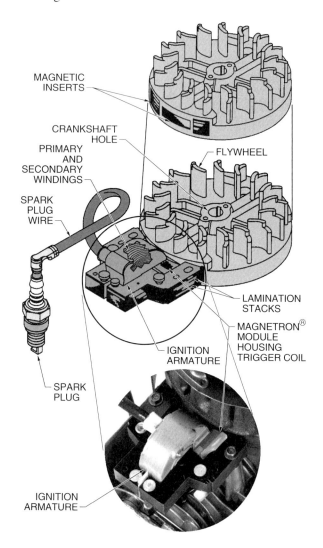

Figure 7-28. As the flywheel magnet insert approaches the Magnetron® trigger coil in the Magnetron® module, a small voltage is induced.

The SCR becomes part of the switching device. When the trigger coil voltage reaches .65 V, the SCR turns ON. The SCR then shorts the current, which controls the transistor pair. When the transistor pair turns OFF, spark is generated in the same manner as the standard Magnetron® ignition system. This changes the timing from retarding as rpm increases to advancing as rpm increases. With increased speed, more voltage is induced faster. The faster the engine operates, the sooner the trigger coil reaches .65 V and spark is initiated. The built-in spark advance is approximately 3° – 5° of crankshaft rotation.

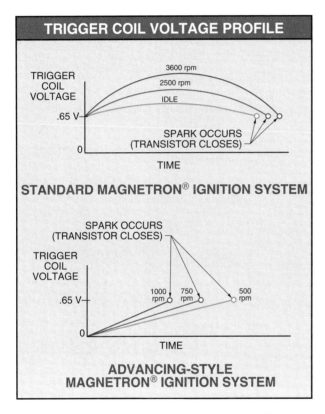

Figure 7-29. The advancing-style Magnetron® ignition system uses a silicon controlled rectifier (SCR) that advances ignition timing as rpm increases.

Ignition systems are designed to operate efficiently with specifically designed components and should not be altered. Original components should be replaced with identical parts. Mixing of ignition system components can cause improper operation.

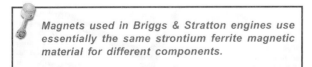

Magnets used in Briggs & Stratton engines use essentially the same strontium ferrite magnetic material for different components.

A Briggs & Stratton Series F engine was used to power the Nickle Bean Picker, manufactured by the Nickle Engineering Works of Saginaw, Michigan in 1927.

Two-Cylinder Ignition System

Some two-cylinder engines use two separate ignition systems. The 180° opposed-twin engine uses a variation of a breaker point ignition or an original Magnetron® ignition system. One ignition armature is used for both cylinders. Each spark plug is attached to opposite ends of the secondary windings. Some two-cylinder engines, such as the Vanguard™ V-Twin, use two separate advancing-style Magnetron® ignition armatures to provide spark.

The Vanguard™ V-Twin engine also uses a special primary circuit harness that is connected to the ground connectors of each Magnetron® ignition armature. Two diodes in the primary circuit harness electrically separate the ignition armatures. The diodes in the harness stop the electrical current from the primary windings of ignition armatures from seeking a path to ground through the other ignition armature.

170 SMALL ENGINES

STARTING SYSTEM

Small engines are equipped with the starting system required for the engine size and application. Starting systems can be a mechanical (rewind) starting system or an electric starting system. See Figure 7-30. A *rewind starting system* is a mechanical starter that commonly consists of a rope, pulley, and return spring used to manually rotate the crankshaft to start an engine. An *electric starting system* is a group of electrical components activated by the operator to rotate the crankshaft when starting an engine. Both rewind starting systems and electric starting systems vary depending on design and application. Larger single-cylinder and multiple-cylinder engines commonly require an electric starting system. Electric starting systems commonly include a starter motor, starter solenoid, ignition switch, and a battery.

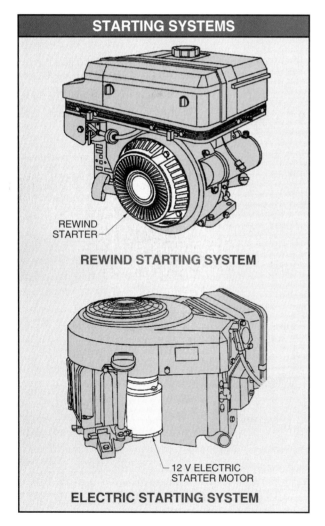

Figure 7-30. Small engine starting systems are designed for the engine size and application.

Starter Motor

A *starter motor* is an electric motor that drives the engine flywheel when starting. A starter motor is commonly a 12 VDC motor, but can also be a 120 V motor. The starter motor rotates the crankshaft and other engine components to engine starting speed (greater than 350 rpm). Starter motors are designed and sized for specific engine applications, but all contain similar components. See Figure 7-31.

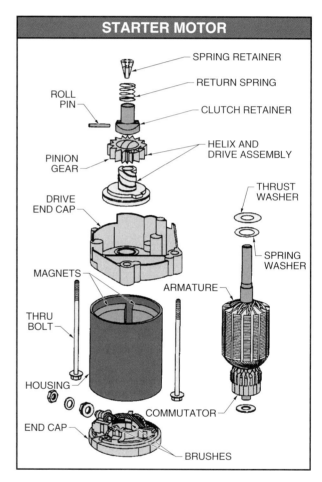

Figure 7-31. Starter motors are designed and sized for specific engine applications, but all contain similar components.

When a starter motor is energized, the armature begins spinning. The spinning motion of the armature drives the pinion gear to the end of the helix. The *pinion gear* is a gear on the starter motor that follows the helix to engage and drive the flywheel ring gear. The *helix* is the component or portion of the armature shaft that has helical grooves to provide axial movement of the starter pinion gear. The helix can be machined from steel or molded from a polymer material. The *flywheel ring gear* is the gear attached to the engine flywheel driven by the pinion gear during engine starting. When the pinion gear reaches the end of the helix, it engages with the flywheel ring gear. When the starter motor is de-energized, armature rpm slows, allowing the return spring (in most designs) to disengage the pinion gear from the flywheel ring gear by pushing it down the helix.

Some starting systems use an automotive-style starter motor with a starter solenoid to engage and disengage the pinion gear with the flywheel ring gear. A *starter solenoid* is an electrical switch with internal contacts opened or closed using a magnetic field produced by a coil. See Figure 7-32. When the ignition switch is turned to the start position, current flows from the battery energizing the solenoid coil to close the internal contacts. The closed contacts provide an electrical path for starting current flow to the starter motor to rotate the crankshaft. Most outdoor power equipment requires a solenoid to allow a low amperage switch to activate the high amperage starter motor circuit.

motor armature. Force from the rotating armature actuates the sliding pinion gear, causing physical contact with the flywheel ring gear, which is attached to the flywheel. The pinion gear is then free to travel along the helix coupled with a break-away clutch.

A *break-away clutch* is an engine component that allows slippage to prevent damage to the pinion gear during a misfire or unexpected reverse rotation. See Figure 7-33. A break-away clutch also provides a release if the starter motor load exceeds the capacity of the starter motor. The break-away clutch then allows free rotation of the armature shaft to prevent overheating of starter motor components.

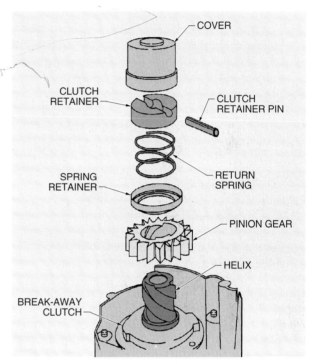

Figure 7-33. A break-away clutch allows slippage to prevent damage to the pinion gear and provides a release if the starter motor is overloaded.

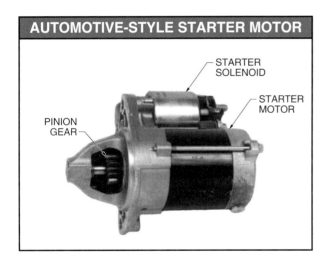

Figure 7-32. Automotive-style starter motors use a starter solenoid to engage and disengage the pinion gear with the flywheel ring gear.

12 V Starter Motor Operation. The 12 V starter motor operates using basic electrical principles. A pair of permanent ceramic magnets are located in the outer shell of the starter motor. These magnets are stationary and produce a magnetic field with a specific polarity affecting the windings on the armature assembly. The pinion gear and helix drive assembly operation is based on the speed of the starter

The components of the 12 V starter motor operate using basic induction theory. The starter motor armature is a shaft containing multiple windings of insulated copper wire around a series of lamination stacks similar to those found in an alternator stator. The commutator is located at the end of the armature shaft farthest from the pinion gear. A *commutator* is a sectional piece of copper that is directly connected to many loops of copper wire in contact with brushes. *Brushes* are carbon components in contact with the commutator that carry battery current to operate the starter motor. Most Briggs & Stratton starter motors contain two pairs of brushes. One pair of brushes is connected together by a copper wire and the positive side of the battery. The other pair of brushes is connected to the starter motor housing attached to ground.

When the starter motor is energized, the battery current flows through the brushes to the commutator. The current then flows through windings on the armature shaft and out the other set of brushes to the starter motor housing. The brushes, windings, and starter motor housing provide the circuit for current flow back to the battery.

Homelite, Inc.
This generator is rated at 5000 continuous watts and is powered by an 11 HP engine with low oil shutoff.

Current flowing through each separate winding of the armature produces a magnetic field having a specific polarity. The magnetic field is the same polarity aligned with the permanent magnets surrounding the armature. The two magnetic fields of the same polarity repel each other. The magnetic fields are timed to produce rotation (torque) as the armature shaft rotates through the magnetic fields.

The sections of the commutator allow current flow to the many separate windings of the armature for smooth operation. As the armature shaft rotates, a new section of the commutator comes in contact with the brushes. The current flow through the different winding produces another magnetic field of the same orientation as the permanent magnets located in the shell. Induction of the magnetic fields in the armature winding creates heat. The greater the load on the starter motor, the more current required to rotate the armature shaft. The windings in the starter motor can become very hot during high load conditions.

Excessive heat can cause the protective varnish insulation on the windings to melt. If the varnish on the windings melts, the windings are no longer insulated from each other and current follows the shortest path, resulting in a dead short. This reduces the magnetic field in a single winding and the torque available to rotate the engine. Other windings must allow more current to flow to maintain starter motor operation. This causes overheating, and increases the problem until the varnish on many windings has melted. A distinctive smell signals that the starter motor has failed.

120 V Starter Motor Operation. A 120 V starter motor operates similar to a 12 V starter motor. The primary difference is that a 120 V motor incorporates a bridge rectifier to convert the 120 VAC from the power source to DC current. The 120 V system also uses a winding of copper wire in place of the permanent magnets used on the 12 V starter motor. A DC current is applied to a winding attached to the starter motor housing. The moving current produces a magnetic field similar to the permanent ceramic magnet in the 12 V starter motor.

A 120 V starter motor obtains power from a 120 V wall outlet instead of a 12 V battery. Special care must be taken to prevent parasitic loads on engines equipped with a 120 V starter motor. A 12 V starter motor system indicates to the operator when there is considerable load on the starter motor by a weakening battery and poor starter motor performance. However, the 120 V starter motor continues to draw current from the 120 V wall outlet regardless of the load. This can lead to severe overheating of the armature windings and eventual starter motor failure.

 Magnets and magnetic inserts used on Briggs & Stratton engines are not magnetized until moments before they are installed on the engine.

COOLING AND LUBRICATION SYSTEMS

CHAPTER 8

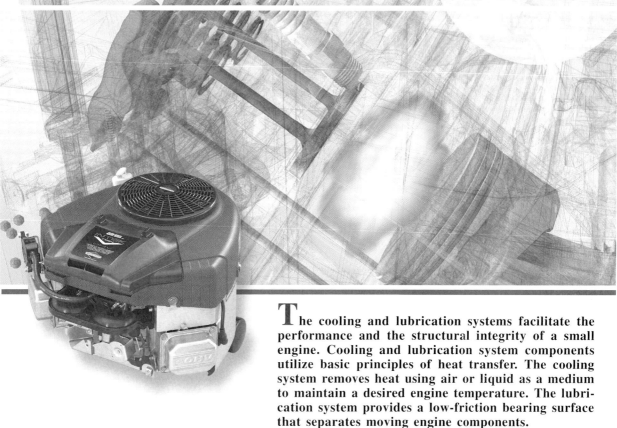

The cooling and lubrication systems facilitate the performance and the structural integrity of a small engine. Cooling and lubrication system components utilize basic principles of heat transfer. The cooling system removes heat using air or liquid as a medium to maintain a desired engine temperature. The lubrication system provides a low-friction bearing surface that separates moving engine components.

ENGINE HEAT

Like other forms of energy, heat is capable of doing work. Heat flows whenever a temperature difference exists in a material. Heat always flows from an area of higher temperature to an area of lower temperature. See Figure 8-1. Heat flow is similar to electron flow when there is a difference of voltage between two points in a circuit.

Matter does not contain heat, rather, it contains kinetic energy. Once absorbed into the material, kinetic energy no longer is considered heat and becomes internal energy. *Internal energy* is the sum of all energy in a substance, including potential and kinetic energy.

Temperature difference and heat transfer direction

When a hand comes in direct contact with a hot engine, heat (kinetic energy) flows from the engine to the hand because of the temperature difference. When a hand comes in direct contact with an ice cube, heat flows from the hand into the ice. The direction of heat transfer is always from a higher temperature to a lower temperature. The rate of heat transfer depends on the thermal conductivity of the material.

174 SMALL ENGINES

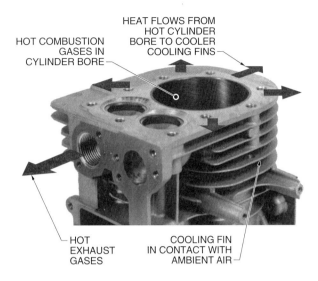

Figure 8-1. Engine heat flows out through hot exhaust gases and from high temperatures in the cylinder bore to cooler temperatures in the cooling fins to control operating temperature.

During the operating cycle of an internal combustion engine, combustion gas temperature varies greatly in a short period of time. Combustion gas temperature is typically lowest toward the end of the intake event. During the compression event, the temperature increases from the compression of the air and fuel molecules. See Figure 8-2. The combustion gas temperature increases until it peaks soon after ignition. The combustion gas temperature decreases during the power and exhaust events. A rapid temperature decrease occurs as the intake valve opens to allow the cool air-fuel mixture into the combustion chamber.

ENGINE MATERIALS AND CHARACTERISTICS

Small engines are commonly manufactured from cast iron alloy or cast aluminum alloy. Cast iron alloy used for small engines contains carbon, phosphorus, and silicon in addition to the base element iron. Cast aluminum alloy contains silicon, copper, and other elements in addition to the base element aluminum. The physical properties of these materials determine the specific characteristics of the cylinder block, crankcase, cylinder head, and internal components of the engine.

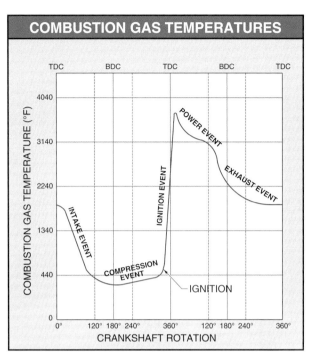

Figure 8-2. Combustion gas temperatures vary with different events in the operating cycle of an internal combustion engine.

Cast iron alloy has long been used for engine blocks and internal components in the small engine industry. Desirable properties of cast iron alloy include:

- greater density resulting in higher compressive strength compared to cast aluminum alloy
- structural integrity when mechanically or thermally stressed
- a porous surface providing small oil reservoirs
- excellent lubrication provided by graphite commonly contained in cast iron
- reduced dimensional changes when placed under thermal stress compared to cast aluminum alloy

Undesirable properties of cast iron alloy include:

- increased weight
- a high propensity for oxidation/corrosion
- difficulty in machining
- relatively poor heat conductivity and dissipation

 Engines equipped with an Oil Gard® system shut down when exceeding a 15° angle of operation.

Cast aluminum alloy is relatively new to the small engine industry compared to cast iron alloy. Desirable properties of cast aluminum alloy include:

- high strength-to-weight ratio
- ease of manufacture
- lower production cost than cast iron alloy
- excellent heat dissipation

Undesirable properties of cast aluminum alloy include:

- higher raw material costs compared to cast iron alloy
- less resistant to wear than cast iron alloy
- greater thermal expansion than cast iron alloy

A common concern with cast aluminum alloy is that aluminum tends to distort when exposed to high temperatures and retains the distortion longer than most cast iron alloys. However, with a proper cooling system, cast aluminum alloy provides a durable, lightweight, and cost-effective alternative to cast iron alloy.

Heat-Induced Expansion

In most cases, matter expands when heated. This expansion is caused by increased vibration of atoms or molecules, causing them to move farther apart from each other. The effects of heat on matter vary. Cast iron alloy engines and components tend to resist the absorption of heat. Cast aluminum alloy engines and components have a higher degree of heat absorption. The heat absorption rate of a material is not always a reliable indicator of its ability to dissipate (release) heat. Heat-induced expansion must be considered when selecting materials for specific components used in small engines. Heat-induced expansion is affected by thermal conductivity, thermal expansion, thermal growth, and thermal distortion.

Thermal Conductivity. *Thermal conductivity* is the ability of a material to conduct and transfer heat. Aluminum has one of the highest rates of thermal conductivity of all common metals. The rate of thermal conductivity for cast iron is considerably lower. Thermal conductivity allows heat to transfer through a mass to the area of lowest temperature. On an air-cooled engine, the area of lowest temperature is the cooling fins. On a liquid-cooled engine, the area of lowest temperature is the radiator fins. See Figure 8-3.

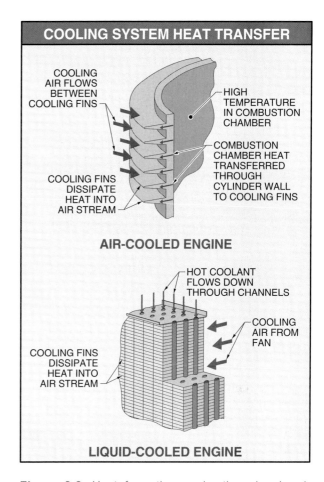

Figure 8-3. Heat from the combustion chamber is transferred to the air stream as it flows between cooling fins or through a radiator.

Thermal Expansion. *Thermal expansion* is the expansion of a material when it is subjected to heat. Thermal expansion is determined by the molecular structure of the material. As the temperature of the material increases, its atoms and molecules vibrate with greater intensity. This causes linear expansion of the material in all directions. Cast aluminum alloy expands faster and with greater magnitude than cast iron alloy, as indicated by the coefficient of thermal expansion. The *coefficient of thermal expansion* is the unit change in dimension of a material by changing the temperature 1°F. The coefficient of thermal expansion for aluminum is .00001244 linear expansion per unit length per °F at 68°F. The coefficient of thermal expansion for cast iron is .00000655 linear expansion per unit length per °F at 68°F.

 Friction is one way of dissipating available energy.

Thermal Growth. *Thermal growth* is the increase in size of a material when heated with little or no change back to original dimensions. Even when subjected to extreme cold, a material that has experienced thermal growth cannot return to its original dimensions. Thermal growth occurs in all directions. See Figure 8-4.

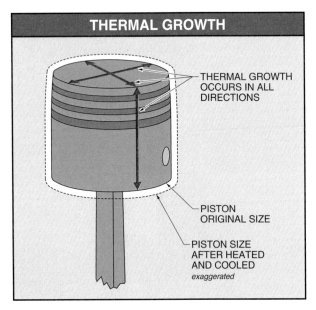

Figure 8-4. Thermal growth increases the size of a material in all directions with little or no change back to original dimensions when heat is removed.

Thermal Distortion. *Thermal distortion* is an asymmetrical or nonlinear thermal expansion of a material. Thermal distortion can result in thermal expansion or thermal growth. All internal combustion engines experience some degree of distortion during operation. The properties of the material, design of the component, and the amount of heat exposure dictate the rate and amount of thermal distortion. In a typical small engine, the amount of heat exposure is determined by operating speed, environment, and load.

Thermal distortion primarily affects the operation and durability of the cylinder bore and valve seat inserts. When an engine is operating under load, the combustion chamber is the main source of heat in the engine. The cylinder bore is a common place for thermal distortion to occur due to combustion chamber location and component friction. Heat produced during engine operation is commonly localized into specific areas of the cylinder bore, combustion chamber, and cylinder head, causing temperature variations. These temperature variations occur regardless of the material. Temperature variations and the structure and design of the cylinder block are the primary causes of thermal distortion. See Figure 8-5.

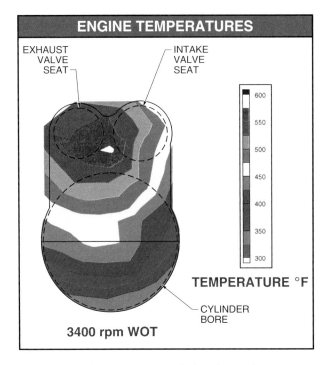

Figure 8-5. Temperature variation in engine components is a primary cause of thermal distortion.

Thermal expansion of the cylinder block is caused by the direct application of heat. The cylinder block responds by attempting to expand in all directions (within the constraints of the structural design to minimize thermal distortion). As expansion occurs, the cylinder bore increases in diameter. The material in the cylinder block surrounding the cylinder bore expands, and the cylinder bore itself increases in diameter as the engine temperature increases and reaches operating temperature.

When a cylinder block is heated, the structural constraints of the cooling fins, valve seat area, valve guide, and intake and exhaust ports cause nonlinear expansion. Localized heating of specific areas of the engine block and combustion chamber also contributes to cylinder bore distortion. The function of any cooling system is to provide a method to transfer heat from the engine to another medium to prevent overheating.

Heat transfer from the combustion gases to the cylinder wall and from the cylinder wall to the cooling medium occurs by conduction, convection, or radiation. *Conduction* is heat transfer that occurs from atom to atom when molecules come in direct contact with each other, and through vibration, when kinetic energy is passed from atom to atom. *Convection* is heat transfer that occurs when heat is transferred by currents in a fluid. *Radiation* is heat transfer that occurs as radiant energy without a material carrier.

Heat transfer through the cylinder wall occurs only by conduction. Although temperatures in the combustion chamber can peak at approximately 3000°F during the first milliseconds of combustion, an overall average temperature range is 1200°F – 1400°F. Cylinder wall temperature is considerably lower because a thin layer of gasoline adheres to the cylinder wall and acts as an insulator. Heat flows from the combustion chamber through a gasoline and oil film, through the cylinder wall, to the coolest portion of the engine block at the cooling fin tips. See Figure 8-6.

AIR-COOLED ENGINE COOLING SYSTEMS

An *air-cooled engine* is an engine that circulates air around the cylinder block and cylinder head to maintain the desired engine temperature. Although air is not the most efficient medium for transferring heat, it is plentiful and usually provides a sufficient temperature difference with the engine block to dissipate heat.

Engine block material transfers heat by conduction. Cooling air circulating around the cylinder block and cylinder head transfers heat by convection. As air passes the engine block and components, it picks up heat, and the atoms and molecules in the air begin to move faster. When the air enters the atmosphere, it releases the heat. See Figure 8-7. Heat is also removed from the engine by exhaust gases, radiant heat emitted from engine components, and the lubrication system.

The maximum angle of operation of a splash lubrication system is the same as that of a full pressure lubrication system. In most applications, carburetor operation determines the maximum angle of operation.

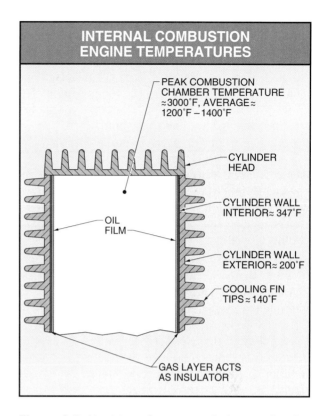

Figure 8-6. Heat transfer occurs during combustion by conduction from 3000°F in the combustion chamber to 140°F at the cooling fin tips.

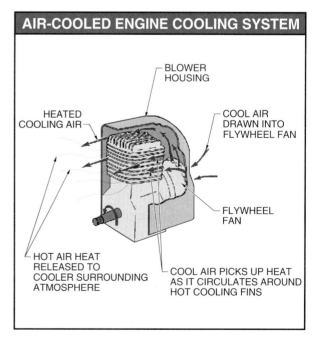

Figure 8-7. An air-cooled engine transfers heat by convection as the air circulates around the cylinder block and cylinder head.

Air-Cooled Engine Cooling System Components

Air-cooled engine cooling system components commonly include a rotating screen, cooling fan or flywheel with cooling fins, blower housing, air guides, cylinder cooling fins, and lubricating oil. A *rotating screen* is an engine component attached to the outer side of the flywheel that prevents harmful foreign matter from entering the path of cooling air to the engine. The rotating screen spins with the flywheel when the engine is operating to block sticks, stones, paper, and other debris from entering the cooling air path.

The rotating screen also serves as a cutting device for any grass or weeds discharged from a lawn mowing application. See Figure 8-8. Grass blown by the wind or the discharge of a mower deck encounters the small holes in the rotating screen. If the grass passes through the holes, it is chopped into smaller pieces. This reduces the size of the debris to make it easier for the cooling air flow to eject debris past the cooling fins of the engine.

Boiling point and temperature change

The boiling point of any liquid is the temperature at which the phase change from liquid to gas occurs. The application of heat when a liquid has reached the boiling point (at one atmosphere) causes liquid to boil. Regardless of the additional heat applied to the liquid, until the phase change (at one atmosphere) from liquid to gas is complete, the temperature of the liquid cannot increase past the boiling point.

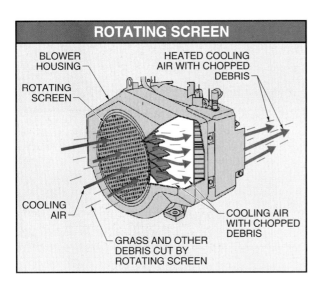

Figure 8-8. A rotating screen prevents harmful foreign matter from entering the path of cooling air and serves as a cutting device for any grass or weeds discharged from a lawn mowing application.

Snapper, Inc.

The 6 HP engine used to power this walk-behind mower has a 20 oz oil capacity.

A *cooling fan* is an engine component that supplies cooling air to the engine when rotated. The cooling fan or flywheel with cooling fins are the most important components of an air-cooled engine cooling system. Rotation of the cooling fan draws ambient air into the engine.

Most Briggs & Stratton engines feature a phase modulated cooling fan. A *phase modulated cooling fan* is a cooling fan that has blades spaced at different distances from each other. Phase modulated cooling fans and cooling fins on flywheels reduce noise by alternating the frequency of the sound of driven air. Air driven by a cooling fan or by cooling fins on a flywheel is a significant contributor to the overall sound produced by a small air-cooled engine. See Figure 8-9.

Figure 8-9. Cooling fins with different sizes and spacings on a flywheel reduce overall engine noise by alternating the sound frequency of driven air.

Ambient air temperature is an important factor in providing proper cooling for the engine. As cooler ambient air passes across hot engine surfaces, kinetic energy is transferred from the engine to the moving air. The greater the difference in temperature between the ambient air and the engine cylinder block, the greater and faster the energy transfer.

Cooling air is routed by the blower housing and air guides. A *blower housing* is a sheet metal or composite material component that encompasses the fan to direct cooling air to the cylinder block and cylinder head. An *air guide* is a sheet metal component used to direct cooling air from the blower housing to the cylinder cooling fins. A *cooling fin* is an integral thin cast strip designed to provide efficient air circulation and dissipation of heat away from the engine cylinder block into the air stream. Cooling fins provide a larger surface area of the cylinder block contacting ambient air to increase cooling efficiency.

Air-cooled engines also indirectly transfer heat from internal components through lubricating oil. As the lubricating oil is splashed around the inside of the crankcase, it transfers accumulated heat to the cylinder block walls. The heat is then transferred to the cooling fins on the outside surface of the cylinder block. The cooling fins complete the cooling process by transferring heat to cooling air.

Engine Ducting

Engine ducting is required when the application requires more control of cooling air. Some applications require a fully-ducted or partially-ducted engine. A *fully-ducted engine* is an air-cooled engine in which cooling air flow routing and rate are controlled by air guides and a sealed blower housing. A *partially-ducted engine* is an air-cooled engine in which cooling air is provided by the blower housing and ambient air flow. See Figure 8-10.

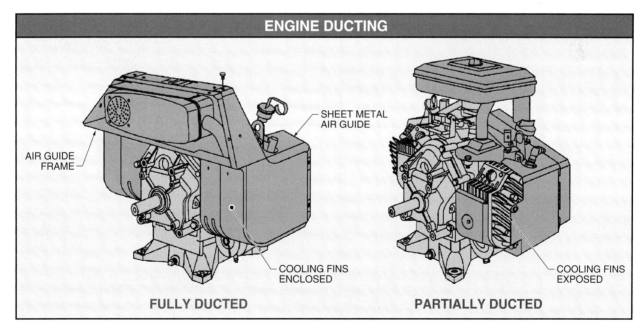

Figure 8-10. Fully-ducted engines provide cooling air using sheet metal air guides and a sealed blower housing. Partially-ducted engines provide cooling air using a blower housing and ambient air flow.

Fully-ducted engines control cooling air flow throughout the entire engine and increase the percentage of heat exchanged between cooling air and the engine. Fully-ducted engines are commonly used on applications that operate in special environments that could affect cooling air flow such as some lawn and garden tractors, asphalt rollers, and carpet cleaning machines.

Partially-ducted engines provide less control of cooling air flow than fully-ducted engines. Most engine application environments do not affect cooling air flow, and only require partial ducting. Partially-ducted engines also allow easier access to engine components than fully-ducted engines.

Cooling Air Plenum

A *cooling air plenum* is a duct made from sheet metal, plastic, or similar materials that provides a specific path for the cooling air to enter the engine cooling system. A cooling air plenum is required when the air flow source and/or direction are not sufficient to meet engine cooling needs. The cooling air plenum draws cooling air from outside the engine enclosure through a duct to prevent cooling air from mixing with hot air surrounding the engine for maximum engine efficiency. See Figure 8-11.

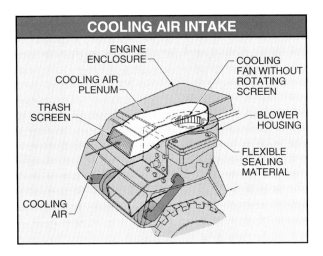

Figure 8-11. A plenum provides a specific path for the cooling air to enter the engine cooling system without mixing with hot air surrounding the engine.

 A liquid-cooled engine has a more consistent engine operating temperature, which results in longer engine life and greater combustion efficiency than an air-cooled engine.

The cooling air plenum is routed to the cooling system air intake opening in the blower housing, where it is sealed with a flexible sealing material such as foam rubber. The cooling air intake duct should be large enough to provide adequate and unrestricted air flow. Typically, the cross-section of the cooling air plenum should be at least 1 sq in. for each cu in. of engine displacement.

Rotating screens on the flywheel should not be used in conjunction with a cooling air plenum intake system. The rotating screen can restrict air flow and cause overheating in some applications. A trash screen ahead of the cooling air plenum with a minimum of 6″ square with 60% open screen for each cu in. of engine displacement is recommended.

The cooling air plenum opening should be positioned away from any implement discharge. If possible, the trash screen should be positioned in a vertical plane to be self-cleaning. The trash screen should be readily visible and accessible to the operator for cleaning efficiency.

Cooling Air Discharge

Cooling air discharge from the engine is collected and routed out of the engine enclosure to avoid overheating the engine compartment. A cooling air discharge duct captures all of the hot air passing by the cooling fins of an air-cooled engine. See Figure 8-12. The cooling air discharge duct must have sufficient cross section to permit hot air to flow without restriction. The cross-sectional size requirement of the cooling air discharge duct is same as the requirement for the cooling air intake duct.

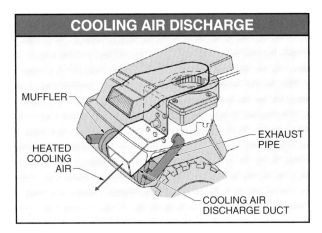

Figure 8-12. A cooling air discharge duct routes hot air away from the engine compartment to avoid overheating.

The exhaust system contains the hottest components attached to the engine, such as the muffler. If possible, the majority of the exhaust system should be located outside the engine enclosure to reduce heat inside the engine compartment. Hot exhaust system components or hot exhaust gases can ignite dead grass or other debris collected around and/or on the exhaust system. Exhaust system components must be located outside to prevent the collection of grass and other debris. The direction of exhaust gas and heated cooling air must also be controlled to prevent undesirable conditions such as brown spots on the grass or extra heat for the operator.

When an engine is shut OFF, cooling air provided to the engine also stops. The engine must have enough ventilation to prevent overheating during hot soak back. *Hot soak back* is the period immediately following the initial shutdown of an engine when cooling air flow has stopped and the engine enclosure temperature increases for a brief time. The most effective enclosure is one that has an opening on the bottom below the carburetor. This allows cool air to flow around the carburetor to minimize the temperature increase of the fuel during hot soak back and reduce hot restart problems.

Professional Chemicals Corporation

This carpet cleaner is powered by a 16 HP engine that heats water using exhaust heat and a catalytic converter.

Without a pressurized cooling system, water in the coolant mixture would reach its boiling point. A liquid-cooled engine can produce more heat than the coolant can absorb. This causes coolant to change from liquid to vapor. A water pump cannot efficiently pump vapor, and the ability to cool the engine would be compromised if a significant amount of coolant changed to vapor.

LIQUID-COOLED ENGINE COOLING SYSTEMS

A *liquid-cooled engine* is an engine that circulates coolant through cavities in the cylinder block and cylinder head to maintain desired engine temperature. The *water jacket* is a series of interconnected cavities cast into the engine block and cylinder head for the circulation of coolant. The primary difference between an air-cooled engine and a liquid-cooled engine is the medium used to transfer engine heat to ambient air. A liquid-cooled engine uses a mixture of water and antifreeze circulated in the engine cylinder block and other engine components. *Antifreeze* is an ethylene glycol chemical mixture used with water to lower the freezing point of engine coolant.

A liquid-cooled system operates most efficiently when coolant in the system is under pressure. The boiling point of a liquid is raised significantly with a relatively small amount of pressure in the system. For example, the cooling system in Briggs & Stratton liquid-cooled engines is pressurized to 11 psi – 13 psi. Coolant in this cooling system commonly reaches temperatures exceeding 212°F under heavy loads.

A pressurized cooling system produces a pressure greater than atmospheric pressure. This allows the coolant to absorb more heat and prevents it from boiling until it reaches a higher temperature. With a pressurized cooling system, the boiling point of the coolant is raised and is not reached under most normal conditions. This same principle allows water to boil at a lower temperature in higher altitudes having less atmospheric pressure. See Appendix.

Liquid-Cooled Engine Cooling System Components

A liquid-cooled engine cooling system continuously circulates coolant throughout the engine. Accumulated engine heat in the coolant is released to the cooling air at the radiator. Coolant is circulated through the engine block and cylinder head at relatively low pressures. Liquid-cooled engine cooling system components commonly include a radiator, radiator cooling fan, water pump, and thermostat. See Figure 8-13. These components are connected to each other using special high-temperature rubber hoses and fittings.

182 SMALL ENGINES

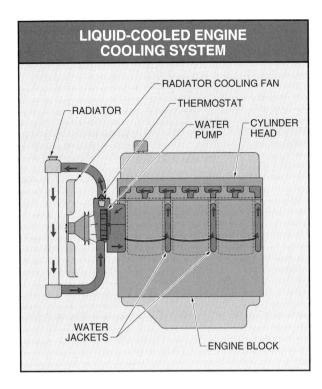

Figure 8-13. A liquid-cooled engine cooling system continuously circulates coolant throughout the engine to remove heat from the engine block and cylinder head.

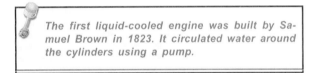

The first liquid-cooled engine was built by Samuel Brown in 1823. It circulated water around the cylinders using a pump.

Cylinder bore diameter and thermal expansion

It is common for small engine service technicians to think that as a cylinder block is heated, the cylinder bore diameter decreases. This is not the case, as every dimension of a heated cylinder bore expands proportionally. For example, if a photograph of a metal flat washer is enlarged, all features of the washer are enlarged, including the hole. Likewise, the cylinder bore in a small engine increases in diameter as the entire cylinder block is heated. This is the same principle used when heat is used for the removal of stubborn or rusted nuts. The hole in the nut expands when heated, allowing the nut to be broken free from rust or other debris.

Radiator

A *radiator* is a multi-channeled container that allows air to pass around the channels to remove heat from the liquid within. Thin metal fins on the radiator channels increase the amount of surface area in contact with cooling air to improve heat transfer efficiency from the coolant in the radiator. An inlet and an outlet on the radiator allow the passage of coolant to and from the engine. Hotter coolant from the engine flows in the radiator through the top inlet. Cooler coolant from the radiator flows to the engine through the bottom inlet. The radiator also serves as a coolant reservoir by storing coolant in the top tank and bottom tank. See Figure 8-14.

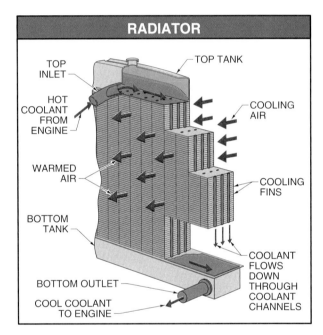

Figure 8-14. A radiator increases the surface area in contact with cooling air to improve heat transfer efficiency from the coolant.

Water Pump

A *water pump* is an engine component that moves coolant through passages of a liquid cooling system. The water pump circulates the coolant through the engine block, usually the cylinder head, and radiator. A water pump is a centrifugal pump consisting of a housing, a sealed bearing, an inlet, an impeller, a shaft, and an outlet. A radiator cooling fan is commonly attached to a pulley bolted to the shaft of the water pump. A *radiator cooling fan* is a device that pulls or pushes cooling air through a radiator. See Figure 8-15.

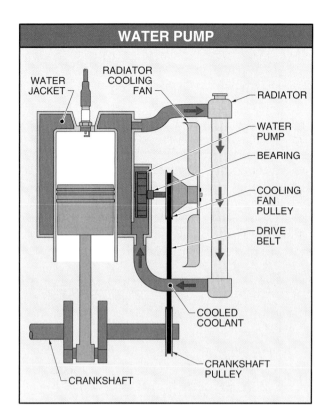

Figure 8-15. A water pump is a belt-driven centrifugal pump that moves coolant through the passages of a liquid cooling system.

In most applications, the water pump impeller, pulley, and cooling fan are driven by a belt that connects the crankshaft pulley to the water pump pulley. This maintains a coolant flow and air flow proportional to engine speed. Some applications utilize a fan driven by an electric motor. This allows greater flexibility when locating the radiator and radiator cooling fan. However, the water pump is still driven by a belt off the crankshaft pulley.

Thermostat

A *thermostat* is a valve placed between the radiator and the engine block on liquid-cooled engines that regulates the flow of coolant. The thermostat is used to control engine temperature and to reduce the engine warm-up period. The thermostat operates using a temperature-sensitive spring and valve that is submerged in the coolant. When the engine temperature is the same or cooler than ambient temperature, the thermostat is closed.

When the engine is started, the water pump circulates coolant throughout the engine block and cylinder head through a bypass hose that typically connects the intake manifold or cylinder head to the water pump. See Figure 8-16. With the thermostat closed, the coolant in the engine block picks up heat from the engine. The coolant in the radiator remains in place until the coolant temperature in the engine block reaches 130°F – 185°F. The increase in temperature causes wax in the wax-filled cylinder to expand to open the valve in the thermostat. This allows coolant to flow from the water pump through the cylinder block and cylinder head and back to the radiator. The thermostat return spring applies pressure to close the valve when the coolant has cooled. The thermostat valve regulates coolant flow by opening and closing to maintain the desired temperature in the cooling system. Normal operating temperatures vary with engine design and commonly range from 175°F – 195°F.

Engine cooling system testing

During engine cooling system testing, the ambient temperature should be within the expected normal ambient temperature range. The engine should have completed the break-in period, and test load should be approximately 75% of maximum power (carburetor throttle position approximately 50% open) for approximately 30 minutes to 60 minutes. Maximum allowable temperatures above ambient temperature of the cooling air, carburetor intake air, fuel in the fuel tank, crankcase oil, and cylinder head during cooling system testing are specified by the manufacturer.

The Moto-Scoot scooter, produced during the 1930s and 1940s, was powered by a Briggs & Stratton engine.

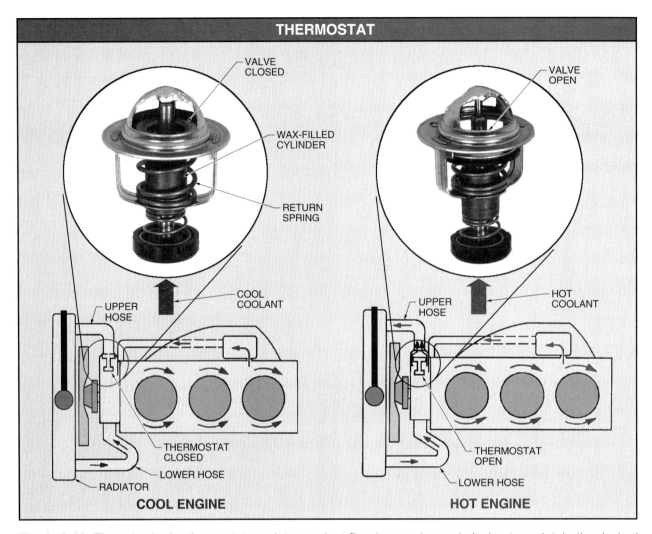

Figure 8-16. The valve in the thermostat regulates coolant flow by opening and closing to maintain the desired engine temperature and to reduce the engine warm-up period.

LUBRICATION

Lubrication of an engine helps to maintain proper operation by reducing friction and cooling internal engine components. *Friction* is the resistance to motion that occurs when two surfaces slide against each other. When two bearing surfaces are in contact, axial and/or radial motion produces friction. Friction in an internal combustion engine is reduced using the physical characteristics of lubricating oil.

All bearing surfaces in an internal combustion engine are susceptible to friction. An integrally machined aluminum bearing surface of a connecting rod, crankcase cover, sump, or cylinder block has a smooth, shiny surface. However, viewing this surface under a microscope reveals asperities. *Asperities* are tiny projections from the machining process which produce surface roughness or unevenness. Under a microscope, asperities resemble a series of mountain peaks and valleys. See Figure 8-17.

Asperities are separated by a film of oil. Only the very tops or peaks are in actual contact with the running surface of the mating component. During the break-in period, the tops of these asperities are worn off by friction. The result of the wear is a series of flat plateaus which provide the actual bearing surface. The remaining valleys act as small reservoirs for additional lubrication and as catch basins for component debris.

 At atmospheric pressure, the maximum temperature that water can reach is 212°F before changing state.

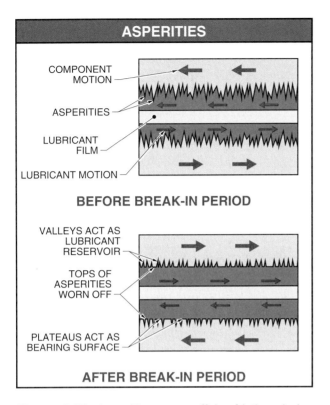

Figure 8-17. Asperities worn off by friction during break-in provide a series of plateaus and valleys for bearing surfaces and lubricant reservoirs.

Oil Characteristics

The primary oil characteristic concerning a small engine service technician is oil viscosity. *Viscosity* is the internal resistance to flow of a fluid. The main property of oil viscosity is thickness. Oil viscosity is important in the ability to protect bearing surfaces in a small engine. Oil viscosity indicates the expected thickness of the oil film separating bearing surfaces. The thickness of the oil film is also directly related to the clearance between any two bearing surfaces.

The ability of the oil to flow at low temperatures is also important in the protection of bearing surfaces. Resistance to flow, or oil thickness, increases as ambient temperature decreases. This usually requires a longer operating time before oil flows smoothly. Premature wear can occur at all bearing surfaces in the engine if an excessive amount of time is required for oil flow.

Oil film thickness decreases with an increase in oil temperature. Oil film thickness can be depleted in high ambient and operating temperatures. Oil specified for an application is a compromise between an oil that flows at low temperatures but still protects the engine at high ambient and/or operating temperatures. Oil recommendations for small engines follow the definition and standards provided by the Society of Automotive Engineers (SAE) and the American Petroleum Institute (API). These two organizations provide standards in both viscosity and additive packages for the majority of lubricants manufactured worldwide.

Oil Viscosity Rating. The SAE assigns a viscosity rating number to engine oil. The *SAE viscosity rating* is a number based on the volume of a base oil that flows through a specific orifice at a specified temperature, atmospheric pressure, and time period. A high viscosity rating results from a small volume of oil flowing through the orifice caused by high resistance to flow. A low viscosity rating results from a large volume of oil flowing through the orifice caused by low resistance to flow. The higher the viscosity rating number, the thicker the oil. The viscosity rating number assigned to an oil does not change, but oil viscosity can change with temperature and ambient pressure. The viscosity rating number or weight for use in internal combustion engines ranges from SAE 10 to SAE 50. Oils recommended for small air-cooled engines are based on the ambient operating temperature range anticipated before the next oil change. See Figure 8-18.

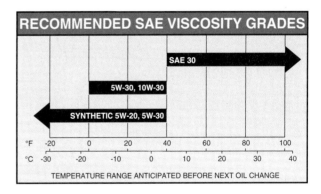

Figure 8-18. Oil recommendations for small air-cooled engines are a compromise between the required oil flow at low temperatures and adequate protection at high ambient and/or operating temperatures.

Oil Standards. The API provides standards that indicate the performance level and quality of engine oil as indicated by the API Engine Oil Licensing and Certification System (EOLCS). The EOLCS is a voluntary program to help the consumer identify products that have satisfied the requirements for licensing and certification. The EOLCS program was developed through a cooperative effort between the petroleum industry, represented by API, and vehicle and engine manufacturers, represented by the American Automobile Manufacturers Association (AAMA). The program is backed by rigorous monitoring and enforcement.

The EOLCS program is designed to define, certify, and monitor engine oil performance deemed necessary for satisfactory equipment life and performance. API-licensed marketers may display the API Service Symbol and the API Certification Mark for use by the consumer. See Figure 8-19. The API Service Symbol indicates oil performance level, oil viscosity, and energy-conserving properties. Oil performance is indicated by service categories for gasoline and diesel engines. The letter "S" indicates an oil suitable for gasoline engines. The letter "C" indicates an oil suitable for diesel engines.

EasyRake®/EverGreen International, Inc.
This chipper/shredder, powered by a 5 HP engine, self-feeds limbs and branches up to 3" in diameter.

The second letter indicates the service category. Service categories are determined by laboratory and engine tests to measure oil performance in protection from wear, sludge, rust, varnish, and other harmful actions. For example, the letters SJ indicate the oil is suitable for gasoline engines. Service categories are listed alphabetically in order of development. The service category SJ was introduced in October 1996 and exceeds the performance properties of SH and the obsolete SA through SG service categories. Most manufacturers have minimum API service category requirements. For example, Briggs & Stratton recommends SAE 30 oil with an API rating of SF or better.

Oil viscosity on the API Service Symbol is the SAE oil viscosity grade. Energy-conserving properties are indicated in the bottom of the API Service Symbol. To be listed as energy conserving, an oil must demonstrate energy-conserving properties when compared with reference oil in an engine test. The API Certification Mark indicates that an oil satisfies the most current requirements of the International Lubricant Standardization and Approval Committee (ILSAC) minimum performance standard for the application.

The API rating establishes standards for minimum performance requirements for specific engine service. Generally, these standards are intended for liquid-cooled internal combustion engines. Although the majority of small engines are air-cooled, the API classification is applicable within certain guidelines.

Multi-Viscosity Oils. A *multi-viscosity oil* is an oil that has the characteristics of two viscosity ratings for the required flow at low ambient temperatures and has adequate oil film protection at high operating and/or ambient temperatures. Multi-viscosity oils offer a low viscosity rating when cold and a high viscosity rating when hot. For example, an oil rated as SAE 10W-30 indicates that at 0°F the viscosity of the oil is low, as designated by the number 10. The letter W indicates that the oil is winter or cold weather rated, and the rating test was performed at 0°F. Without the W rating, the oil meets the specifications of SAE 10 oil at 212°F. The 30 in the rating number indicates the oil provides the viscosity of a SAE 30 oil at normal operating temperatures.

Multi-viscosity oils contain complex polymers to achieve the desired characteristics. A *polymer* is a molecule consisting of repeating structural units that have been chemically formulated to perform in a specific manner. Polymers used in multi-viscosity oils are heat-sensitive. When a multi-viscosity oil is cold, polymer molecules act like molecular springs and retract in length, causing little if any change in the viscosity of the base oil. The oil flows and lubricates like the base oil. As the oil reaches operating temperature (200°F – 300°F), the polymer molecules expand in length and increase in size to raise oil viscosity in proportion to heat applied.

OIL STANDARDS

SERVICE CATEGORY: GASOLINE ENGINES*

SA – *Obsolete* – For older engines, no performance requirement. Use only when specifically recommended by the manufacturer.

SB – *Obsolete* – For older engines, use only when specifically recommended by the manufacturer.

SC – *Obsolete* – For 1967 and older engines.

SD – *Obsolete* – For 1971 and older engines.

SE – *Obsolete* – For 1979 and older engines.

SF – *Obsolete* – For 1988 and older engines.

SG – *Obsolete* – For 1993 and older engines

SH – *Obsolete* – For 1996 and older engines. Valid when preceded by current C categories.

SJ – *Current* – For 2001 and older automotive engines.

SL – *Current* – For all automotive engines presently in use. Introduced July 1, 2001. SL oils are designed to provide better high-temperature deposit control and lower oil consumption. Some of these oils may also meet the latest ILSAC specification and/or qualify as Energy Conserving.

* S = oil suitable for gasoline engines

SERVICE CATEGORY: DIESEL ENGINES**

CA – *Obsolete* – For light duty engines (1940s and 1950s).

CB – *Obsolete* – Moderate duty engines for 1949 – 1960.

CC – *Obsolete* – For engines introduced in 1961.

CD – *Obsolete* – Introduced in 1955. For certain naturally aspirated and turbocharged engines.

CD-II – *Obsolete* – Introduced in 1987. For two-stroke cycle engines.

CE – *Obsolete* – Introduced in 1987. For high-speed, four-stroke, naturally aspirated and turbocharged engines. Can be used in place of CC and CD oils.

CF – *Current* – Introduced in 1994. For off-road, indirect-injected and other diesel engines including those using fuel with over 0.5% weight sulfur. Can be used in place of CD oils.

CF-2 – *Current* – Introduced in 1994. For severe duty, two-stroke cycle engines. Can be used in place of CD-II oils.

CF-4 – *Current* – Introduced in 1990. For high-speed, four-stroke naturally aspirated and turbocharged engines. Can be used in place of CD and CE oils.

CG-4 – *Current* – Introduced in 1995. For severe duty, high-speed, four-stroke engines using fuel with less than 0.5% weight sulfur. CG-4 oils are required for engines meeting 1994 emission standards. Can be used in place of CD, CE, and CF-4 oils.

CH-4 – *Current* – Introduced in 1998. For high-speed, four-stroke engines designed to meet 1998 exhaust emission standards. CH-4 oils are specifically compounded for use with diesel fuels ranging in sulfur content up to 0.5% weight. Can be used in place of CD, CE, CF-4, and CG-4 oils.

CI-4 – *Current* – Introduced September 5, 2002. For high-speed, four-stroke engines designed to meet 2004 exhaust emission standards implemented in 2002. CI-4 oils are formulated to sustain engine durability where exhaust gas recirculation (EGR) is used and are intended for use with diesel fuels ranging in sulfur content up to 0.5% weight. Can be used in place of CD, CE, CF-4, CG-4 and CH-4 oils.

** C = oil suitable for diesel engines

API CERTIFICATION MARK
- ADMINISTRATING ORGANIZATION
- SPECIFIC APPLICATION
- MEETS ALL CERTIFICATION REQUIREMENTS

API SERVICE SYMBOL
- SAE OIL VISCOSITY RATING NUMBER
- OIL PERFORMANCE LEVEL
- SERVICE CATEGORY
- ENERGY-CONSERVING PROPERTIES

API MARKS

American Petroleum Institute

Figure 8-19. The API provides standards that indicate the performance level and quality of engine oil.

Homelite, Inc.
This centrifugal pump has a priming lift of 25′ at sea level and pumps up to 17,600 gallons per hour (gph).

Oil Selection. Most small air-cooled engines operate with oil temperatures that can exceed 275°F. Multi-viscosity oil does not perform as well as single-viscosity (straight grade) oil in small air-cooled engines. Higher oil temperatures in small air-cooled engines cause polymer molecules and other oil additives in multi-viscosity oil to vaporize rapidly. This reduces oil viscosity and can lead to increased oil consumption and a reduction of oil film thickness required for bearing surfaces. Multi-viscosity oils perform better in liquid-cooled engines where oil temperatures rarely reach 230°F.

Briggs & Stratton recommends SAE 30 for temperatures at or above 40°F. Briggs & Stratton recommends SAE 10W-30 multi-viscosity oil when the ambient temperature is below 40°F. At lower temperatures, a multi-viscosity oil provides easier starting, quicker lubricant flow to engine components during startup, while maintaining required oil viscosity after normal operating temperatures are reached.

Air-cooled small engines operate hotter than liquid-cooled or automobile engines. Use of multi-viscosity oils such as SAE 10W-30 at or above 40°F results in high oil consumption and possible engine damage. The oil level should be checked more frequently if multi-viscosity oil is used. See Figure 8-20. Use of SAE 30 oil below 40°F can result in hard starting and possible cylinder damage from inadequate start-up lubrication.

Figure 8-20. Use of multi-viscosity oils such as SAE 10W-30 above 40°F may result in high oil consumption and require more frequent checking than SAE 30 oil.

Lubrication Systems

A lubrication system provides oil to appropriate areas of the engine to maintain a film of oil to separate bearing surfaces. Mechanical components are used to move oil from the reservoir to the bearing surfaces. Lubrication systems commonly used on small engines include the splash lubrication system, pressure filtration lubrication system, or pressure lubrication system.

Splash Lubrication System. A *splash lubrication system* is an engine lubrication system in which oil is directed to moving parts by a splashing motion. Splash lubrication systems are simple in design and are commonly used on horizontal and vertical shaft engines. Oil is directed to bearing surfaces with a dipper or slinger. See Figure 8-21. A *dipper* is an engine component attached to the connecting rod which directs oil from the oil reservoir to bearing surfaces. The dipper enters and exits the oil reservoir as the piston travels to and from BDC to splash and distribute oil throughout the crankcase.

 In 1885, the spark plug similar to those used today was invented by Etienne Lenoir.

Cooling and Lubrication Systems **189**

Pressure Filtration Lubrication System. A *pressure filtration lubrication system* is an engine lubrication system in which a pump is used to circulate oil in a limited area of the engine. Oil is directed to bearing surfaces by a gerotor oil pump used in conjunction with a slinger to provide the necessary lubrication to the engine. A *gerotor oil pump* is an oil pump that consists of a multiple-lobed inner rotor meshing with an outer rotor to discharge oil under pressure. See Figure 8-22. The gerotor oil pump inner rotor contains four to six lobes at the circumference. The outer rotor has one more lobe space than the inner rotor. The gerotor oil pump body has a passageway that connects to the oil reservoir. The inner rotor is rotated by the pump shaft without any radial or axial motion. As the inner rotor is rotated, the outer rotor rotates from the meshing of the lobes. A low-pressure area or vacuum is created within the pump body near the passageway to the oil reservoir from the extra lobe on the outer gear during rotation. A groove in the cover commonly houses an O-ring or gasket to seal the pump body.

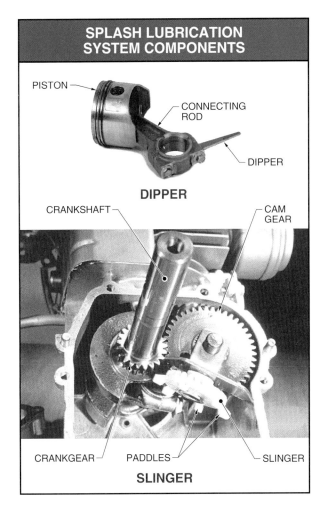

Figure 8-21. A dipper and slinger are splash lubrication system components that direct oil to moving parts with a splashing motion.

A *slinger* is a splash lubrication system component used on vertical crankshaft engines consisting of a spinning gear with multiple paddles cast into the plastic gear body. The slinger distributes the oil throughout the crankcase. The slinger meshes with the crankgear or cam gear. Approximately one-third of the slinger gear is located above the oil level in the crankcase. As the slinger gear spins, oil is discharged by the paddles and splashed throughout the crankcase.

Splash lubrication systems require the recommended oil level in the crankcase for proper oil distribution. Splash lubrication systems provide an effective, reliable, and low-cost method of engine lubrication and are used on virtually every application. The simplicity and dependability of the splash lubrication system makes it the standard lubrication system for most small air-cooled engines.

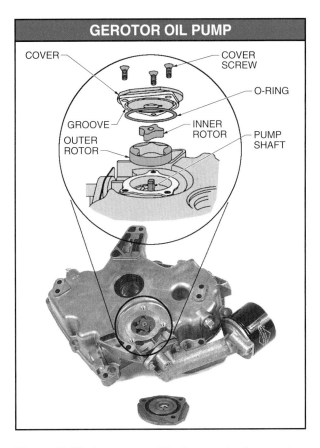

Figure 8-22. A pressure filtration lubrication system uses a gerotor oil pump to circulate oil to a limited area of the engine.

Pressure filtration systems are used on engines and applications where it is desirable to filter the oil. The primary function of the gerotor oil pump is to move oil from the oil reservoir through the oil filter and return it to the oil reservoir. The oil filter contains pleated paper filtering material to trap dirt, metal particles, carbon, and other foreign matter from oil drawn from the oil reservoir. See Figure 8-23. A rubber sealing gasket provides a positive seal to prevent leakage. A spring-loaded bypass valve opens if oil cannot pass through a clogged filter. This allows oil to continue to be routed through lubrication passages.

Pressure Lubrication System. A *pressure lubrication system* is an engine lubrication system in which a pump is used as the primary component to circulate oil throughout the entire engine. This type of lubrication system provides consistent pressure and volume of oil to all bearing surfaces in the engine. A malfunction in the pressure lubrication system causes a disruption of oil to engine components and can result in severe damage to the engine. A pressure lubrication system consists of a gerotor oil pump, a pick-up screen, an oil filter, and oil passages throughout the engine. See Figure 8-24.

Figure 8-23. The oil filter traps dirt, metal particles, carbon, and other foreign matter in pleated paper filtering material from oil drawn from the oil reservoir.

In addition to the ability to filter the oil, most pressure filtration systems provide a method of lubricating the magneto side of the main crankshaft bearing surfaces. This is an important feature as the magneto side main bearing on a vertical crankshaft engine is sensitive to oil supply delivery fluctuation. If the engine is operated for a period of time under a low-oil condition, this bearing surface is usually the first to show signs of insufficient lubrication. This occurs because of the distance of the magneto side main bearing from the oil reservoir and oil slinger. A pressure filtration system pumps the required amount of oil directly from the oil reservoir to the magneto side main bearing even in a relatively low oil level condition.

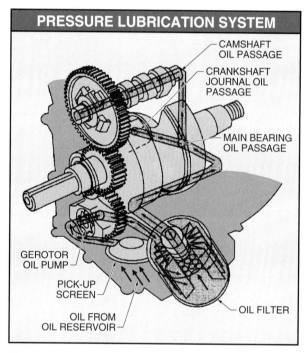

Figure 8-24. A pressure lubrication system circulates oil to all bearing surfaces throughout the engine.

A typical pressure lubrication system uses a gear drive for the gerotor oil pump. The gear meshes with the crankgear or cam gear, which provide the power to rotate the inner rotor of the gerotor oil pump. Although there is relatively low pressure in the crankcase of an operating engine, the gerotor oil pump produces a lower pressure in the oil pump body. This pressure difference pushes oil into the gerotor oil pump lobes.

Once the oil reaches the pump body and lobes, the squeezing action of the rotating rotor lobes pressurizes the oil. Based on oil temperature, oil viscosity, and engine speed, a pressure of up to 70 psi is

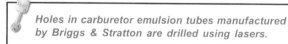

Holes in carburetor emulsion tubes manufactured by Briggs & Stratton are drilled using lasers.

commonly achieved. The pressurized oil is then forced through oil passages to the oil filter and bearing surfaces of the engine. Oil is then returned to the oil reservoir in the sump or crankcase of the engine.

The lubrication of the cylinder walls and some internal gears in the engine is accomplished by the spraying of high pressure oil as it exits a precisely-drilled orifice in the connecting rod. The orientation of the connecting rod must be correct to ensure that oil spray is directed to bearing surfaces in the engine. In addition, the calibration of the orifice provides a predictable pressure release system which provides some degree of pressure regulation.

Oil Pressure Regulation

Oil pressure produced by the oil pump in a pressure lubrication system is regulated by an oil pressure relief valve consisting of a spring, check ball, and dump orifice. As oil pressure builds in the system, the spring and check ball move within a cylinder machined into the pump body. When the oil pressure overcomes the spring pressure, the check ball moves to uncover a dump orifice to allow some oil to return to the oil reservoir. See Figure 8-25. As the oil is discharged into the oil reservoir, oil pressure decreases. The pressure decrease is proportional to the size of the orifice and flow of oil. In normal operation of a pressure lubrication system, pressure is regulated whenever the engine is operating.

A pressure lubrication system commonly increases engine oil temperature approximately 5% – 10%. Some of the heat in the oil is dissipated through the oil filter. The remaining heat is dissipated by the engine cooling system. The benefits of a pressure lubrication system greatly outweigh the slight increase in oil temperature.

Low Oil Level Warning System

Some engines are equipped with a low oil level warning system to alert the operator to a low oil level condition in the engine. For example, the Briggs & Stratton Oil Gard® system prevents engine damage by shutting down an engine operating with insufficient oil. The Oil Gard® system also shuts down the engine if the angle of operation causes a low oil level condition. See Figure 8-26. An Oil Gard® can be a float type or spark gap type system. In a float type Oil Gard® system, a float connected to a switch is inserted into the oil reservoir of the crankcase. When oil is at the correct level, the float is in a raised position to open the switch. When the oil level drops below the minimum recommended oil level, the float drops to close the switch. The closed switch completes a circuit from the primary windings of the ignition coil through the oil sensor. This causes the warning light to flash and directs primary ignition voltage to ground to stop the engine. The engine cannot be started until the oil level is restored to the correct level.

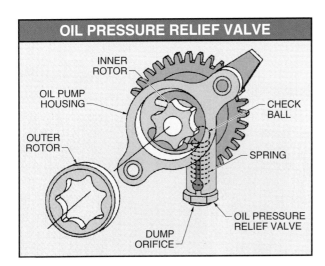

Figure 8-25. An oil pressure relief valve uses a spring and check ball to regulate oil pressure in a pressure lubrication system.

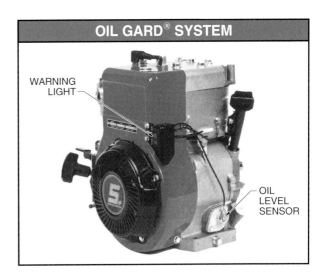

Figure 8-26. The Briggs & Stratton Oil Gard® system prevents engine damage from a low oil level condition in the engine.

In a spark gap type Oil Gard® system, an oil level sensor with a spark gap submerged in oil is connected to a lead from the ignition armature. When the oil level is sufficient, the submerged spark gap is filled with oil, preventing a spark from jumping the gap. When the oil level drops below the minimum recommended oil level, the spark gap is no longer submerged and is exposed to air in the crankcase. The resistance of the oil sensor spark gap in the crankcase is lower than the spark plug gap. This causes the armature to fire across the oil sensor spark gap and bypass the spark plug to stop the engine. The engine cannot be started until the oil level is restored to the correct level.

 Overall engine temperature varies with ambient air temperature. The amount of heat transferred for engine cooling is determined by the temperature difference between overall engine temperature and ambient air temperature.

COOLING AND LUBRICATION SYSTEM SERVICE PROCEDURES

Cooling and lubrication system service procedures commonly performed include checking the oil, changing the oil and oil filter, and removing debris. The oil and oil filter should be changed according to the application manufacturer's recommendations. On air-cooled engines, debris may pass through the rotating screen and become lodged between the cooling fins. This can reduce the cooling capacity of the engine and cause overheating. Cooling fins should be cleaned every 100 hours of engine operation. On liquid-cooled engines, the coolant and fan belt should be checked and changed as required. Maintenance schedules provided by the manufacturer detail specific maintenance operations. Service should be performed more frequently if the engine is operated in dirty or dusty conditions, under heavy load or in high ambient temperatures. See Figure 8-27.

VANGUARD™ THREE-CYLINDER GASOLINE ENGINE MAINTENANCE SCHEDULE							
System	Maintenance Operation	Daily	Every 50 Hours	Every 100 Hours	Every 200 Hours	Every 600 Hours	Yearly
Cooling System	Check coolant	✓					
	Change coolant						✓
	Check fan belt			✓			
Lubrication System	Check oil level	✓					
	Check for oil leaks	✓					
	Change oil		✓ [1]		✓ [2,3]		
	Change oil filter		✓ [1]		✓ [2,3]		

[1] Perform first maintenance operation after 50 hours
[2] Then perform maintenance operation at this interval
[3] Service more often when operating under heavy load or in high ambient temperatures

Figure 8-27. Maintenance schedules detail maintenance operations and intervals for optimum engine performance.

MULTIPLE-CYLINDER ENGINES

CHAPTER 9

Multiple-cylinder engines are commonly used to power applications with reduced vibration and have greater efficiency than single-cylinder engines. Multiple-cylinder engines require the same basic systems for operation, but vary in complexity with engine design. Common multiple-cylinder small engine designs include the 180° opposed-twin, the in-line, and the 60° V-twin.

MULTIPLE-CYLINDER ENGINE DESIGN

A *multiple-cylinder engine* is an engine that contains more than one cylinder. Multiple-cylinder engines produce less vibration during operation than single-cylinder engines. Most single-cylinder engines have 50% – 60% counterweighting using crankshaft counterweights to balance approximately one-half of piston motion forces. In a multiple-cylinder engine, piston motion forces can be counterbalanced by the inertia of another moving piston in addition to crankshaft counterweights.

Power strokes in multiple-cylinder engines occur more frequently, providing smoother operation. The increased frequency of power strokes reduces the duration of deceleration between power strokes. This results in greater power consistency across the rpm range and allows more of the power produced to be applied to the application. Common multiple-cylinder small engine designs include the 180° opposed-twin, the in-line, and the 60° V-twin. A *180° opposed-twin engine* is an engine that has two horizontal cylinders opposite each other. An *in-line engine* is an engine that has two or more parallel cylinders adjoining each other. A *60° V-twin engine* is an engine that has two cylinders forming a V-shaped angle at 60° to a horizontal plane. These engine designs are represented by the Briggs & Stratton 180° opposed-twin, Vanguard™ three-cylinder, and Vanguard™ V-Twin. See Figure 9-1.

 Briggs & Stratton is the largest manufacturer of automotive locks and keys.

194 SMALL ENGINES

Figure 9-1. Common multiple-cylinder small engine designs include the 180° opposed-twin, the in-line, and the 60° V-twin.

Each multiple-cylinder engine design has certain characteristics that offer advantages for various applications. For example, a 180° opposed-twin engine provides the best balance for minimal vibration of two-cylinder engine designs. Piston motion in one cylinder is offset by piston inertia in the other cylinder. The engine has a broad power range that is suitable for a variety of applications ranging from lawn and garden tractors to industrial/commercial applications. This has made the 180° opposed-twin engine a popular small engine design.

An in-line engine houses adjoining cylinders in a compact space. In-line gasoline or diesel engines are commonly used in medium to large industrial and commercial applications. In-line engine design and operating characteristics are similar to the characteristics of a liquid-cooled automotive engine.

The Vanguard™ V-Twin engine design combines the advantages of the compact size of an in-line engine with the smoothness of a 180° opposed-twin engine. The relatively large bore and short stroke provide the horsepower at 3600 rpm commonly required for many outdoor power equipment applications.

Multiple-Cylinder Engine Displacement

Multiple-cylinder engine displacement is based on the combined displacement of all cylinders. This allows greater horsepower ratings in a smaller package when compared to single-cylinder engines. *Displacement (swept volume)* is the volume that a piston displaces in an engine when it travels from TDC to BDC during the same piston stroke. Generally, the greater the displacement of the engine, the more power the engine can produce. Multiple-cylinder engine displacement is determined by multiplying the number of cylinders by the displacement of a single cylinder. See Figure 9-2. When bore and stroke are known, the displacement of a multiple-cylinder engine is found by applying the formula:

$$D = N \times .7854 \times B^2 \times S$$

where
N = number of cylinders
D = displacement (in cu in.)
.7854 = constant
B^2 = bore squared (in sq in.)
S = stroke (in in.)

DETERMINING MULTIPLE-CYLINDER ENGINE DISPLACEMENT

VANGUARD™ V-TWIN ENGINE

2.83″ BORE
2.75″ STROKE

What is the displacement of a Vanguard™ V-Twin engine that has a 2.83″ bore and a 2.75″ stroke?

$D = N \times .7854 \times B^2 \times S$
$D = 2 \times .7854 \times (2.83 \times 2.83) \times 2.75$
$D = 2 \times .7854 \times 8.0089 \times 2.75$
$D = 34.596044 = $ **34.60 cu in.**

Figure 9-2. Multiple-cylinder engine displacement is determined by multiplying the number of cylinders by the displacement of a single cylinder.

For example, what is the displacement of a three-cylinder engine that has a 2.68″ bore and a 2.52″ stroke?

$D = N \times .7854 \times B^2 \times S$
$D = 3 \times .7854 \times (2.68 \times 2.68) \times 2.52$
$D = 3 \times .7854 \times 7.1824 \times 2.52$
$D = 42.646388 = $ **42.65 cu in.**

MULTIPLE-CYLINDER ENGINE SYSTEMS

Multiple-cylinder engines, like single-cylinder engines, include compression, fuel, governor, electrical, cooling, and lubrication systems. These systems vary with the specific engine design and application. Unique system characteristics may provide additional challenges when servicing multiple-cylinder engines. Always refer to the engine manufacturer's repair manual for specific engine system information.

Compression System

The compression system directs, contains, and compresses the air-fuel mixture and discharges exhaust gases. Compression system components in a multiple-cylinder engine function in the same way as compression system components in a single-cylinder engine. Compression system component design varies with engine design and number of cylinders. See Figure 9-3. For example, crankshafts used for multiple-cylinder engines are longer and have multiple crankpin journals to accommodate multiple connecting rods. Longer crankshafts require more support for the increased span and multiple stresses from combustion forces. Crankshafts for multiple-cylinder engines commonly require more than two main bearings to counteract stresses applied to the crankshaft.

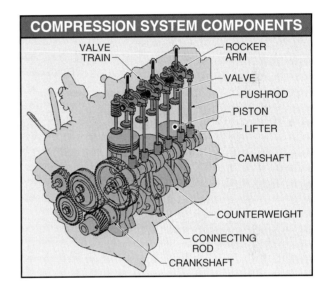

Figure 9-3. Compression system component design in a multiple-cylinder engine varies with engine design and number of cylinders.

Camshafts must also be longer and contain additional cam lobes to actuate the valves in the multiple-cylinder engine. Radial forces applied to the camshaft are increased with the increased number of cylinders and valve train components. This requires multiple camshaft bearing surfaces to counteract stresses applied to the camshaft. A larger capacity lubrication system is required to provide sufficient lubrication to bearing surfaces.

Cylinder heads vary for different multiple-cylinder engine designs. The 180° opposed-twin and the

Vanguard™ V-Twin engines use individual heads similar to those used for single-cylinder engines. In-line engines utilize a single common cylinder head for all cylinders. See Figure 9-4. A common cylinder head seals all cylinders in the engine and can affect compression test results. For example, if a head gasket is damaged or blown, air from the cylinder leakdown tester could fill an adjoining cylinder during the compression test. This can cause inaccurate test results as the adjoining cylinder does not have the correct piston and/or valve orientation. The cooling system on an in-line engine can also affect compression test results as a damaged head gasket or warped head can leak air into the water jacket, causing bubbles to erupt in the radiator. Compression testing of a Vanguard™ V-Twin engine is performed on one cylinder at a time and is much easier than compression testing an in-line engine.

Valve clearance adjustment and camshaft position

Engine design affects routine service procedures such as the valve stem to tappet clearance adjustment. When adjusting valve clearance on a 180° opposed-twin engine, all valve springs must be installed in the engine to provide an accurate measurement. Failure to install valve springs on the valves in the opposite cylinder results in an inaccurate adjustment. The force applied by the valve springs ensures that the camshaft is located in a consistent position for both cylinders during the adjustment process. Without the valve springs installed in the opposite cylinder, the camshaft may not be positioned properly within the bearing surfaces at the camshaft ends, resulting in a possible error of .001″ – .003″ in tappet clearance.

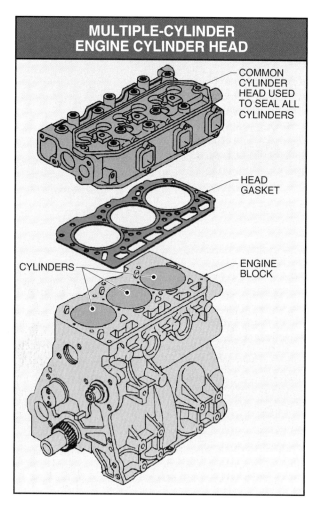

Figure 9-4. A common cylinder head and head gasket are used to seal all cylinders in an in-line engine.

The Vanguard™ V-Twin engine has compression components designed for specific engine operating characteristics. For example, the Vanguard™ V-Twin engine experiences significant thermal expansion when operated under load. This requires a steel pushrod on the intake side valve and an aluminum alloy pushrod on the exhaust side valve. As the Vanguard™ V-Twin engine reaches operating temperature, the cylinder exhibits thermal expansion. Overall dimensions of the engine increase, including the distance from the cam lobe to the rocker arm. This increase causes an increase in the valve clearance where the rocker arm contacts the valve stem.

The compression release system on the Vanguard™ V-Twin engine is located on the exhaust side cam lobes. A steel pushrod does not exhibit the same thermal expansion rate as other aluminum engine components. This causes the compression release system to become progressively less effective as engine temperature increases, resulting in kick back or hard starting. An aluminum pushrod has a thermal expansion rate approximately proportional to the engine and maintains the required valve stem to rocker arm clearance.

 At peak production, Briggs & Stratton can manufacture 50,000 engines a day.

Fuel System

Fuel systems used for multiple-cylinder engines differ slightly from those used on single-cylinder engines. A single carburetor is commonly used with an intake manifold. An *intake manifold* is an engine component that distributes the air-fuel mixture from the carburetor to more than one cylinder. See Figure 9-5. On diesel engines, the intake manifold is used to distribute air from the air cleaner to each cylinder.

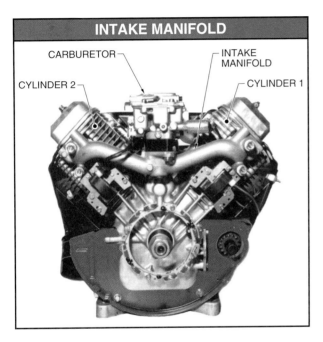

Figure 9-5. An intake manifold on a multiple-cylinder engine distributes the air-fuel mixture from the carburetor to more than one cylinder.

The intake manifold for both gasoline and diesel engines is designed to provide equal, unrestricted flow to each of the cylinders. Unequal flow to the cylinders can cause an unbalanced condition in which each cylinder does not carry an equal portion of the load. This causes premature wear of the engine components carrying the majority of the load. Restriction of flow in the intake manifold can cause turbulence, reducing overall engine efficiency.

Multiple-cylinder engines commonly use bowl-style carburetors. A *bowl-style carburetor* is a carburetor that has a fuel reservoir (bowl) located in the carburetor. See Figure 9-6. A float attached to a needle in the fuel bowl maintains a constant fuel level inside the fuel bowl. The float changes position with the fuel level and moves the needle in the seat to allow fuel to enter into the fuel bowl. Bowl-style carburetors are used in applications where the fuel tank must be located away from the carburetor. An electric or vacuum fuel pump is used to supply fuel from the fuel tank to the fuel bowl.

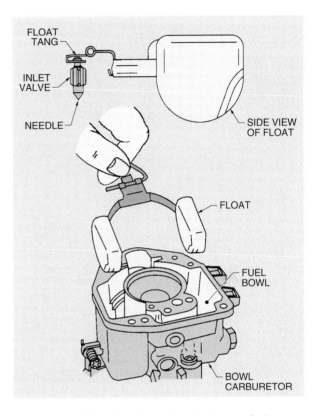

Figure 9-6. A bowl-style carburetor has a fuel reservoir (bowl) located in the carburetor and can be used in locations away from the fuel tank.

The Toro Company

This greens mower has a hydraulic cutting unit drive and is powered by a liquid-cooled three-cylinder engine.

Intake events without ignition during coastdown in a multiple-cylinder engine can cause a fuel vapor buildup in the muffler which can lead to afterfire. *Afterfire* is an engine condition that occurs when the engine continues to operate after the ignition switch is shut OFF. Multiple-cylinder engines commonly have an anti-afterfire solenoid. An *anti-afterfire solenoid* is a device that shuts OFF the fuel at the carburetor to prevent the engine from receiving fuel after the ignition switch is shut OFF. Anti-afterfire solenoids are installed on engines having a battery-powered electrical system.

Carburetor function on a multiple-cylinder engine

A carburetor on a four-stroke cycle single-cylinder engine provides air-fuel mixture for approximately 25% of engine operation time. This is normal, as the air-fuel mixture is only drawn into the cylinder during the intake event. A single carburetor used on a multiple-cylinder engine requires the carburetor to function a greater percentage of each operating cycle, depending on the number of cylinders.

Governor System

Governor systems used for multiple-cylinder engines are usually mechanical or electronic. Pneumatic governor systems are not used on multiple-cylinder engines. Mechanical or electronic governor systems offer more precise control of torque produced and engine speed. Multiple-cylinder engine governor systems may appear more complex, but they operate using the same fundamental principles as single-cylinder engine governor systems.

Some multiple-cylinder engine governor systems have unique operating characteristics. For example, the governor system on later model 180° opposed-twin cylinder engines has a governed idle lockout mechanism. Linkage used on this governor system limits the travel of the governor arm when the throttle control is in the slow or idle position. Most two-cylinder engines produce a moderate amount of power at slower speeds compared to single-cylinder engines. The governed idle lockout linkage limits the travel of the governor arm and throttle plate so the governor system can only respond to light loads in the idle position. When the speed control cable or linkage is moved beyond the idle position, the lockout device becomes inactive and the governor arm is allowed to move unrestricted.

The Vanguard™ three-cylinder engine can be equipped with either a mechanical governor system or an electronic governor system. An electronic governor system uses a limited angle torque (LAT) motor in place of the governor spring and speed-sensing device used in a mechanical governor system. See Figure 9-7. The governor return spring applies force against the LAT motor linkage in an attempt to close the throttle plate. This is a safety mechanism designed to prevent overspeeding by returning the throttle plate to idle position if an electrical failure occurs. The electronic governor provides the most precise engine governing capabilities of all Briggs & Stratton governor systems. Some applications require the engine be equipped with an electronic governor system that is adjustable to four different speed ranges, including idle, fast idle (for warm-up), operation, and transport.

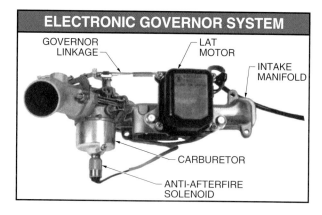

Figure 9-7. An electronic governor system uses a limited angle torque (LAT) motor to provide precise engine speed control.

The Vanguard™ V-Twin engine uses a mechanical governor system that applies more force at the governor shaft than any other Briggs & Stratton engine. A unique feature of this governor system is the location of the main governor spring when used on a generator or other application requiring close speed regulation. Most other governor systems used on generator applications require the governor spring to be located closest to the governor shaft. On the Vanguard™ V-Twin engine, the combination of linkage geometry, spring rate, and high forces generated at the governor shaft allows the governor spring location to be moved away from the governor shaft for improved engine speed control. The governor spring hole closest to the governor shaft is used for all other normal duty applications. A governor spring with a higher spring rate is used for other nongenerator applications to offset the increased forces produced by this governor system.

Electrical System

Multiple-cylinder engine electrical systems include electrical components required for starting and operating the engine and equipment accessories. Multiple-cylinder engines commonly require an electric starting system. The battery size or capacity should be large enough to power the starter motor above the minimum speed for starting. Multiple-cylinder engines commonly have more sophisticated engine components for ignition distribution to the cylinders. Multiple-cylinder engine electrical systems vary with the engine design and application requirements.

Ignition System. Most multiple-cylinder engine ignition systems require more ignition components than single-cylinder engines. An exception is the 180° opposed-twin cylinder gasoline engine. In this ignition system, the ignition armature provides spark to both cylinders simultaneously. Spark occurs in one cylinder near the end of the compression event, igniting the air-fuel mixture for the power stroke.

As spark occurs in one cylinder, a corresponding spark occurs simultaneously in the other cylinder during the exhaust event. The spark that occurs during the exhaust event has no effect on the operation of the engine. The spark occurs toward the end of the exhaust event, when the piston is discharging exhaust gases from the combustion chamber and there is no air-fuel mixture left to ignite. This type of ignition eliminates the need for an ignition timing and distribution system.

The Vanguard™ three-cylinder engine and the Vanguard™ V-Twin engine require ignition timing components to provide spark to the cylinder at the proper intervals. The Vanguard™ three-cylinder engine uses three ignition coils in conjunction with an ignition module for timing of the spark to the appropriate cylinder. See Figure 9-8. Engine timing is provided by the trigger. A *trigger* is a magnetic pick-up located near the crankshaft pulley that senses and counts crankshaft rotation.

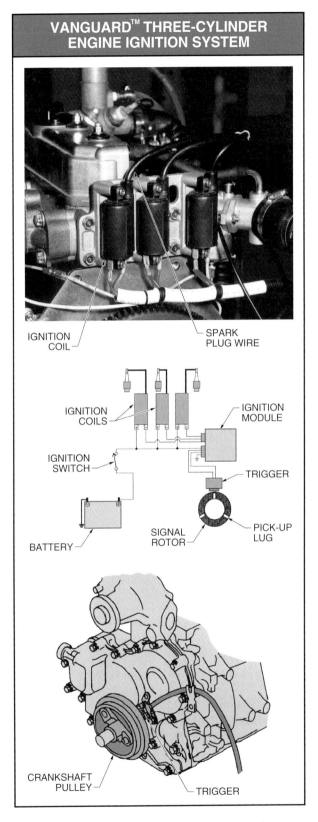

Figure 9-8. The Vanguard™ three-cylinder engine uses three ignition coils in conjunction with an ignition module for timing of the spark to the appropriate cylinder.

200 SMALL ENGINES

Ransomes America Corporation
This utility vehicle is powered by a 16 HP engine that produces 24 lb-ft of torque at 2200 rpm.

A signal rotor mounted on the crankshaft pulley rotates. Pick-up lugs on the signal rotor pass through the magnetic field of the trigger. The signal is sent from the trigger to the ignition module. A signal is then sent to the ignition coil, which sends high voltage to the specific cylinder spark plug. Firing order is determined by the location of the lugs on the signal rotor.

The Vanguard™ V-Twin engine uses two separate ignition armatures to provide a properly timed spark to each cylinder. The ignition armatures are electrically separated by two diodes installed in the wire harness that connects the ignition armatures to a single ignition switch. The diodes provide the independent ignition armatures a common path to ground to eliminate spark to the engine.

Two independent ignition armatures are required because of the firing intervals required by the Vanguard™ V-Twin engine. Diodes are used to isolate the ignition armatures from each other. See Figure 9-9. Without diodes in the circuit, the primary circuit of either ignition armature would have a path to ground through the opposite ignition armature. This would result in a no-spark condition. Engine startup, operation, and shutdown problems can be caused by a faulty diode.

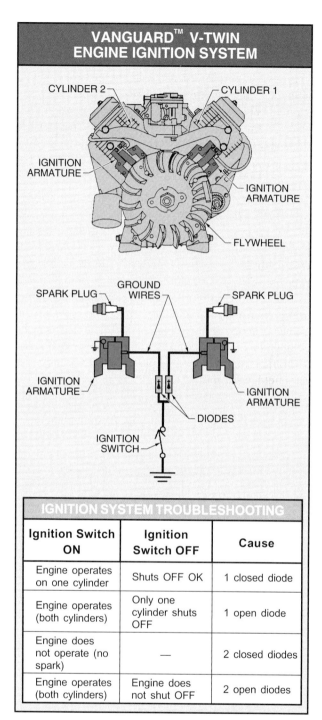

Figure 9-9. The Vanguard™ V-Twin engine uses two separate ignition armatures to provide a properly timed spark to each cylinder.

 The Vanguard™ V-Twin engine features a vented air cleaner assembly that draws air for the carburetor from the flywheel cooling fan and ejects air and any debris out through a series of holes in the air cleaner base or cover.

IGNITION SYSTEM TROUBLESHOOTING

Ignition Switch ON	Ignition Switch OFF	Cause
Engine operates on one cylinder	Shuts OFF OK	1 closed diode
Engine operates (both cylinders)	Only one cylinder shuts OFF	1 open diode
Engine does not operate (no spark)	—	2 closed diodes
Engine operates (both cylinders)	Engine does not shut OFF	2 open diodes

Charging System. The charging system on a multiple-cylinder engine replaces the electricity drawn from the battery by starting and accessories. An alternator, commonly used in conjunction with a regulator-rectifier, maintains the battery at full charge and supplies electricity for accessories such as lights and electric lifts.

Virtually all existing Briggs & Stratton alternator systems can be used in 180° opposed-twin engines or Vanguard™ V-Twin engines. However, the Vanguard™ three-cylinder engine has an automotive-style alternator driven by a fan belt. This alternator is a 14 A or 40 A automotive-style alternator which utilizes a voltage regulator and rectifier. Operation of the alternator on the Vanguard™ three-cylinder engine is similar to other Briggs & Stratton alternators.

The alternator is driven by a fan belt and uses a field winding that creates the magnetic field instead of permanent magnets found on the inside of a typical Briggs & Stratton flywheel. See Figure 9-10. The alternator has electrical connections to the battery, the field wire which supplies current to the field winding in the body of the alternator, and the indicator system, which displays a loss of charge.

Murray Inc.

The engine used to power this garden tractor is a 180° opposed-twin engine design.

Starting System. A starting system consists of a starter motor, a starter solenoid, an ignition switch, and a battery. Multiple-cylinder engines commonly have more displacement than single-cylinder engines and require a significant amount of starting effort if a rewind starter is provided. Most multiple-cylinder engines use a 12 V electric starter motor powered by a battery to start the engine.

For example, the 180° opposed-twin engine uses a heavy-duty 12 V starter motor and unique mounting extension. The heavy-duty starter includes larger magnets, stronger armature windings, and stronger bearings at each end of the armature shaft. This provides the power required to overcome loads from larger engine displacement, engine component mass, and connected application components.

The Vanguard™ three-cylinder engine uses an automotive-style starter to supply the required starting power. When the ignition switch is turned to start position, a path is created from the battery to the starter solenoid. See Figure 9-11. The starter solenoid is energized, and closes internal contacts to create another current path from the battery to the starter motor. The starter motor is energized and the starter pinion gear meshes with the flywheel ring gear. The flywheel is then driven by the starter motor to rotate the crankshaft. The main differences in the starting systems on multiple-cylinder engines are the power, size, and durability of the starter motor.

Figure 9-10. The Vanguard™ three-cylinder engine has an automotive-style alternator driven by a fan belt.

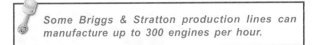

Some Briggs & Stratton production lines can manufacture up to 300 engines per hour.

202 SMALL ENGINES

Figure 9-11. When the starter motor is energized, the starter pinion gear meshes with the flywheel ring gear to rotate the crankshaft.

Starter motor power required is based on application parasitic load, duty cycle, and fuel; and is rated as 800 W, 1000 W, or 1200 W. *Parasitic load* is any load applied to an engine that is over and above the frictional load of an engine, such as a lawn mower blade. *Duty cycle* is the length of time (expressed as a percentage) that equipment can operate continuously at its rated output within a given time period. For example, the duty cycle for a typical Briggs & Stratton starter motor is 25% (15 sec) in a 60 sec time period. Engine fuel affects starter power required as diesel engines have higher compression ratios and require more starter power than gasoline engines.

Automotive-style starter motors use a solenoid and a drive lever. See Figure 9-12. The solenoid moves the drive lever to engage the starter pinion gear with the flywheel ring gear. Some starter motors use a clutch to reduce wear on the starter pinion gear from initial engagement with the flywheel. The Vanguard™ V-Twin electric starter motor is also designed and manufactured for specific applications. For example, an optional steel starter pinion gear and flywheel ring gear can be used in heavy service applications.

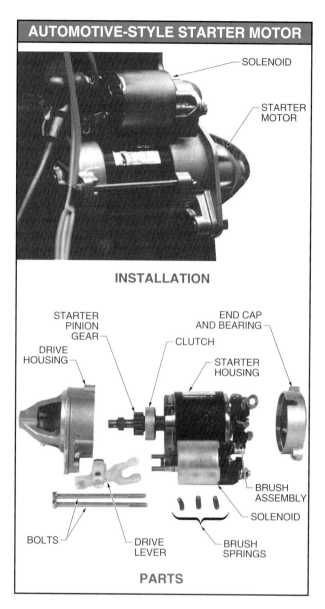

Figure 9-12. In an automotive-style starter motor, the solenoid moves the drive lever to engage the starter pinion gear with gear teeth on the flywheel.

Cooling System

Multiple-cylinder gasoline or diesel engines can be either air-cooled or liquid-cooled. Air-cooled engines are more compact in size and weigh less than liquid-cooled engines. Liquid-cooled engines require a radiator, radiator cooling fan, water pump, thermostat, and hoses. See Figure 9-13. These components are not required on an air-cooled engine.

An improper coolant mixture of water and antifreeze can freeze and expand. In some cases, enough pressure is exerted to crack a cylinder head or cylinder block. All liquid-cooled engines use freeze plugs to prevent possible damage from freezing coolant. A *freeze plug* is a concave-shaped metal plug pressed into a hole at the water jacket used to provide a release for pressure from freezing coolant in a liquid-cooled engine. Freeze plugs are easy to replace and inexpensive when compared with the cost of replacing other engine components damaged from freezing coolant.

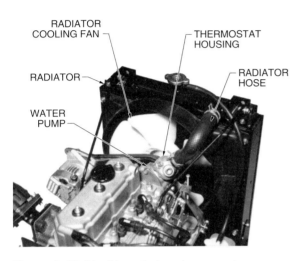

Figure 9-13. Liquid-cooled engines require more components than air-cooled engines, but have a more consistent engine operating temperature to increase overall engine life.

Liquid-cooled engines provide a more consistent operating temperature and have reduced operating noise and vibration. A more consistent engine operating temperature modifies radical temperature variations from ambient conditions and loads to increase overall engine life. The 180° opposed-twin and Vanguard™ V-Twin designs are air-cooled, similar to other single-cylinder engines.

A significant amount of engine heat is removed from the engine in the engine exhaust. Engine exhaust is routed from the combustion chamber on multiple-cylinder engines using an exhaust manifold. An *exhaust manifold* is an engine component that collects and directs exhaust gases from each cylinder to the muffler. Exhaust gases must be safely routed away from the operator, engine components, and application components. See Figure 9-14. The exhaust manifold must also minimize the restriction of the exhaust gas flow. Exhaust flow restriction causes unnecessary back pressure and reduces volumetric efficiency, resulting in a loss of horsepower.

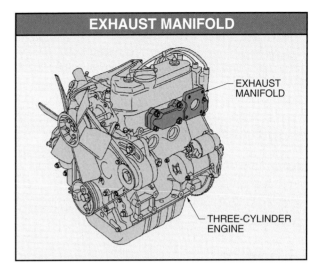

Figure 9-14. An exhaust manifold directs exhaust gases from each cylinder to the muffler for safe discharge into the atmosphere.

Some multiple-cylinder engine designs use a header pipe. A *header pipe* is a separate exhaust pipe used for each cylinder. A header pipe allows each cylinder to evacuate exhaust gases from the combustion chamber. Header pipes may be attached to allow the use of a common muffler.

MTD Products Inc.
A 16 HP engine is used to power this utility vehicle.

Lubrication System

Pressure lubrication systems are commonly used on multiple-cylinder engines. An oil pump provides the pressure required to deliver the necessary lubricant to bearing surfaces at the main bearings, connecting rod, and camshaft. The oil pump is commonly driven by an oil pump gear that meshes with the crankgear or cam gear. Gerotor oil pumps are commonly used on multiple-cylinder engines.

Multiple-cylinder engines can also be equipped with an oil filter to trap dirt, carbon, and/or other foreign matter in the oil as it circulates in the lubrication system. The main difference in the lubrication system of multiple-cylinder engines compared to single-cylinder engines is the number and location of oil passageways leading to and from bearing surfaces. Multiple-cylinder engines with more bearing surfaces require a lubrication system capable of maintaining the required oil film on bearing surfaces in all operating conditions. Some 180° opposed-twin cylinder engines are manufactured with splash lubrication systems.

Wacker Corporation

This walk-behind trowel, powered by an 8 HP engine, has a 48″ diameter and a trowel speed range of 70 rpm – 100 rpm.

MULTIPLE-CYLINDER DIESEL ENGINES

In the past, diesel engines have been used as a power source for trucks, ships, and trains. Small multiple-cylinder diesel engines are now commonly used to power medium- to large-size lawn and garden equipment and other outdoor power equipment. Multiple-cylinder diesel engines generally have a longer life span and are often the preferred choice over comparable gasoline engines. Advantages of a diesel engine include:
- greater fuel economy
- lower fuel volatility
- lower exhaust gas emissions
- longer life span

Disadvantages of a diesel engine include:
- more difficult cold weather starting
- higher engine noise
- greater manufacturing costs
- exhaust gas particulates (soot) and odor
- variable diesel fuel quality and availability
- heavier engine components

Diesel engines offer greater fuel efficiency, require no electrical ignition system, and have safer fuel characteristics. Diesel fuel provides more potential energy, and has more Btu/gal. than gasoline. Diesel fuel is rated at approximately 138,000 Btu/gal., compared to gasoline at approximately 115,000 Btu/gal. Diesel fuel is less volatile and has a flash point of 125°F, compared to –36°F for gasoline, reducing the risk of explosion or fire.

Diesel engines have a higher compression ratio than gasoline engines, which increases fuel vaporization during the compression event to increase fuel efficiency. For example, the compression ratio on the Vanguard™ three-cylinder gasoline engine is 8.6 : 1. The compression ratio on the Vanguard™ three-cylinder diesel engine is 22 : 1. Diesel engines often produce more torque than gasoline engines due to a relatively longer stroke and higher compression ratio. The higher compression ratio and combustion pressure requires heavier, higher-strength engine components in diesel engines.

Diesel engines typically exhibit lower horsepower-to-weight ratios than gasoline engines due to heavier engine components and air-fuel ratios during operation. For example, the Vanguard™ DM 700 G gasoline three-cylinder engine has a .18 HP/lb ratio. The Vanguard™ DM 700 D diesel three-cylinder engine has a .12 HP/lb ratio. At idle, some diesel engines exhibit an air-fuel ratio as high as 100 : 1. Under load, the air-fuel ratio is rarely less than 20 : 1. This results in high fuel efficiency, but less horsepower is produced in comparison with overall engine weight.

Compression Ignition

Diesel engines are compression ignition engines and do not require an electric spark to detonate the fuel. However, like the electric spark in an ignition system, diesel fuel must be delivered to the correct cylinder at the correct time for proper engine operation. See Figure 9-15. The injection pump is mechanically timed with the crankshaft on diesel engines. This simplifies engine operation by eliminating the need for ignition components such as the ignition coil and spark plugs. Timing is provided by the injection pump indirectly connected to the crankshaft.

The injection pump is similar to the distributor cap used on multiple-cylinder spark-ignition engines. Instead of providing a spark to a specific cylinder at the precise time, the injection pump provides a high-pressure fuel spray to the cylinder. One injection pump may supply more than one cylinder, or a separate injection pump may be required for each cylinder. Fuel pressure must exceed the compressed air pressure in the cylinder to be fed into the cylinder.

The injector functions as an ON/OFF valve to introduce fuel into the cylinder. Multiple-cylinder engines require an injector for each cylinder. The injector is hydraulically activated by the pressurized fuel delivered from the injection pump. Diesel fuel is more difficult to ignite at lower temperatures than gasoline.

A glow plug mounted in the cylinder head preheats air inside the combustion chamber to facilitate initial ignition of diesel fuel during cold startup. The glow plug begins to glow and release heat from current passing through an internal heating coil. See Figure 9-16. Diesel engines are shut off by stopping fuel flow to the engine using mechanical linkage or a fuel solenoid. A *fuel solenoid* is an electrically actuated component that controls the flow of fuel in the injection pump.

MULTIPLE-CYLINDER ENGINE SERVICE PROCEDURES

Multiple-cylinder engine service is similar to single-cylinder engine service with differences in ignition, engine filters, and cooling system engine component service. Ignition service involves engine components providing and distributing spark to more than one cylinder. On multiple-cylinder engines, it is possible for the engine to operate without the proper spark to all cylinders. This can complicate engine service and troubleshooting procedures. For example, a bad ignition coil on a three-cylinder engine affects engine performance, but does not stop the engine. A separate ignition coil is used for each cylinder. However, a bad ignition coil stops a 180° opposed-twin engine because there is only one ignition coil. Poor ignition-related performance could also be caused by a problem with spark plugs or spark plug wires.

Figure 9-15. Compression ignition engines require diesel fuel to be delivered to the correct cylinder at the correct time for proper engine operation.

 The carburetor on a multiple-cylinder engine operates the same as a carburetor on a single-cylinder engine.

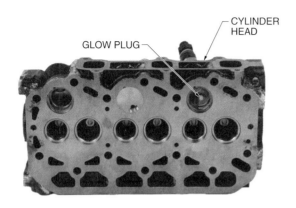

Figure 9-16. A glow plug mounted in the cylinder head preheats air inside the combustion chamber to facilitate ignition of diesel fuel during cold startup.

Engine filter service includes the changing of air, oil, and fuel filters as required. For example, on a Vanguard™ three-cylinder gasoline engine, air filter service is recommended every 50 hours. Oil filters should be replaced every 200 hours. Fuel filters should be changed every 600 hours. More frequent service is required when operating in dirty or dusty conditions, under heavy load, or in high ambient temperatures. See Appendix. Failure to replace filters at the recommended intervals can result in premature wear, poor engine performance, and/or engine failure.

Air-cooled multiple-cylinder engines require routine inspection of the cooling fins and the blower fan. Liquid-cooled multiple-cylinder engines require routine maintenance to the cooling system. Coolant level should be checked daily. The fan belt driving the cooling fan and water pump should be checked for any unusual signs of wear and the proper tension every 100 hours of service. The fan belt should be replaced if there is any visible cracking on the bottom or glazing on the sides. A newly-installed fan belt must be adjusted to the correct tension.

Diesel Engine Service Procedures

Diesel engines require similar service as gasoline engines. A diesel engine requires air, fuel, and compression to operate, but most required service is usually related to fuel system problems. Engine components supplying fuel to the engine are more complex and require special maintenance. The fuel filter removes water and foreign matter from the diesel fuel and must be changed at the recommended intervals. For example, Briggs & Stratton recommends that the Vanguard™ three-cylinder diesel engine fuel filter be changed every 800 hours of service. Lack of fuel is a common diesel engine problem and is often caused by a faulty fuel solenoid and/or injection pump, plugged fuel lines, a faulty/plugged injector, or air in the fuel system.

The easiest way to determine if fuel is supplied to the cylinder is to disconnect the exhaust system at the exhaust manifold and observe the exhaust gases discharged as the engine is turned over. If dense white smoke pours out of the exhaust manifold, the engine is getting fuel. Another method is to remove the injector from the engine and reconnect it to the fuel line. The injector should be suspended in air and pointed safely away so injector nozzle function can be observed. As the engine rotates, fuel should spray freely from the injector at the point of ignition.

Warning: Never place a hand or any part of the body in the direct discharge path of an injector nozzle. Diesel fuel discharged from an injector is highly pressurized and can penetrate through the skin to cause serious injury.

Trapped air in any part of the diesel engine fuel system prevents the engine from starting. Air is commonly removed from the fuel system by cracking open (loosening) fuel line fittings when the engine is turning over. Fuel line fittings are cracked at the injection pump and/or at each injector. The fuel line fitting is left open until a steady flow of fuel is discharged to purge any air. The fuel line fitting is then closed and tightened.

Diesel engines rely on compression to ignite the fuel. Any loss of compression can be detrimental to the operation of the engine. The lack of compression, or low compression, is commonly caused by damaged and/or worn piston rings, a blown head gasket, leaking valves, and/or a damaged cylinder head.

Diesel engines and air volume

A diesel engine is sometimes referred to as a constant-volume engine. During each intake event, theoretically the same volume of air is introduced into the combustion chamber. The displacement and compression ratio is constant. With gasoline engines, the displacement and compression ratio is also constant. However, the amount of air introduced into the combustion chamber is variable based on the throttle plate position inside the carburetor.

TROUBLESHOOTING

CHAPTER 10

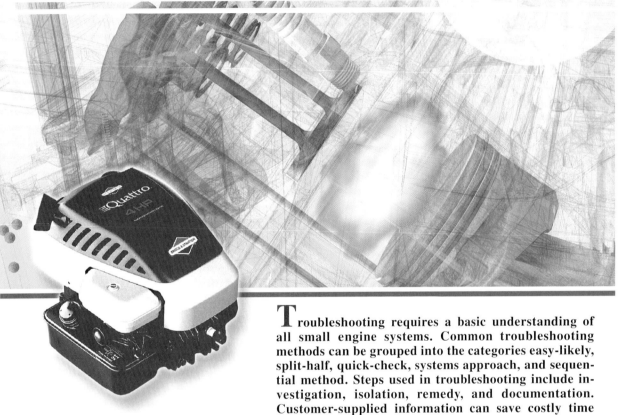

Troubleshooting requires a basic understanding of all small engine systems. Common troubleshooting methods can be grouped into the categories easy-likely, split-half, quick-check, systems approach, and sequential method. Steps used in troubleshooting include investigation, isolation, remedy, and documentation. Customer-supplied information can save costly time when investigating an engine problem.

TROUBLESHOOTING METHODS

Troubleshooting is the systematic elimination of the various parts of a system or process to locate a malfunctioning part. A *process* is a sequence of operations that accomplishes desired results. A *malfunction* is the failure of a system, equipment, or part to operate as designed. When troubleshooting, the small engine service technician:

- works systematically
- never makes assumptions
- isolates the cause of the problem as quickly as possible
- analyzes all affected factors related to the condition
- remedies the cause of the problem, not just the effect

EasyRake®/EverGreen International, Inc.
This pull-behind dethatcher unit uses 100 rotating tines powered by a 5 HP engine.

Disorganized or hit-or-miss methods usually result in high repair costs and dissatisfied customers. Common troubleshooting methods can be grouped into the categories easy-likely, split-half, quick-check, systems approach, and sequential method.

Gasoline blending and fuel system troubleshooting

Gasoline is blended to provide optimum engine performance within a certain ambient temperature range. When troubleshooting fuel system related performance problems, verify that the gasoline is fresh. Leftover gasoline may be the cause of starting and performance problems. Using gasoline blended for summer use in the winter results in hard starting because the gasoline vaporizes at a higher temperature. Using gasoline blended for winter use in the summer can result in the fuel vaporizing too quickly, causing vapor lock.

Easy-Likely Method

The *easy-likely method* is a troubleshooting method that isolates the cause of a malfunction by grouping possible causes as easy, difficult, likely, and unlikely. Once the possible causes of the problem have been grouped into these four areas, the problem is solved in sequence by identifying the easy-likely cause, the easy-unlikely cause, the difficult-likely cause, and the difficult-unlikely cause. The easy-likely method of troubleshooting relies on a depth of engine operation knowledge and past repair experience.

For example, the easy-likely method can be applied to a lawn tractor with a no-start condition. See Figure 10-1. When the ignition key is turned to the start position, there is no engine motion associated with the starting system. The easiest and most likely cause, such as a problem with battery connections and/or battery condition, is checked first. An easy-unlikely cause is the safety switch, which is checked next. A difficult-likely cause, such as problems with the starter solenoid and/or the starter motor, is checked next. A difficult-unlikely cause, such as a severely bent crankshaft restricting engine rotation, is checked last.

EASY-LIKELY TROUBLESHOOTING METHOD		
NO-START CONDITION	EASY	DIFFICULT
LIKELY	EASY-LIKELY • BATTERY CONNECTIONS • BATTERY CONDITION	DIFFICULT-LIKELY • STARTER MOTOR • STARTER SOLENOID
UNLIKELY	EASY-UNLIKELY • SAFETY SWITCH	DIFFICULT-UNLIKELY • SEVERELY BENT CRANKSHAFT

Figure 10-1. The easy-likely troubleshooting method groups possible causes into four areas. The cause is then isolated by progressing from simple to complex.

Split-Half Method

The *split-half method* is a troubleshooting method that isolates the cause of a malfunction by splitting parts of a system in half until the cause is isolated. The split-half method is traditionally used on electrical circuits when there is an input but no output. It is also effective on mechanical systems. A test for power is performed at the halfway point of the system. If power is found at the halfway point, the problem must be downstream. If power is not found at the halfway point, the problem must be upstream. The next step is to split the problem half of the system in half again to eliminate another portion of the system. The process continues until the cause of the problem is isolated.

For example, an engine with a no-start condition can be diagnosed by using the split-half method. See Figure 10-2. The engine does not turn over when the ignition switch is in the start position. The starting circuit is first divided in half by measuring voltage at the ignition switch. Correct voltage at the ignition switch indicates the problem is downstream. Incorrect voltage at the ignition switch indicates the problem is upstream. Voltage is correct, and the problem area is divided in half by measuring voltage at the starter solenoid. If voltage is not present, the problem is narrowed to the area between the ignition switch and the output side of the starter solenoid. The problem area is divided in half by measuring voltage at the starter solenoid. Voltage is present, which isolates the problem to the starter solenoid since no voltage was measured going out of the starter solenoid. The split-half troubleshooting method is often used by experienced small engine service technicians unfamiliar with a specific system.

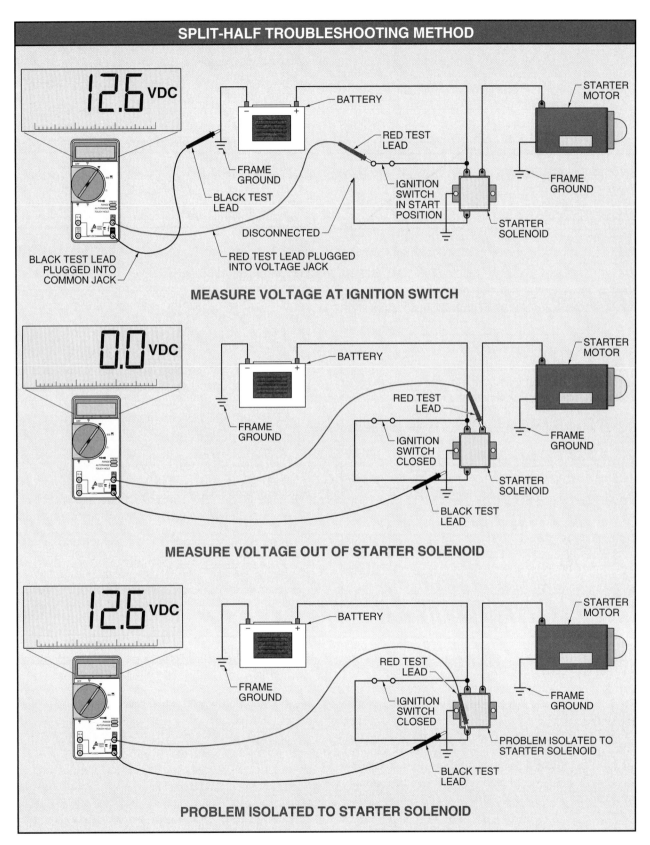

Figure 10-2. The split-half troubleshooting method isolates the cause of a malfunction by splitting the system and system parts in half and testing for proper function.

210 SMALL ENGINES

Quick-Check Method

The *quick-check method* is a troubleshooting method that isolates the cause of a malfunction by focusing on common problems identified by the manufacturer, product history, and/or service technician experience. The manufacturer frequently provides this information to service personnel. For example, engine overspeeding is commonly caused by a governor system malfunction. Information provided by the manufacturer can allow isolation of the problem. See Figure 10-3. Like the easy-likely method, the quick-check method can be performed while a customer is present and waiting. The quick-check method provides immediate information that can be invaluable later during service.

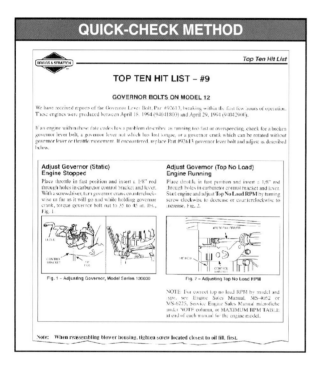

Figure 10-3. The quick-check method commonly uses information provided to service personnel by the engine manufacturer.

Systems Approach Method

The *systems approach method* is a troubleshooting method that isolates the cause of a malfunction by dividing the engine into separate systems and subsystems. Small engines can be divided into compression, fuel, electrical, cooling, and lubrication systems. Each individual system is then divided into specific subsystems. For example, an overheating problem is usually a cooling system problem.

The cooling system on an air-cooled engine includes such subsystems as ducting, flywheel, cooling fins, and engine speed. The problem is isolated using a sequential process from a system, to a subsystem, to an individual component. See Figure 10-4. The systems approach method of troubleshooting is frequently used by experienced technicians and is often used by the manufacturer to organize troubleshooting information distributed to small engine service technicians.

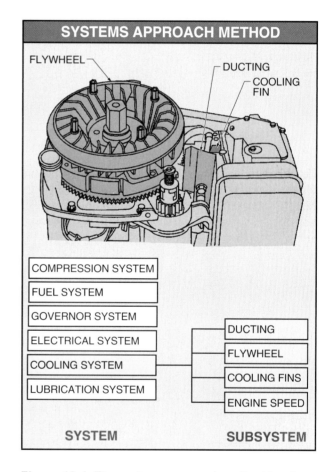

Figure 10-4. The systems approach method is a troubleshooting method that isolates the cause of a malfunction by dividing the engine into compression, fuel, electrical, cooling, and lubrication systems and subsystems.

 A 180° opposed-twin engine produces spark at each spark plug every time the flywheel magnet passes the ignition armature. As one spark plug ignites the charge in cylinder one, the other spark plug fires in an empty cylinder.

Sequential Method

The *sequential method* is a troubleshooting method that isolates the cause of a malfunction by starting at one end of a system and progressing to the other end with sequential checks. For example, on a lawn tractor that does not propel itself, troubleshooting starts at the beginning of the drive train system, the engine PTO. If there is output from the engine, the next step is to verify that power is going into the transmission. Testing of transmission output, differential input, differential output, and drive wheel rotation is sequentially performed. The sequential method requires broad knowledge of the system and testing skills. Experienced technicians usually avoid this method because it requires more time than other troubleshooting methods.

TROUBLESHOOTING STEPS

Troubleshooting steps vary depending on the troubleshooting method used. The differences commonly relate to the number of steps, terminology used, or both. However, all troubleshooting requires a systematic and logical sequence using knowledge, testing procedures, past experience, and reasoning. Troubleshooting steps can be grouped into the areas of investigation, isolation, remedy, and documentation.

Investigation

Investigation of the problem is the first step of troubleshooting. Investigation is based on inspection and/or information obtained through testing, equipment age, equipment operating history, customer-supplied information, and past history of similar problems such as previous work orders. A small engine service technician begins an investigation using steps similar to those a medical doctor takes to diagnose a patient. Patient history is referenced, and tests are ordered based on the specific problem after consulting with the patient. Likewise, the small engine service technician references equipment history. Customer-supplied information is critical to determining the next course of action.

Customer-Supplied Information. Customer-supplied information can save costly time when investigating an engine problem. Information supplied varies with the technical expertise of the customer. Initial conversation allows the service technician to assess customer knowledge about the equipment. Information is gathered about how the problem started, how the equipment normally functions, if the equipment has a history of problems, what action was taken to remedy the problem in the past, and if manufacturer's operating procedures were followed. See Figure 10-5.

Figure 10-5. The small engine service technician obtains valuable information from the customer about the equipment to reduce troubleshooting time and service cost.

Obtaining valuable troubleshooting information from the customer requires utilizing interpersonal skills. *Interpersonal skills* are strategies and actions which allow a person to communicate effectively with other individuals in a variety of situations. Occasionally, small engine service technicians may have to work with customers who are upset. Equipment failure is frustrating, and at times this frustration is vented on the small engine service technician.

Communication with an individual who is upset is best accomplished by relaxing and remaining calm. Listening carefully to an uninterrupted explanation of the problem allows time for individuals to collect their thoughts. Focusing on the engine problem and taking clear notes enhances the possibility of isolating the cause of the problem. Specific service procedures required and an estimated time frame should be clearly stated and documented. Clear communication eliminates misunderstandings and reduces repair time. After customer-supplied information is gathered, an inspection of the equipment is performed.

Inspection. During inspection, equipment is checked for any unusual conditions such as leaks, broken or bent parts, or obvious deficiencies. If the problem occurs during operation, the engine is started by following manufacturer's recommended start-up procedures. Check fluid levels, guard positions, and loose parts before starting the engine. See Figure 10-6.

Figure 10-6. During inspection, sight, sound, and touch are used to determine engine condition. Fluid levels, guard positions, and loose parts are checked before starting the engine.

Warning: Equipment brought in for service may have been previously worked on by an unqualified person. Always perform a safety check of any engine components that could cause a safety hazard during engine startup.

During inspection, sight, sound, and touch are used. Start the engine and test one part of the system at a time. Perform all hand-operated tests first. Operate the equipment long enough to obtain normal operating rpm, temperature, and pressures.

Isolation

Isolation of the problem begins with a list of possible causes derived from the investigation of the problem. When isolating the problem, it is helpful to broadly classify engine operations into three modes: startup, operation, and shutdown. Certain engine components operate during each of these modes and can be investigated as required. *Startup* is the engine operation mode in which the engine cranks, the air-fuel mixture is drawn into the cylinder and compressed, and ignition occurs. A common problem that occurs during startup is that the engine does not turn over, or the engine turns over but does not start.

Operation is the engine operation mode in which the engine has completed the startup mode and operates normally. Some common problems that occur during operation mode are erratic problems such as hunting and surging, misfiring, and overspeeding. *Shutdown* is the engine operation mode when the engine is shut down after operation. Gasoline engines are shut down by terminating the electrical pulse to the spark plug. Diesel engines are shut down by terminating the fuel injected into the cylinder.

To isolate the problem, a specific troubleshooting method is required. Experienced small engine service technicians commonly use the easy-likely method, unless they are unfamiliar with the device. Regardless of the method used, a logical troubleshooting sequence progressing from simple to complex is used to isolate the problem. All possible causes should be checked, no matter how simple. This includes checking the condition of a battery or the amount of fuel in the fuel tank. Problems similar to those that have occurred in the past must be considered.

 Ethanol is contained in over 10% of all gasoline sold in the United States.

Repair manuals from the manufacturer include recommended troubleshooting and repair procedures on specific engines using flow charts and troubleshooting charts. A *flow chart* is a diagram that shows a logical sequence of steps for a given set of conditions. See Figure 10-7. Flow charts help small engine service technicians follow a logical path when trying to isolate a problem.

Flow charts may use symbols and interconnecting lines to provide direction. Symbols used with flow charts vary depending on the manufacturer. Common symbols include an ellipse, rectangle, diamond, and arrow. An ellipse indicates the beginning and end of a flow chart. A rectangle contains a set of instructions. A diamond contains a question stated so that a yes or no answer is achieved. The yes or no answer determines the direction to follow through the flow chart. Arrows throughout the flow chart indicate the direction of flow. A *troubleshooting chart* is a logical listing of problems and recommended actions. See Figure 10-8.

When troubleshooting, it is easy to become focused on the symptoms or effects rather than the cause of the problem. If a dead end is reached, the problem should be reconsidered with all possible causes listed and checked. Other small engine service technicians experienced with the problem should be consulted. The most effective small engine service technician uses all resources available to isolate a problem quickly and economically.

Remedy

After isolating the problem, all possible ways to remedy the problem must be considered. The best remedy is determined by considering cost, time, quality, safety, and liability. The age of the engine has a great effect on how to best remedy the problem. Most small engines follow a typical life expectancy curve, including a break-in period, useful life, and wear-out period. See Figure 10-9.

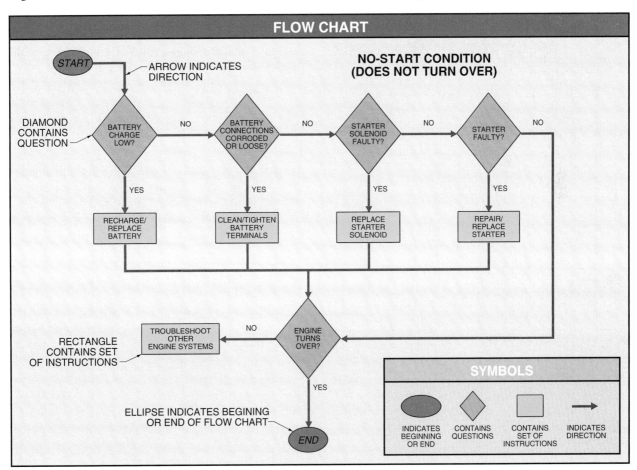

Figure 10-7. A flow chart shows a logical sequence of steps to assist a small engine service technician when attempting to isolate a problem.

TROUBLESHOOTING COMPRESSION SYSTEMS

Problem Prior to Testing	Test Result	Solution
Lack of power or starting problem	Reading is green – minimal air flow	Investigate for other (non-compression) problems
Lack of power or starting problem	Reading is green – minimal (audible) air flow with small amount coming through the head gasket	Replace head gasket
Lack of power or starting problem	Reading is green – all the air escaping from one component	Check that the piston is at TDC on the compression stroke, rotate 720°, lock the crankshaft, and retest. If the reading is correct, investigate the possible problem with that component
Lack of power or starting problem	Reading is red – all the air escaping from one component	Check that the piston is at TDC on the compression stroke, rotate 720°, lock the crankshaft, and retest. If the reading is correct, investigate the possible problem with that component
Lack of power or starting problem	Reading is red – the air escaping is from several components	Check that the piston is at TDC on the compression stroke, lock the crankshaft, and retest. If the reading is correct, investigate the possible problems starting with the component that appeared to leak the greatest volume of air

Figure 10-8. A troubleshooting chart provided by the manufacturer lists problems and recommended actions.

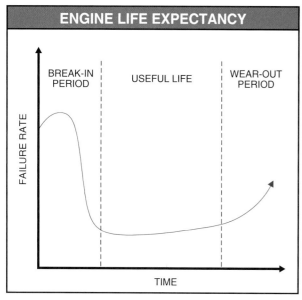

Figure 10-9. The age of the engine has an effect on service and whether to repair or replace the engine.

Break-in period is the period of time required for the running surfaces of piston rings and the surface of the cylinder bore to conform to one another after initial startup. *Useful life* is the period of time after the break-in period when most small engines operate as designed. *Wear-out period* is the period of time after the useful life of the engine when normal wear failures begin to occur. If an engine problem occurs during the break-in period, the cause could be an improper installation or operating procedure. If the problem occurred during useful life, repair rather than replacement may be economically feasible. If a major failure occurred during the wear-out period, the engine may not be worth repairing.

After the problem has been remedied, the engine is monitored for proper operation. Sufficient time is allowed for the engine to reach normal operating levels and load conditions. The engine is inspected again for proper operation.

Documentation

All service procedures and replaced and/or repaired parts should be listed on the shop service/repair ticket. These documents are used for maintaining inventory and identifying recurring service problems. Small engine service technicians should notify engine manufacturers about recurring problems. Briggs & Stratton uses information provided by service technicians in the field to improve product performance and service. Annual technical update seminars for small engine service technicians inform the dealer network of common problems and actions required. See Figure 10-10.

TROUBLESHOOTING COMPRESSION SYSTEMS

Engine compression system components include the valves, valve seats, pistons, piston rings, cylinder bore, and head gaskets. An engine with little or no compression does not operate. Compression loss is usually the result of a blown or leaking cylinder head gasket, valves sticking or not seating properly, or piston rings not sealing.

A quick method for checking compression in an engine is the rebound compression test. The *rebound compression test* is a test for engine compression in which the flywheel is spun counterclockwise (backward) and checked for the amount of rebound force. See Figure 10-11. If the flywheel rebounds sharply, compression is satisfactory. Slight or no rebound indicates low compression. The rebound compression test has limited effectiveness when determining engine compression. The amount of rebound force is not quantified and varies greatly with a warm engine and a cold engine. The rebound compression test is best used as a preliminary test before performing the more accurate cylinder leakdown test.

Figure 10-10. Annual technical update seminars for small engine service technicians provide information about specific engine problems.

For example, Briggs & Stratton received several reports from the field that the Model 40-42 180° opposed-twin engine was experiencing excessive oil consumption. After extensive testing, a service repair kit was designed and made available that eliminated the oil consumption symptom. Information about the problem, remedy, and service repair kit were included in the technical update material provided to small engine service technicians.

Repairs performed while an engine is under warranty require that a warranty claim be submitted to the manufacturer in order to receive reimbursement. Warranty claim forms must be properly filled out in order to receive reimbursement. Warranty claim information is also used by the manufacturer to identify recurring service problems in the field. See Appendix.

 A plastic sandwich bag placed under a fuel tank cap before tightening eliminates fuel spillage from the fuel tank when equipment is tipped for service.

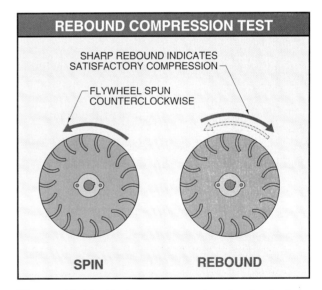

Figure 10-11. Engine compression is checked by hand before performing a cylinder leakdown test.

A *cylinder leakdown test* is a test that checks the sealing capability of compression components of a small engine using compressed air. A cylinder leakdown test provides greater accuracy and should be performed before the engine is disassembled to help isolate the faulty compression component.

Testing Compression Systems

The cylinder leakdown tester is placed in series with a compressed air source and the engine being tested. The crankshaft must be locked at TDC between the compression and power stroke before compressed air is introduced into the cylinder leakdown tester. The overall condition of compression components is determined by the amount of leakage after the combustion chamber is filled with compressed air. Air continues to fill the combustion chamber until air pressure in the chamber equals the regulated supply air going into the tester. Engines normally have some combustion chamber leakage, and some air may continue to flow. An engine having satisfactory compression displays a reading in the green range on the leakdown gauge with a minimum of audible leakage. A reading in the red/green or red range, along with high audible leakage, indicates a problem with compression components.

All internal combustion engines leak some air when tested with a cylinder leakdown tester. The best way to determine the location of a leak is the audible noise from the source as it passes through any engine opening. The engine is inspected for noise with the air cleaner removed at the muffler and the oil fill dipstick or breather tube. Head gaskets, hose, and fittings should also be checked for leaks.

Compression loss can present different symptoms. For example, an exhaust valve leak can cause an engine to appear to have little or no fuel for cold starting. When the piston moves toward the crankshaft, maverick air can enter the combustion chamber through the exhaust valve. The maverick air takes the place of the air-fuel mixture drawn through the carburetor. Without the proper air-fuel mixture, the engine has an insufficient amount of fuel vapor and does not start. See Figure 10-12.

With a leaking exhaust valve, an engine with less than optimal compression can sometimes be started by priming the combustion chamber. Less fuel is required to keep an engine operating than is required for initial cold starting. The engine may operate acceptably when warm, but the leaking exhaust valve still affects engine operation. Maverick air continues to enter the combustion chamber each time the piston tries to draw air-fuel mixture into the combustion chamber through the carburetor. The lean air-fuel mixture produces higher combustion chamber temperatures, which can cause damage to the valve train and other compression system components.

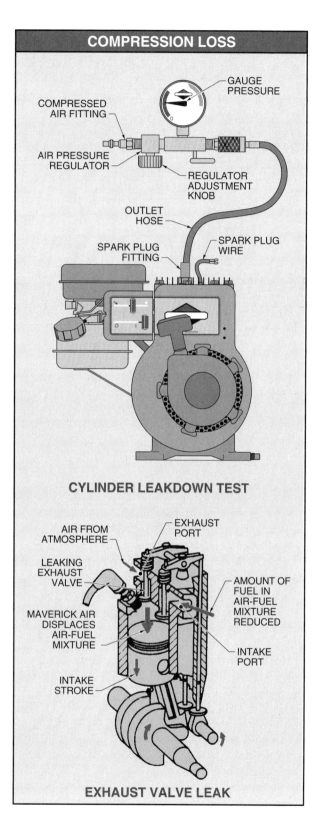

Figure 10-12. Compression loss from a leaking exhaust valve results in maverick air entering the combustion chamber, which reduces the amount of fuel present.

TROUBLESHOOTING FUEL AND GOVERNOR SYSTEMS

The fuel and governor systems function together in a small engine to provide desired engine performance. The fuel system includes the fuel tank, fuel lines, fuel filter, fuel pump, and carburetor. Proper governor system operation is affected by fuel system components, engine design, load, and ambient operating conditions. These variables can individually or collectively cause engine operation problems. In some cases, the same symptom may be the result of causes ranging from improper air-fuel mixture to clogged fins on a flywheel. This makes troubleshooting difficult, and many other engine problems are mistakenly diagnosed as governor system malfunctions. Common fuel system problems include starting problems, performance problems, and carburetor leakage. Common governor system problems are performance problems related to hunting and surging and overspeeding.

Starting Problems

Starting problems occur when the engine is cranking but does not operate. For an engine to operate there must be air, fuel, compression of the charge, and a properly timed spark. If compression and ignition test good, the engine is not starting because of a fuel system problem. More fuel is required during cold startup than during normal operation. Starting problems are isolated by starting at the fuel supply.

Fuel Supply. The required fuel supply to the carburetor is verified by clamping off the fuel line in order to stop fuel flow and removing the fuel line from the inlet fitting of the carburetor. With the fuel cap installed, release the clamping device and allow the fuel to drain into an approved container. The fuel flow should be as wide as the inside diameter of the fuel line. See Figure 10-13. If the engine is equipped with a fuel pump, remove the spark plug wires from the spark plugs. Turn the engine over and perform the same test. The fuel should be delivered with distinct pulses at cranking speeds. If pulses are not present, the fuel pump or fuel pump vacuum connections may be defective and/or the fuel line or fuel filter may be clogged.

Figure 10-13. The fuel flow should be as wide as the inside diameter of the fuel line.

Primer Bulb System and Choke. Some small engines are equipped with a primer bulb system to force fuel into the venturi to assist in starting the engine. The amount of fuel forced into the carburetor is directly related to how hard and how many times the primer bulb is depressed. To verify that the primer bulb is working, remove the air cleaner assembly and look directly into the throat of the carburetor. A pulse of fuel should be present when the primer bulb is depressed.

Starting problems are often caused by an improperly adjusted choke on engines not equipped with a primer system. To verify that the choke is functioning, look inside the throat of the carburetor for fuel. The nature of the starting problem provides possible causes. For example, a customer complaint of hard starting when the engine is cold but normal starting when hot indicates a possible choke malfunction. The choke assembly, cables, and linkage should be checked completely for proper function.

Another method to verify primer bulb system or choke operation is to remove the spark plug and examine the spark plug electrode. A wet spark plug electrode indicates that fuel is being fed to the combustion chamber. A dry spark plug electrode indicates that fuel is not being fed to the combustion chamber. An extremely wet spark plug indicates that the choke may be stuck closed or the engine has been excessively primed or choked.

218 SMALL ENGINES

Anti-Afterfire Solenoid. An *anti-afterfire solenoid* is a device that shuts OFF the fuel at the carburetor to prevent the engine from receiving fuel after the ignition switch is shut OFF. Anti-afterfire solenoids are installed on engines that have a battery-powered electrical system. If the anti-afterfire solenoid is faulty or there is a bad connection, the anti-afterfire solenoid cannot be energized and fuel flow is stopped. See Figure 10-14. To test the anti-afterfire solenoid, connect one pole of a 9 V battery to the spade terminal and the other pole to the solenoid case. A sharp distinct movement should be seen. If not, the solenoid is defective. Care should be taken to prevent damage caused by overextension of the plunger during removal and installation.

When testing some anti-afterfire solenoids, the plunger must be touched or nudged to initiate movement. When installed in the carburetor, the anti-afterfire plunger contacts the orifice in the jet before reaching the maximum extended position. When removed from the carburetor, the plunger is in maximum extended position. A nudge may be required because power from the 9 V battery may not be sufficient power to retract the plunger from the maximum extended position.

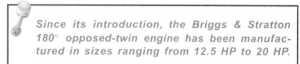

Since its introduction, the Briggs & Stratton 180° opposed-twin engine has been manufactured in sizes ranging from 12.5 HP to 20 HP.

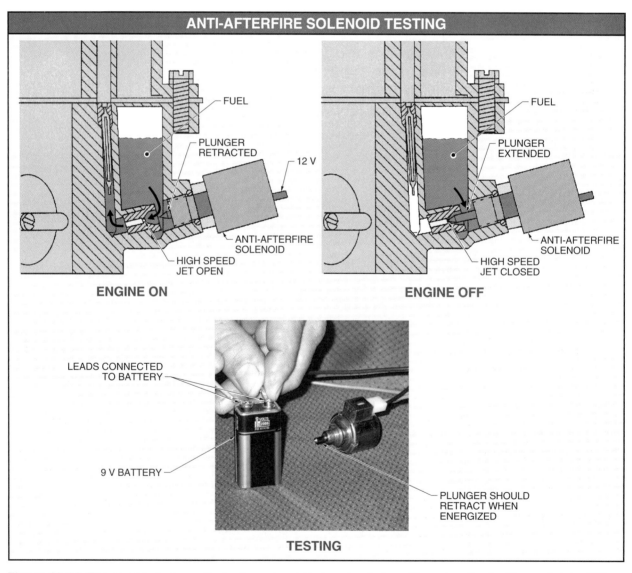

Figure 10-14. The plunger in the anti-afterfire solenoid is in extended position when the ignition switch is OFF.

Debris in Carburetor. Debris lodged in the main jet of the carburetor causes the engine to perform similar to an engine in an under-choked condition. In some cases, debris in the main jet can completely stop all fuel flow into the throat of the carburetor. To check for debris requires examination of the main jet assembly.

Performance Problems

Performance problems occur when the engine is operating. Common performance problems directly related to fuel system and governor system function involve hunting and surging. *Hunting* is the undesirable quick changing of engine rpm when set at a desired speed. *Surging* is the undesirable slow changing of engine rpm in a cyclical pattern when set at a desired speed. Hunting and surging are similar malfunctions and are terms commonly used together in the field when describing undesirable engine rpm variation. Hunting and surging problems occur at idle, top no-load speed, or under load. Most hunting and surging performance problems are related to an interruption of fuel flow.

The first step in troubleshooting a fuel flow problem is to determine whether the problem is in the carburetor or the governor system. To begin troubleshooting, the governor system, linkage, and springs must be in proper position and must appear to be operating correctly with no apparent binding. A static governor adjustment should also be performed. Some engines have multiple positions for the governor spring. Any deviation from the correct position can cause symptoms that are difficult to accurately diagnose and correct.

Hunting and Surging at True Idle. Engine hunting and surging at true idle is caused by a fuel delivery problem or an air leak. *True idle* is the carburetor setting when the throttle plate linkage is against the idle speed adjusting screw after idle mixture adjustment. Because the throttle plate is held stationary during true idle, hunting and surging must be caused by an improper air-fuel mixture related to an air leak or an obstruction in the idle circuit. At true idle, the governor spring applies no force on the throttle plate and has no effect on the idle characteristics of the engine.

A specialized gear-type transmission for marine use was introduced for Briggs & Stratton inboard motor applications in 1940.

Hunting and Surging at Governed Idle. An engine hunting and surging only at governed idle and equipped with an idle mixture adjustment has a governor system or carburetor problem. The idle mixture must be adjusted correctly. Hold the throttle plate linkage against the idle speed adjusting screw and increase the idle speed to the specified governed idle speed. See Figure 10-15. If the engine operates without hunting or surging, the problem is the governed idle spring or linkage. If the engine continues to hunt and surge, the problem is in the carburetor. After testing, return the engine to the correct idle speed.

Hunting and Surging at Top No-Load Speed. Troubleshoot hunting and surging at top no-load speed using the same sequential steps used to isolate a governor system or carburetor problem during true idle and governed idle. Once the idle mixture is adjusted and the engine idles smoothly, increase the engine speed using the idle speed adjusting screw. Hold the throttle plate linkage against the idle speed adjusting screw until the engine reaches the specified top no-load speed. See Figure 10-16. Without any appreciable load, fuel is provided by the idle circuit.

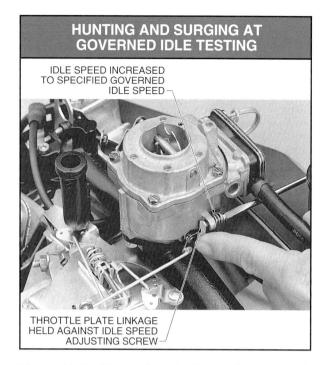

Figure 10-15. Troubleshoot hunting and surging only at governed idle by holding the throttle plate linkage against the idle speed adjusting screw and increasing the idle speed to the specified governed idle speed.

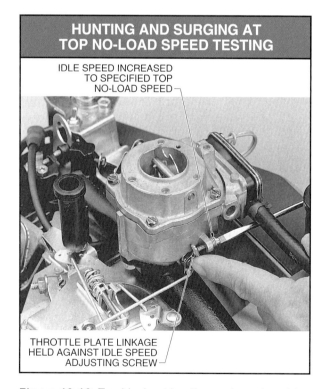

Figure 10-16. Troubleshoot hunting and surging at top no-load speed by holding the throttle plate linkage against the idle speed adjusting screw and increasing the idle speed to the specified top no-load speed.

If the engine continues to hunt and surge, the carburetor is the probable cause. In this test condition, the governor system has no effect on engine speed. The idle speed adjusting screw controls the throttle plate position, which affects engine speed. If the engine operates properly when controlled by the idle speed adjusting screw but hunts and surges when controlled by the governor spring, the governor system is suspect. Check the governor linkage for resistance and binding, and replace the governed idle spring and the main governor spring.

Hunting and Surging Under Load. Hunting and surging under load usually indicates a carburetor or fuel delivery system problem rather than a governor system problem. Fuel fed under load is primarily fed through the main jet and emulsion tube. Most loads are constant enough to maintain the rpm of the engine. The governor system has very little additional effect on the performance of an engine under load except for applications with sizable varying loads such as a generator.

Poor Performance Under Load. Poor performance under load requires first eliminating compression component problems. If the problem is isolated to the fuel system, the cause is usually debris in the main jet or air bleeds. To isolate the problem component, examine the exhaust gas when the engine is under load. If black smoke is present, there is an excess of fuel. This condition may be caused by an incorrect float level setting, a partially clogged main air bleed, or debris lodged between the needle and the seat. If black smoke is not present, and there is no black residue on the muffler deflector, the main air bleed or main jet is probably obstructed.

A primer bulb system or choke can also be used to quickly troubleshoot carburetor problems related to a lean or rich condition. See Figure 10-17. After making carburetor adjustments, slowly close the choke plate when the engine is operating poorly. If performance improves, the air-fuel mixture is too lean. If performance worsens, the air-fuel mixture is too rich. A primer bulb system can also be used to test for a lean or rich condition. Depressing the primer bulb injects extra gas into the carburetor. If the engine air-fuel mixture is too lean, the injection of the extra fuel should improve engine performance. If the engine air-fuel mixture is too rich, the injection of the extra fuel should worsen engine performance.

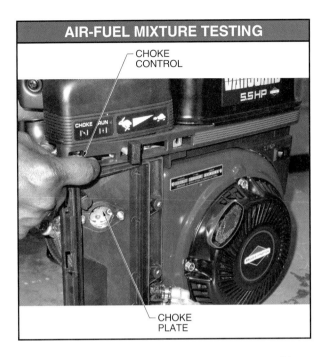

Figure 10-17. Troubleshoot air-fuel mixture problems by closing the choke plate to simulate a rich air-fuel mixture condition.

Engine manufacturers commonly have a variety of problem solving charts available that highlight typical engine problems, probable causes, and recommended solutions. Some troubleshooting charts are specific to engine systems or components. See Figure 10-18.

Overspeeding. Overspeeding problems are commonly caused by binding, missing, or inoperative governor system linkage or components. See Appendix. Overspeeding problems are checked by using the procedure:

1. Move the throttle control lever through its range of motion to ensure smooth operation of the linkages. Place the throttle control lever in the idle position to relieve all tension from the governor spring. Perform a static governor adjustment.

2. Continue the check of the linkage by actuating the governor system components to identify any binding or bent linkage or pivot points.

3. If the linkage moves freely, place the throttle plate and linkage in the idle position.

4. Remove the governor spring and start engine.

5. If the engine goes immediately to idle, the internal governor gear or external governor vane and pivot are operational. If the engine goes to WOT and engine speed progressively increases, the speed-sensing device is the cause of the overspeeding problem.

6. If the engine stays at idle, the governor spring and/or linkage controls are the cause of the problem.

Warning: Before removal of the governor spring and starting of the engine, a positive ignition shutoff procedure must be prepared. If the internal governor gear is the cause of the overspeeding, there must be a procedure for quickly shutting OFF the ignition system before the engine reaches dangerous speeds.

Harmonic Hunting and Surging. *Harmonic hunting and surging* is the undesirable quick and/or slow changing of engine rpm in a cyclical duration caused by excessive governor spring vibration. Testing for harmonic hunting and surging can be performed by operating the engine at the speed where harmonic hunting and surging is most evident. Observe the motion of the governor spring. If the governor spring vibrates or oscillates excessively, dampen the vibration by placing a pencil eraser against the vibrating spring. If the hunting and surging disappears, replace the governor spring.

If the application is not speed-sensitive, adjust to a different top no-load speed. If the harmonic hunting and surging problem cannot be solved by speed adjustment, install a support bracket from the engine to the application that decreases the natural vibration frequency of the application. Selection and installation of a support bracket is usually done by trial and error. Usually, a bracket installed from an engine head bolt to the frame of the application changes the natural vibration frequency.

Multiple Symptoms. Multiple symptoms tend to follow a pattern in which common combinations are found. For example, a common combination is that the engine hunts and surges at idle and top no-load speed but operates properly under load. The circuit feeding fuel to the engine during idle and top no-load speeds is primarily the low speed/transitional circuit. If the engine operates properly and produces acceptable power under load, then the main jet, main fuel supply, and main air bleed are presumed good. The problem must lie in the idle/transition circuit and the pilot jet (if equipped). Transitional holes should be checked for debris.

CARBURETOR PROBLEM SOLVING CHART		
Problem	Cause	Solution
Flo-Jet carburetor leaking after being transported	Float bounce	Use the fuel shutoff valve
Flo-Jet carburetor leaks during operation	Fuel tank too far above carburetor causing excessive pressure at the needle valve	Lower the tank to a maximum of 45″ above the carburetor
	Loose, missing, incorrectly assembled/adjusted, or damaged parts	Correct parts problem
	Contaminated fuel	Clean system/replace fuel
Flo-Jet carburetor leaks shortly after engine is shut OFF	Long coast-down period	Lower engine speed to idle before shutting down
	Fuel leaking past main nozzle	Clean/replace emulsion tube/nozzle
	Loose, missing, incorrectly assembled/adjusted, or damaged parts	Correct parts problem
Engine dies at idle, runs normal at high speed-full load, but surges when running at high speed-no load	Idle passage is blocked	Clean the passage
Engine runs normal at high speed with or without load, but at idle it runs rough with a rhythmic idle	Idle air bleed is blocked	Clean air bleed
Engine runs normal at high speed and idle with no load, but at high speed under load there is a severe loss of power and the engine dies	High speed pick-up tube partially blocked	Clean obstruction
At idle, engine is running slightly fast. At high speed under load it is very rich, blowing black smoke, and top speed does not exceed 2200 rpm. At top speed with no load it is also slightly rich	High speed air bleed is partially blocked	Clean obstruction

Figure 10-18. Troubleshooting efficiency can be improved with specific engine component troubleshooting charts supplied by the manufacturer.

Carburetor Leakage

Carburetor leakage in float-style carburetors is most commonly caused by dirt or debris in the fuel system. True factory defects are rare. However, there is always the small possibility of a defect in a casting or a component in the carburetor body that can cause leakage. No part of a carburetor should be overlooked when performing a troubleshooting procedure or test. Float-style carburetors may have a static leak and/or dynamic leak.

Static Leaks. A *static leak* is an undesirable discharge of gasoline which occurs when the engine is not operating. A static leak can occur immediately after the engine is stopped or days after the last operation. A pressure test can be used to determine the source of the static leak. Remove the carburetor from the engine. Remove the float bowl and invert the carburetor with float and needle installed. Wet the needle and seat area with a small amount of gasoline. Use a thumb pump with a clear nonrubber hose to pressurize the inlet needle to a minimum of 8 psi. See Figure 10-19. The inlet needle should pop off and reseat at 2 psi or greater. If the needle does not seal, clean or replace and test again. If the needle seals and holds pressure at 2 psi or greater for at least 5 min, it is acceptable. If the carburetor does not hold 2 psi for 5 min or continues to exhibit a leak when the engine is in service, additional testing is required.

Caution: Excess finger pressure can cause damage to the needle and/or seat surface. The Viton® portion of the needle and seat combination should be replaced after the pressure test.

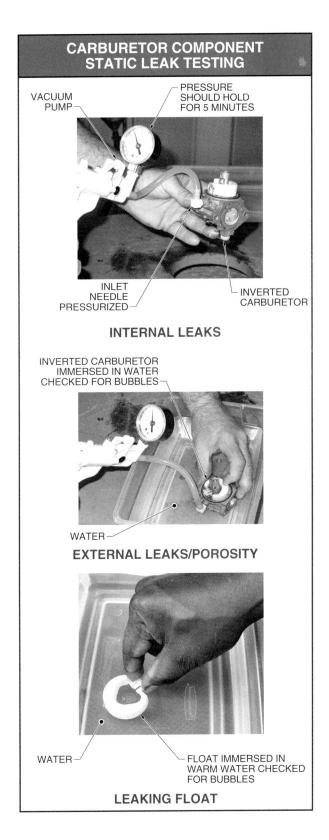

Figure 10-19. Factory defects causing carburetor leakage are rare, but no part of a carburetor should be overlooked when testing.

To test a carburetor for possible porosity or external leakage, invert the carburetor and immerse in water. Hold the float with a finger to ensure a seal between the needle and the seat. Pressurize the inlet needle to minimum of 8 psi. Porosity can cause bubbles to appear from the casting between the brass seat insert and the metal casting or from the fuel inlet fitting. If the pressure test reveals no bubbles from any component parts, check the float by removing it from the carburetor body and immersing it in warm water, and watching for bubbles emerging from a seam.

If all carburetor components test good, the leakage problem was probably caused by dirt or debris. Dirt is the most common cause of leakage after an engine is not used for a period of time. A small particle of dirt on a needle or seat surface can allow an engine to perform properly, but can still cause leaking after sitting idle over a period of time. Most dirt and debris is removed from the fuel by the fuel filter. All fuel filters are rated by the size of the particles measured in microns (μ) allowed to pass through. If the filter size is too large, dirt capable of plugging the jets can enter. If the filter size is too small, fuel flow may be restricted, causing a lean air-fuel mixture.

When an engine is not in use, the fuel in the fuel bowl is exposed to the atmosphere through the fuel bowl vent. Fine particles of dirt or debris normally suspended in the fuel that pass through the carburetor settle and collect at the lowest point between the fuel inlet fitting and the inlet needle. As the fuel slowly evaporates, the float responds by lifting the inlet needle slightly off its seat. Dirt in the fuel can then lodge itself between the needle and seat. A slow but steady flow continues into the fuel bowl, eventually causing a static leak.

Static leak testing a carburetor is performed with a 2 qt freestanding tank filled with gasoline. Position the tank 12″–18″ above the carburetor test bench measured from the bottom of the fuel tank. Attach the fuel line to the carburetor to be tested and open the in-line fuel shutoff valve. See Figure 10-20. Place a piece of paper under the carburetor and let stand in a large drain pan. The carburetor should not leak on the paper overnight.

The fuel supply system must be completely checked to eliminate the cause of the carburetor leak. Inspect the fuel tank cap. An improperly vented fuel tank cap can allow pressure to build up in the tank, which could overcome the float assembly and result in a intermittent leak. The fuel tank should be drained completely and inspected for debris, dirt, or scaling if the tank is metal. Clean or replace as required. After service to the tank, remove and replace all fuel lines and fuel filters. Install the engine manufacturer's recommended fuel filter for the specific application. Complete disassembly of every removable carburetor part and careful inspection and service can eliminate recurring problems. On recurring leakage problems, remove the inlet seat in the carburetor body if possible. Refer to manufacturer's procedures. Dirt can lodge in small crevices and carburetor components. If disturbed, this can cause leakage in the future.

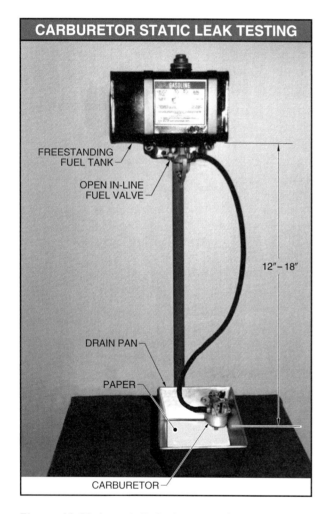

Figure 10-20. In a static leak test, carburetor leakage is indicated by gasoline present on a sheet of paper when left overnight.

 In 1893, Karl Benz designed a carburetor that used a throttle plate for regulating the air-fuel mixture to the engine.

 Over 95% of carburetor performance problems are caused by dirt in the fuel system.

When a fuel filter is removed, a small amount of the inner rubber hose is removed as the barbed end of the filter is pulled out. Small rubber particles can enter into the carburetor and cause obstructions as well as leaks. These rubber particles are very small and difficult to see with the naked eye. When a fuel filter is removed, the fuel line should be replaced. See Figure 10-21.

Caution: Never use a threaded fastener to plug a fuel line during service. The threads cut the rubber fuel line and can result in small rubber particles entering the fuel system.

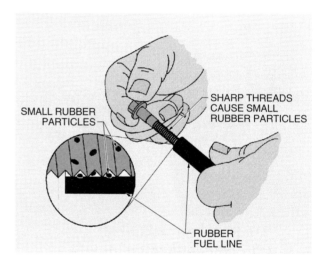

Figure 10-21. Use of a threaded fastener to plug a fuel line during service can result in small rubber particles entering the fuel system.

Dynamic Leaks. A *dynamic leak* is an undesirable discharge of gasoline which occurs when the engine is operating. Most dynamic leaks are caused by the same problems that cause static leaks. Dynamic leaks can be external or internal. External dynamic leaks are commonly caused by deterioration of the fuel tank and strap assemblies and/or improperly installed, damaged, or missing gaskets. Internal dynamic leaks may be indicated by a leak through the brass drain insert in the throat of the carburetor or fuel wetting of the air filter element. Internal leaks can be caused by:

- loose or damaged emulsion tube

- clogged air filter element (externally vented carburetors only)

- excessive vibration
- incorrect fuel inlet seat size
- improper float level
- clogged muffler
- porosity in the carburetor casting
- improper valve clearance
- loose intake valve seat

TROUBLESHOOTING ELECTRICAL SYSTEMS

The electrical system of a small engine includes the starting system, charging system, and ignition system. Starting system components include the battery, starter motor, starter solenoid, starter switch, and battery cables. Engines with rewind starters do not require electrical starting components, and problems are primarily related to mechanical component failure. Charging system components include the alternator, diodes, flywheel, and regulator/rectifier. Ignition system components include the ignition armature, spark plugs, spark plug wires, and Magnetron® ignition module (or breaker points and condenser).

 OPEI estimates that up to 50,000,000 acres of lawn area are mowed in the United States annually.

Starting System

Troubleshooting a starting system requires consideration of related components and systems and is commonly performed by isolating separate components and testing each from easiest to most difficult. If the starter motor does not energize when the ignition switch is turned to the start position, there may be several different causes that may or may not be associated with the starting system. For example, the cause may be a malfunctioning starter interlock mechanism. When the application contains a starter interlock mechanism, troubleshooting should begin at the battery.

Testing an installed battery is best performed with a battery loading device. A *battery loading device* is an electrical test tool that applies an electrical load to the battery while measuring amperage and voltage. The rate of electrical discharge and voltage in each cell of the battery is monitored throughout the test. See Figure 10-22.

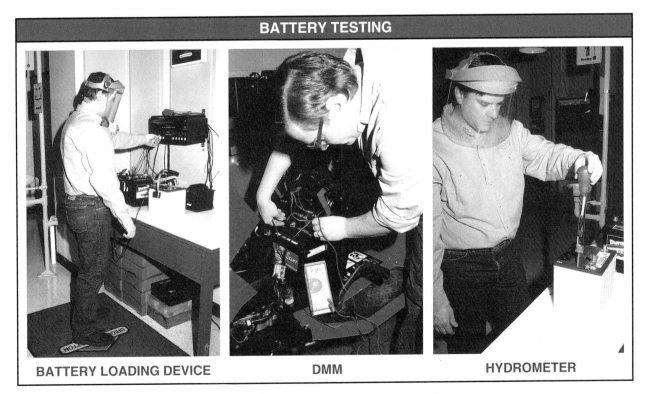

Figure 10-22. Troubleshooting a starting system should begin by testing the battery.

One of the biggest customers of Briggs & Stratton in the 1920s was Bolens, Inc. of Port Washington, Wisconsin, which produced the Bolens garden cultivator.

Most small engine service technicians quickly test a battery using a digital multimeter (DMM). The DMM is connected in parallel with the battery. The DC voltage should read over 12 V. If the battery is completely charged, the voltage reading may be as high as 12.6 V. The potential difference of the battery is sufficient if the voltage reading is greater than 12 V. The battery capacity or charge must also be checked.

The battery capacity is tested by checking the specific gravity of the electrolyte. *Specific gravity* is a comparison of the mass of a given sample volume compared to an equal volume of water. A *hydrometer* is an instrument used to measure the specific gravity of a liquid. As a battery is discharged, the electrolyte loses sulfuric acid and gains water. This allows the measurement of the specific gravity of electrolyte to indicate the relative charge of the battery. Generally, the specific gravity reading of the electrolyte in the battery should be greater than 1.225. If the specific gravity reading of the electrolyte from any one cell varies more than .050 from the others, the battery should be replaced.

After battery voltage and capacity test good, the next step in troubleshooting the starting system is to verify that the engine is capable of turning over by rotating the engine 360° by hand. If the engine rotates, the next step is to apply 12 V from the battery to the starter solenoid. A heavy gauge jumper wire is used to connect the positive battery cable end and the starter solenoid terminal. This directs 12 V from the battery to the starter motor. See Figure 10-23. If the starter motor does not energize, the starter motor is faulty. Faulty starter motors are usually replaced rather than repaired. However, some minor repair procedures such as brush replacement can be economical. Procedures for testing 12 V electric starter motors are usually specified in the manufacturer's repair manual. If the starter motor energizes and turns the engine over at 350 rpm or greater, the problem is caused by some other starting system component.

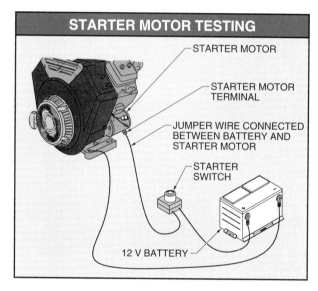

Figure 10-23. A jumper wire is used when troubleshooting the starting system to determine if the starter motor is faulty.

On engines equipped with a separate starter solenoid, a jumper wire can be used from the battery to the spade connector on the starter solenoid that is connected to the ignition switch. If the starter solenoid has a single spade connector, connect the jumper wire. If the starter solenoid has two spade connectors, remove the wire connector that is not directly connected to the ignition switch. Run a jumper wire from that spade connector on the starter solenoid to ground. The attached jumper wire is carrying 12 V from the battery to the other spade connector. This directs 12 V from the battery to the starter solenoid and eliminates any ground side safety switches that may be on the application. With the 12 V directed to the starter solenoid, the starter motor should energize. If this test produces no starter motor rotation, or the starter solenoid makes a loud clicking noise followed by intermittent starter motor action, the starter solenoid or the ground connection is defective and should be repaired or replaced. If the starter motor and starter solenoid test good, test for 12 V at the wire from the battery, through the ignition switch to the small spade terminal connected to the starter solenoid. With the ignition switch in the start position, it should read battery voltage at the wire connection. If there is no voltage at the wire, the starter system safety interlock and the ignition switch should be tested following starter system interlock test procedures in the manufacturer's repair manual.

Charging System

Troubleshooting charging systems includes finding the cause of a dead battery or overcharging (excessive boiling of the electrolyte). Testing the output of a charging system isolates the battery and connections from the alternator system. The two basic tests commonly used for all Briggs & Stratton alternator systems are the AC voltage test and the DC amperage test. The reference values listed in the manufacturer's repair manual determine the minimum output of the charging system in question. If both AC volt and DC amp reference values are listed, the AC voltage test should be performed first.

AC Voltage Test. An *AC voltage test* is a test that uses a DMM to indicate the voltage potential of the alternator stator. Set the selector switch to AC voltage. Plug the red test lead into the voltage jack. Plug the black test lead into the common jack. Connect the red test lead to the output connector lead of the stator. Connect the black test lead to a good engine ground. Start the engine and bring it up to 3600 rpm. See Figure 10-24. If AC voltage output is below specifications, the stator probably contains a short in the windings. If the AC voltage output is equal to or greater than the specifications, the stator is good. Two exceptions to this test procedure are the 10-13-16 A and 20 A alternator systems. When testing these systems, the black test lead is connected to the second stator connector pin on the output lead of the stator instead of a good engine ground.

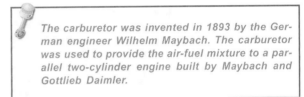

The carburetor was invented in 1893 by the German engineer Wilhelm Maybach. The carburetor was used to provide the air-fuel mixture to a parallel two-cylinder engine built by Maybach and Gottlieb Daimler.

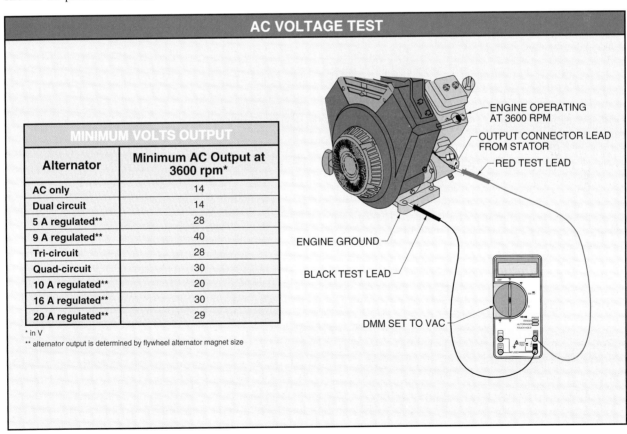

Figure 10-24. The AC voltage test uses a DMM to determine the voltage potential of the alternator stator. If the AC voltage output is equal to or greater than the specifications, the stator is good.

DC Amperage Test. A *DC amperage test* is a test that uses a DMM to indicate the current that should enter the battery if all connections to the battery are good. The test is performed the same way regardless of the charging system. Set the selector switch to DC amperage. Plug the red test lead into the 10 A jack. Plug the black test lead into the common jack. Connect the red test lead to the output connector lead from the stator or regulator/rectifier. Connect the black test lead to the positive side of the battery. See Figure 10-25.

The engine should be started only after the DMM connections are complete. This reduces the chance of a voltage spike entering the regulator/rectifier and possibly causing component failure. With the engine operating at 3600 rpm, the DMM reads the actual current delivered to the battery through the charging circuit. In this test, the battery is used in series to provide a resistive load. A *resistive load* is an applied load that reduces the possibility of the alternator system delivering full amperage through the circuit.

Proper test lead connections to the stator

A 10-13-16 A or 20 A Briggs & Stratton stator does not provide a ground connection to the stator lamination stacks. Each winding end is fed out from under the flywheel as an output lead. If this stator is tested by connecting the black test lead to ground as with most other Briggs & Stratton alternator systems, the AC volts reading is slightly more than one-half of the expected and satisfactory reading. The voltage measurement proves that a complete electrical circuit is not required to measure voltage. With the test leads connected incorrectly, the stator voltage reading is purely a potential quantity, since there is no electrical flow without a completed circuit.

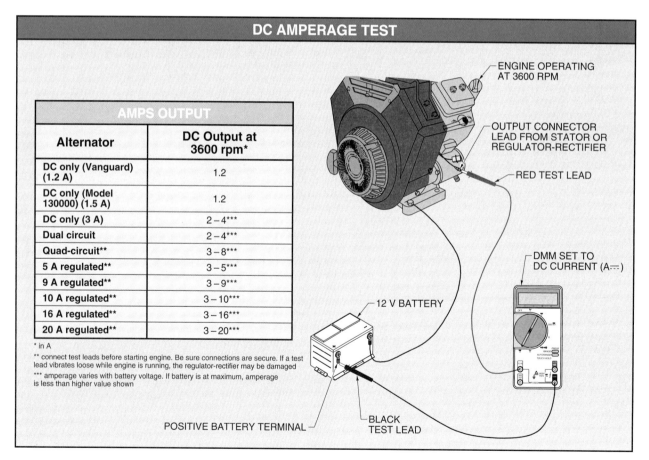

DC AMPERAGE TEST

AMPS OUTPUT

Alternator	DC Output at 3600 rpm*
DC only (Vanguard) (1.2 A)	1.2
DC only (Model 130000) (1.5 A)	1.2
DC only (3 A)	2–4***
Dual circuit	2–4***
Quad-circuit**	3–8***
5 A regulated**	3–5***
9 A regulated**	3–9***
10 A regulated**	3–10***
16 A regulated**	3–16***
20 A regulated**	3–20***

* in A
** connect test leads before starting engine. Be sure connections are secure. If a test lead vibrates loose while engine is running, the regulator-rectifier may be damaged
*** amperage varies with battery voltage. If battery is at maximum, amperage is less than higher value shown

Figure 10-25. A DC amperage test indicates the value of the current that should enter the battery if all of the connections to the battery are good. The engine should be started only after the DMM connections are complete.

If the DC amperage test was performed with the black test lead connected to ground, the alternator system would be subjected to a dead short. This causes the alternator system to deliver the maximum current available from the circuit. Without the battery in series with the circuit, the current produced by the stator would blow the fuse in the DMM or overheat the windings.

Ignition Systems

A spark tester is used to troubleshoot the ignition system. A *spark tester* is a test tool used to test the condition of the ignition system on a small engine. See Figure 10-26. Electricity flows to the spark tester and must jump across an air gap. Air is a poor conductor of electricity, and the .166″ air gap in the spark tester simulates the minimum voltage requirements of an ignition armature under operating conditions. If the ignition armature is capable of producing sufficient voltage to achieve a spark in the spark tester window, it is capable of producing the voltage necessary to operate the engine, even under load.

The *ionization gap* is the distance between the ignition armature pole and the secondary pole in the spark tester. Electricity jumps across the ionization gap, then the large gap in the spark tester. The ionization gap is required to maintain a consistent atmosphere within the spark tester. When the spark jumps the ionization gap, the humidity in the air is reduced within the enclosed window, providing a consistent environment for testing. Testing the ignition system is the same for both a magneto ignition system and a Magnetron® ignition system.

To test the ignition system, the spark tester is connected in series with the spark plug wire and ground. Pulling the rewind starter rope or actuating the electric starting system should produce a visible spark jumping between the two main posts inside the spark tester body. If a visible spark is produced, regardless of the color, the ignition system should be considered good.

If no spark is produced, remove the stop switch wire from the ignition armature to the stop switch and retest. If no spark is produced, remove the blower housing, and inspect the spark plug wire(s) for cuts and/or other damage. Disconnect the stop switch wire directly from the ignition armature and retest. This eliminates any possibility of grounding of the primary or secondary coil in the armature.

To test for a heat-related ignition problem, the spark tester is connected in series between the spark plug wire and the spark plug connector. See Figure 10-27. The engine is started and allowed to reach operating temperature. If the engine malfunctions during operation, the ignition system can be isolated or eliminated as the problem by watching the spark in the spark tester window. An ignition system malfunction caused by losing spark when the engine is hot is indicated by the loss of the spark during coastdown. If the spark is still visible during coastdown until the engine reaches 350 rpm, the ignition system is not the cause.

Many small engine technicians test the ignition system by removing the spark plug and laying it against the cylinder head or engine block to provide a ground for the ignition system. This testing method does not provide accurate results and can be dangerous. The voltage required to jump a .030″ gap in the spark plug outside of the combustion chamber can be as low as 2000 V. It takes a minimum of 8000 V to ignite the charge inside a typical combustion chamber. Testing the spark plug outside the combustion chamber does not provide minimum voltage re-

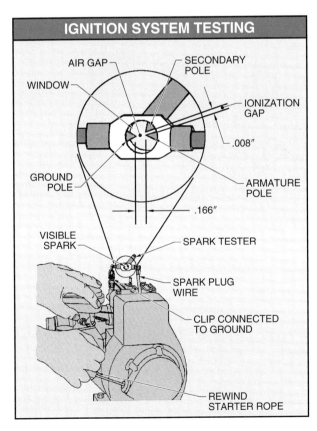

Figure 10-26. A spark tester tests the condition of the ignition system on a small engine by simulating the electrical load on the ignition system.

quirements of an operating engine. In addition to being inaccurate, the open spark across the spark plug gap outside the combustion chamber can create a fire hazard if there are flammable vapors present.

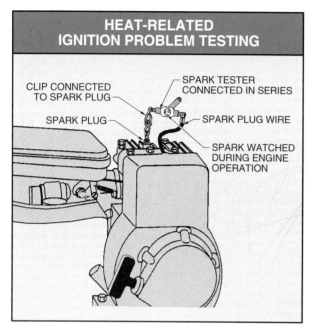

Figure 10-27. Testing for a malfunction caused by losing spark when the engine is hot requires connecting the spark tester in series between the spark plug lead and the spark plug connector.

Magneto Ignition System. If the performance of the points and condenser in the magneto ignition system are suspect, additional testing is required. Specific test procedures for points and condensers are specified in the manufacturer's repair manual. A magneto ignition armature can be tested using a test engine consisting of a horizontal crankshaft engine block, crankshaft, blower housing with rewind assembly, and flywheel. See Figure 10-28.

To test a magneto ignition armature, remove the armature from the engine and temporarily install a Briggs & Stratton Magnetron® conversion kit. Install the armature on the test engine and pull the rewind starter rope with the spark tester attached. If there is spark in the window and the ignition system problem is not heat-related, the armature is considered good. This test engine can also be used to test all Magnetron® ignition system armatures. When testing the 180° opposed two-cylinder engine ignition system armature, the second spark plug wire must be grounded before performing the test.

Figure 10-28. An ignition armature can be quickly tested using a test engine to see if a spark in the spark tester is produced.

TROUBLESHOOTING COOLING SYSTEMS

Troubleshooting the cooling system commonly consists of identifying and solving air and coolant flow and leak problems. Air-cooled engines require a constant, unrestricted flow of fresh air directed around specific engine parts. Liquid-cooled engines require a constant, unrestricted flow of coolant through the radiator and engine cavities. Cooling system problems are commonly caused by recirculation of hot exhaust gases and inadequate coolant flow.

Air-Cooled Engines

Improper flow of hot exhaust gases from the muffler and from engines equipped with an exhaust deflector can cause recirculation of hot exhaust gases into the cooling air intake of the engine. Exhaust recirculation is indicated by dark or sooty deposits on the flywheel and/or rotating screen, sooty streaking along body panels leading to the cooling air intake, and/or missing or damaged air plenums and seals leading to the cooling air intake area.

Liquid-Cooled Engines

Troubleshooting a cooling system on a liquid-cooled engine is often related to problems of overheating and leakage. Overheating is indicated by a temperature gauge reading in excess of 220°F, boiling over of coolant, and/or cooling system leakage. An excessive temperature reading or boiling over of coolant usually indicates a lack of sufficient air flow through the radiator or coolant flow throughout the engine.

Generally, troubleshooting steps on a liquid-cooled engine are similar for high temperature readings and boiling over problems. If the engine has high temperature readings, check the coolant level when the radiator is cool. The coolant level should be no more than 1″ below the neck of the radiator fill hole when cool. See Figure 10-29. Carefully examine the radiator cap for signs of pressure leakage, such as a cut, distorted, or missing rubber washer seal. Inspect the fan belt and fan to verify there is no slippage between the belt and pulleys. Check the fan for proper direction. On most applications, the fan should pull air through the radiator and toward the engine. Remove any debris on the outside of the radiator that could restrict air flow through the fins of the radiator core.

Start the engine with the radiator cap removed. Check the coolant flow as it passes by the filler neck opening as the engine reaches operating temperature. The coolant should be flowing freely when the temperature reaches 200°F. If the coolant in the radiator shows no sign of circulation by the time the temperature gauge reaches 200°F, shut down the engine. Examine the lower radiator hose carefully. If the cooling system has a defective thermostat, the lower radiator hose may have a section that is partially or completely collapsed. Allow the engine to cool before continuing the troubleshooting process.

Warning: Stand away from the engine when checking coolant flow with the radiator cap removed. Hot engine coolant could discharge abruptly if the thermostat is stuck closed.

Remove and test the thermostat following the procedure in the manufacturer's repair manual. If the thermostat tests good, the water pump should be removed and checked. The water pump impeller may show signs of corrosion that can inhibit coolant flow.

A liquid cooling system that leaks can be tested using an automotive-style cooling system pressure tester. Attach the tester and pressurize the system to 11 psi for a minimum of 10 min. Examine all hose connections, radiator fins, tank seams, freeze plugs, and the sight hole (if equipped) in the water pump below the shaft seal. If there are no signs of external leakage, the problem may be a blown head gasket or a cracked water jacket in the engine.

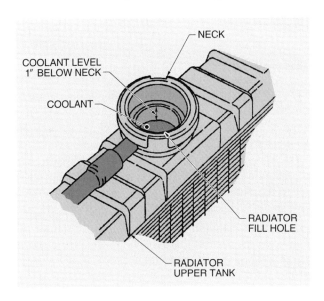

Figure 10-29. High temperature readings and boiling over problems are prevented by maintaining the proper coolant level when the radiator is cool.

TROUBLESHOOTING LUBRICATION SYSTEMS

When troubleshooting lubrication systems, the oil level should be checked using a sight hole (if equipped) to verify the level of oil in the reservoir. If the engine is not equipped with a sight hole, the correct length and style of dipstick for the engine should be verified. Engine failure can be caused by an inadequate or excessive amount of oil. Engine damage caused by both conditions is similar.

Splash Lubrication Systems

The splash lubrication system is a simple design used on horizontal and vertical shaft engines, and there is a limit to what can go wrong. The splash lubrication system functions properly as long as the dipper is correctly attached to the connecting rod and the required oil level is maintained. The oil hole in the connecting rod and aluminum main bearings should be checked for debris and proper assembly orientation on the connecting rod.

The vertical shaft splash lubrication system also must be checked for the same conditions as the horizontal shaft splash lubrication system and the oil slinger. The oil slinger and the governor gear work together. If the governor gear teeth on the slinger wear or fail, the engine will overspeed. Splash lubrication system problems are predicted by monitoring splash system components and affected bearing surfaces.

Pressure Filtration Lubrication Systems

Pressure filtration lubrication systems utilize a gerotor pump to circulate oil in a limited area of the engine. The gerotor pump is used in conjunction with an oil slinger splash system to provide the necessary lubrication of the engine. The gerotor pump is not regulated and may in some cases provide additional oil to the main bearing on the flywheel side by several delivery methods.

The most common pressure filtration system problem is the failure of the driven member of the gerotor pump. Failure can be caused by excessive shock loads to the engine, lack of proper clearance between the pump and cover, or foreign particles in the oil pump.

A failure in the pressure filtration system typically does not cause a lubrication failure of any bearing surface in the engine. The pressure filtration system is designed to provide a method to circulate and filter oil in the reservoir and supply additional, but not primary, lubrication to the main bearing surface on the flywheel side. An engine equipped with a pressure filtration system can perform properly even with the gerotor pump completely removed.

Pressure Lubrication Systems

Pressure lubrication systems provide all of the benefits of the pressure filtration system and act as the primary method of lubrication for the entire engine. A malfunction in the pressure lubrication system can have a catastrophic effect on the engine. Troubleshooting a pressure lubrication system commonly includes testing for high oil pressure, low oil pressure, and lack of oil pressure conditions.

High oil pressure is commonly caused by a defective or stuck pressure relief valve in the oil pump and/or using the incorrect viscosity oil. High oil pressure can lead to significant external oil leaks. Another possible symptom of high oil pressure is cracking of the oil filter. Testing for the proper oil pressure is performed by removing a pipe plug at the oil filter adapter or at the crankcase and taking a measurement with an automotive-style oil pressure gauge. Testing specifications for oil pressure based on engine rpm are listed in the engine manufacturer's repair manual.

Low oil pressure can be caused by several problems such as loose bearing clearances, damaged gerotor oil pump gears or mating surfaces, damaged teeth on the oil pump drive gear, or too much clearance between the gerotor pump gears and the oil pump casting. Although dirt and debris in the passages of the lubrication system may cause oil delivery problems and starvation of one or more bearing surfaces, reducing the flow of oil in the passages does not result in any oil pressure abnormalities.

Lack of oil pressure and delivery can be caused by all of the problems causing low pressure. The most common cause of lack of oil pressure is the incorrect assembly or damage to an internal component part. All oil passages must be in alignment during the reassembly of the engine. See Figure 10-30. The cause of lack of oil pressure is best determined by beginning the inspection of the engine at the oil pickup and following the oil passages through the rest of the engine.

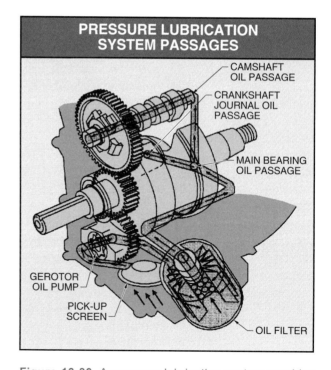

Figure 10-30. A pressure lubrication system provides the primary method of lubrication for the entire engine. A malfunction in any of the oil passages can have a catastrophic effect on the engine.

FAILURE ANALYSIS

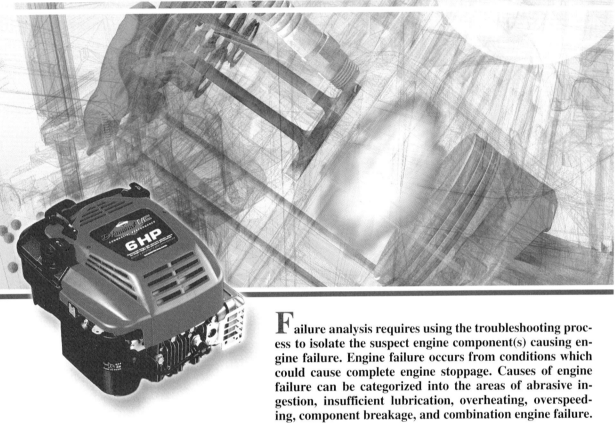

Failure analysis requires using the troubleshooting process to isolate the suspect engine component(s) causing engine failure. Engine failure occurs from conditions which could cause complete engine stoppage. Causes of engine failure can be categorized into the areas of abrasive ingestion, insufficient lubrication, overheating, overspeeding, component breakage, and combination engine failure.

ENGINE FAILURE

Engine failure is the complete stoppage of the engine caused by the failure of one or more engine components. Determining the cause of engine failure is one of the most challenging tasks of a small engine service technician. Most engine problems can be solved by referring back to similar cases and engine service experiences. However, engine failure causes vary and often do not provide historical information. For example, two engines that have failed from the same cause under normal circumstances may not exhibit the same degree of damage. This can cause confusion when determining the primary cause of a similar engine failure.

Although the terms troubleshooting and failure analysis are often used interchangeably, there is a distinct difference between the two. *Troubleshooting* is the systematic elimination of the various parts of a system or process to locate a malfunctioning component. Troubleshooting is commonly performed on an operating engine. *Failure analysis* is the analysis of the engine component or components related to the cause of an engine failure (complete engine stoppage). Troubleshooting steps can be used to systematically isolate the cause(s) of the engine failure. See Appendix. Engine failure causes can be categorized into the areas of abrasive ingestion, insufficient lubrication, overheating, overspeeding, component breakage, and combination engine failures.

 A head gasket leak in a liquid-cooled engine may result in oil in the coolant or coolant in the oil, depending on the leak location.

Abrasive Ingestion

Abrasive ingestion is a cause of engine failure through the undesirable introduction of abrasive particles into a small engine. An *abrasive particle* is a particle with enough hardness to cause the grinding or wearing away of material through friction. Abrasive ingestion is the most common cause of small engine failure. The most common abrasive particle affecting small engines is silica. *Silica* is a compound of the elements silicon (Si) and oxygen (O_2). Silica is commonly found in sand and dirt and is the main ingredient in the mineral quartz. Only the minerals topaz, corundum (sapphire), and diamond in ascending order are rated harder than quartz. Silica particles have sharp edges. The sharp edges are capable of removing a significant amount of material through impact when accelerated by moving air. The hardness and shape of the abrasive particle is a major factor in damage leading to engine failure from abrasive ingestion.

Abrasive particles can be ingested through air leaks in the air intake system, oil fill plug, extended oil dipstick, rubber seals and gaskets, and/or any other breach in engine component sealing integrity. In addition, abrasive particles can be ingested and routed through normal engine operation. For example, airborne abrasive particles are accelerated by moving air and enter the engine with the air-fuel mixture burned in the combustion chamber. Abrasive particles commonly leave evidence of the path taken inside the engine. Most abrasive ingestion effects are similar, but damage varies depending on the characteristics of the affected component. See Figure 11-1.

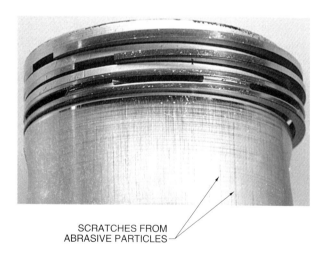

Figure 11-1. Scratches in the piston are caused by abrasive particles embedded in the cylinder wall.

Abrasive Ingestion Effects. Moving air is capable of retaining a large volume of abrasive particles in suspension. As the velocity of the air is increased, the potential quantity and mass of the particles suspended also increases. When traveling through an engine, abrasive particles encounter changes in direction or nonlinear paths. These nonlinear paths can occur at a curve, turn, or around a projection. Abrasive particles with greater mass than the surrounding air tend to maintain a linear path. For example, when air and abrasive particles make a turn in the intake system, the air and some abrasive particles follow the curve, but abrasive particles with greater mass than the surrounding moving air impact and are deposited on the outer surface of the curve. See Figure 11-2.

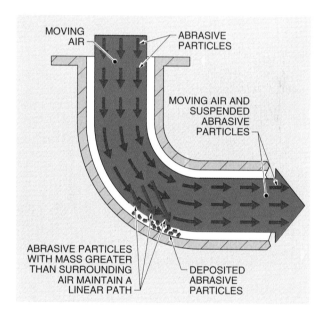

Figure 11-2. Abrasive particles with greater mass than the surrounding moving air maintain a linear path and impact at the outer surface of a curve.

Velocity and hardness cause ingested abrasive particles to act similar to a sandblaster being used to remove paint. Air entering a typical Briggs & Stratton engine can reach speeds up to 35 mph. This velocity, in conjunction with the sharp hard edges of abrasive particles, causes a signature wear pattern. A *signature wear pattern* is an area impacted by abrasive particles having specific appearance and dimensional characteristics. A signature wear pattern in an engine block appears as a satin finished (usually gray) streak or as a teardrop shape on a curved metal or polymer engine component. Signature wear patterns are most evident in areas where abrasive particles have changed direction, such as intake manifolds and intake valve ports.

Analysis of abrasive ingestion requires identification of the path of the abrasive particles. Evidence left by abrasive particle wear can lead directly to the primary cause of the ingestion problem. Repairing an engine contaminated with abrasive particles requires careful examination of all possible effects. This includes all engine components and bearing surfaces, regardless of the degree of abrasive ingestion, to prevent future problems. See Figure 11-3. For example, main bearing damage from abrasive ingestion is often overlooked during engine evaluation. The damage may be difficult to discriminate from normal marks left from machining.

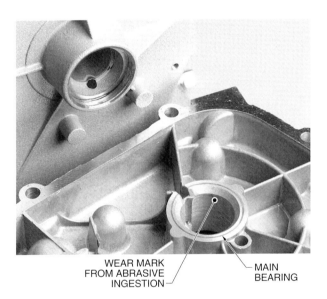

Figure 11-3. Damage to the main bearing from abrasive ingestion is often overlooked during engine evaluation.

Carburetor. Abrasive particles entering the engine through the carburetor can be detected by particles lodged on or around the choke plate. As air passes through the venturi and draws fuel from the emulsion tube, the next component or obstacle in the air stream path is the throttle shaft. Gasoline vapor in the carburetor exhibits some adhesive qualities, and vapor droplets attract abrasive particles which collect on the throttle shaft near the bushing surface. The abrasive particles cause severe premature wear of the throttle shaft and throttle shaft bushing surface. See Figure 11-4. This can allow unfiltered maverick air into the carburetor. *Maverick air* is undesirable, unaccounted air entering the engine through leaks caused by worn, loose, or failed engine components.

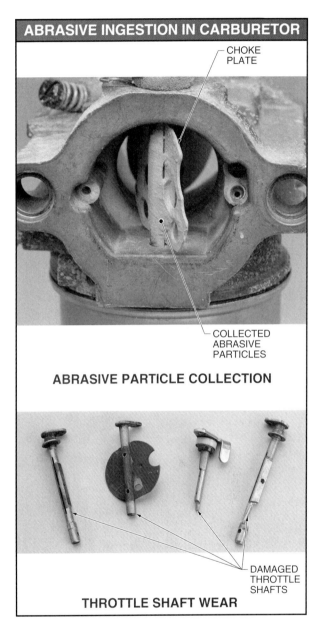

Figure 11-4. Abrasive particles entering through the carburetor collect on the choke plate and choke shaft and cause premature wear of the throttle shaft.

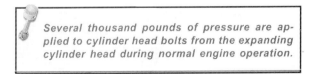

Several thousand pounds of pressure are applied to cylinder head bolts from the expanding cylinder head during normal engine operation.

Maverick air can enter the carburetor after the venturi through a worn throttle shaft bushing. The maverick air contains no fuel when it enters the combustion chamber, resulting in a leaner air-fuel mixture. The lean air-fuel mixture can lead to overheating. The maverick air also enters the combustion chamber unfiltered, allowing abrasive particles that have bypassed the air filter element to enter the engine.

Valve and Valve Guide. Evidence of abrasive ingestion affecting the valve and valve guide is commonly located at the intake valve port. The *intake valve port* is the portion of the engine that provides a path from the carburetor or intake manifold to the intake valve head. The intake valve port contains the intake valve stem and valve guide. As in the throttle shaft area of the carburetor, the air-fuel mixture entering the intake valve port carries abrasive particles. The abrasive particles are deposited and collect wherever there is a nonlinear change in direction. The majority of abrasive particles are deposited on the valve stem and valve guide. A signature wear pattern is created on the intake port and the intake valve stem. The oil film and the reciprocating motion of the intake valve draw abrasive particles into the valve guide, causing premature wear. See Figure 11-5.

Abrasive particles that pass through the intake valve port also produce effects on the valve face and the sealing surface of the valve seat. In extreme cases, a signature wear pattern is evident on the valve seat, consisting of a wide, rounded sealing surface nearest to the cylinder. In most cases, evidence of abrasive ingestion is indicated by an asymmetrical signature wear pattern on 70°–90° of the sealing area on the valve seat. This pattern is caused by abrasive particles that must pass over and contact the valve seat sealing surface before entering the cylinder. This results in an uneven valve seat sealing surface. In addition to abrasive wear, each time the intake valve closes, abrasive particles are trapped between the valve face and valve seat sealing surface. The force of the valve spring and the velocity of the closing valve cause small abrasive particles to be crushed, resulting in excessive wear on the valve seat and an impression pattern on the valve face. An *impression pattern* is a uniform circular signature wear pattern on the intake valve face caused by crushed abrasive particles. See Figure 11-6.

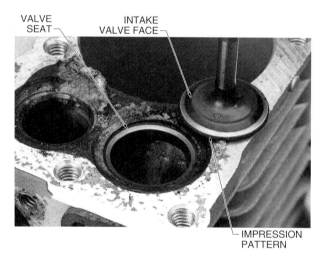

Figure 11-6. Abrasive particles trapped between the intake valve face and the valve seat sealing surface cause excessive wear and produce an impression pattern.

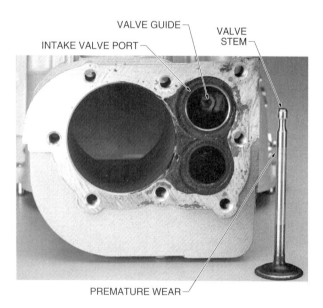

Figure 11-5. Abrasive particles ingested through the intake valve port cause premature wear to the valve guide and valve stem.

Cylinder Bore. Abrasive particles may enter the cylinder bore suspended in the air-fuel mixture. A number of abrasive particles may become embedded in the side of the cylinder wall opposite the intake valve. Some of the abrasive particles flow back to the cylinder wall below the intake valve and adhere to the oil film, becoming embedded in the cylinder wall. Although cast aluminum alloy engines are more susceptible to abrasive particles becoming embedded into the cylinder bore, this also occurs in cast iron cylinder bores. Some of the abrasive particles begin to remove metal on the cylinder wall as the piston and piston rings move up and down in the cylinder bore. The first sign of abrasive ingestion in the cylinder bore is the premature loss of the cross-hatched pattern. See Figure 11-7.

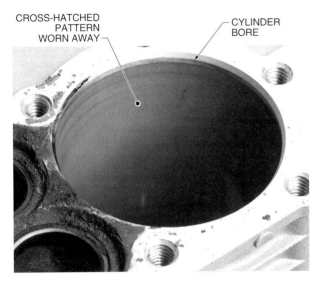

Figure 11-7. The first sign of abrasive ingestion in the cylinder bore is the premature loss of the cross-hatched pattern.

Continued wear from abrasive particles in the cylinder bore can also increase cylinder bore diameter. Cylinder bore wear is checked by taking two measurements (90° to each other) at the top, center, and bottom locations of piston ring travel. The cylinder bore diameter should be the same. The majority of wear from abrasive particles occurs at the top of the cylinder bore nearest to the entry point. The wear profile of cylinder bore wear from abrasive ingestion is similar to an inverted bell. See Figure 11-8.

Piston and Piston Rings. Abrasive particles on the cylinder wall are commonly larger than the oil film separating the piston and piston rings from the cylinder bore. Large abrasive particles tend to scratch the piston, and small abrasive particles tend to polish the piston. The greatest amount of wear from abrasive particles occurs on the oil ring. The oil ring exerts the greatest amount of pressure in a small area of the sealing edges as it moves in the cylinder bore. As the oil ring wears, it becomes wider, reducing wiping and sealing effectiveness. This allows more oil to pass by the oil ring to the compression ring and results in increased oil consumption. See Figure 11-9.

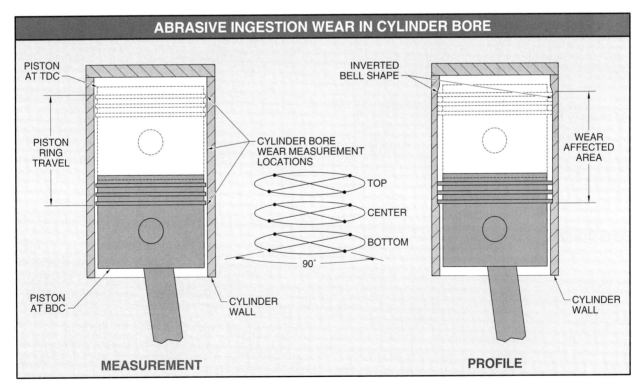

Figure 11-8. Wear from abrasive particles in the cylinder bore commonly occurs at the top of the cylinder bore nearest to the entry point.

238 SMALL ENGINES

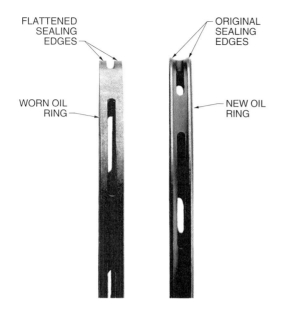

Figure 11-9. Abrasive ingestion has the greatest effect on the oil ring, resulting in increased oil consumption.

Snapper, Inc.
The 14.5 HP engine used to power this lawn tractor features overhead valves and a splash lubrication system.

Abrasive particle deposits in the cylinder bore can also lead to piston ring sticking and uneven piston ring wear. Piston ring sticking is normally caused by excessive oil consumption and carbon deposits in the piston ring lands, which restrict or stop piston ring movement. Piston rings which do not rotate are exposed to uneven piston ring wear caused by abrasive particles embedded in the cylinder wall.

Crankcase Components. A properly operating engine has a negative pressure in the crankcase that can draw in abrasive particles through various entry points. The signature wear pattern for abrasive ingestion affecting crankcase components is a satin gray finish on the surface of the connecting rod and main bearings, crankshaft wear, and camshaft bearing journal wear. See Figure 11-10. Abrasive ingestion also affects bearing surfaces on rotating, sliding, or oscillating counterweights used to balance the engine. Abrasive particles entering the cylinder bore through the crankcase cause increased wear near the crankcase. Less wear occurs near the cylinder head, resulting in a wear profile similar to a bell.

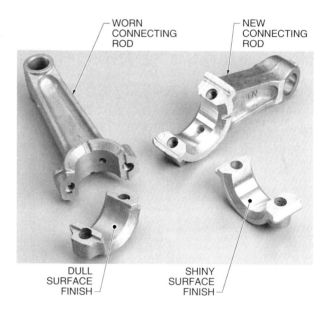

Figure 11-10. Evidence of abrasive particles drawn through the crankcase is indicated by a dull surface finish on the crankshaft bearing surface of the connecting rod.

Abrasive Ingestion Entry. Abrasive ingestion entry is determined by tracing the evidence and the signature wear patterns back to the source. The entry point is located between the component exhibiting no signs of abrasive ingestion and the first sign of abrasive ingestion. For example, abrasive ingestion is suspected, and evidence is not apparent until the intake valve port. The carburetor, throttle shaft, and intake manifold show no evidence of abrasive ingestion. The entry point of abrasive particles must have occurred in the path from the intake manifold to the intake port gasket. The probable entry point is a missing or damaged intake manifold to the intake port gasket. Determining the exact locations where abrasive ingestion has and has not occurred helps to identify the entry point and possible effects on engine components.

Insufficient Lubrication

Insufficient lubrication is a cause of engine failure from the absence, loss, or degradation of the oil film between two bearing surfaces. Insufficient lubrication results in increased friction and heat. The degree of oil film degradation is proportional to the evidence of damage to the failed engine. Engine failures from insufficient lubrication are not always easy to identify. Variables such as engine load, operating conditions, maintenance schedules, and engine tolerance may change the severity of an engine failure. For example, an engine that operates under a consistent light load exerts less combustion pressure against the piston rings, which reduces the unit pressure applied against the cylinder bore. Reduced unit pressure reduces possible damage from an inadequate oil film.

All internal combustion engines produce friction during operation. *Friction* is the resistance to motion that occurs when two surfaces slide against each other. The amount of friction depends on the temperature, surface smoothness, presence of a lubricant, and the force applied to the surfaces in contact. Friction can be static or kinetic (sliding) friction. *Static friction* is the force needed to accelerate or initiate the movement of a stationary mass. *Kinetic friction* is the friction exhibited by a moving mass. Kinetic friction is almost always less than the force needed to initiate motion. The product of any friction is heat.

The functions of oil in an engine include:
- reducing wear by preventing metal-to-metal contact and corrosion
- cooling the engine by transferring heat from internal components to the cylinder block
- sealing the engine and reducing deposit formations by providing a fluid oil film between engine components
- cleaning the engine by suspending dirt and debris and reducing sludge formation

Oil provides a lubricating film to separate moving surfaces and to reduce metal-to-metal contact and resulting friction and heat at bearing surfaces. Bearing surfaces commonly requiring an adequate oil film include crankshaft bearing journals, camshaft bearing surfaces, connecting rod bearing surfaces, the cylinder bore surface, and rotating, sliding, or oscillating counterweight bearing surfaces.

All machined bearing surfaces have asperities. *Asperities* are tiny projections from the machining process which produce surface roughness or unevenness. Oil introduced between any two surfaces should have sufficient thickness to prevent the two surfaces from coming in contact. The level of protection provided by oil is dependent on the speed of the two surfaces, forces acting on the surfaces, thickness of the oil film, and profile of the asperities on the machined surfaces. Engine failures caused by insufficient lubrication are generally classified as low-oil conditions, no-oil conditions, or excessive-oil conditions.

Oil and internal friction

Oil is an incompressible liquid that has the ability to separate moving parts. An oil film between two bearing surfaces adheres to the surfaces of the components. If one surface is moved, the corresponding oil film travels at the same velocity. The oil film on the stationary component remains stationary. This causes some internal friction at the boundary between the moving and stationary layers of the oil. The increase in internal friction and heat is minimal and varies with the oil viscosity.

Low-Oil Conditions. A *low-oil condition* is a condition when the amount of oil in the engine is less than the amount required to maintain the proper oil film to lubricated surfaces. Without proper oil film between moving and stationary bearing surfaces, heat generated by additional friction can result in temperatures high enough to damage the engine component.

With an inadequate oil film between bearing surfaces, the asperities begin to make contact. Friction from this contact causes the asperities on each bearing surface to weld themselves together. The force of the moving engine components (determined by the torque produced) causes the newly formed welds to break away, causing additional damage. Oil remaining on the bearing surface breaks down from the high friction temperatures, the oil loses its viscosity, and additional metal-to-metal contact occurs, causing scoring and/or seizure. *Scoring* is the result of scratching an engine component surface caused by a foreign object or undesirable transfer of metal from metal-to-metal contact between bearing surfaces. See Figure 11-11. *Seizure* is the joining of engine components at the bearing or running surface caused by excessive heat, pressure, and/or friction. The degree of scoring varies with the engine material, component, and load.

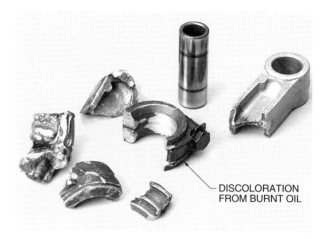

Figure 11-12. Discoloration surrounding an affected bearing surface is caused by oil that has overheated and carbonized.

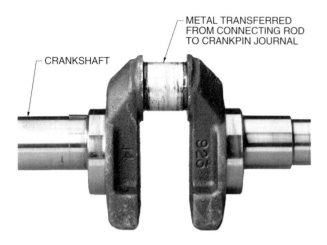

Figure 11-11. Friction heat from an inadequate oil film can result in the transfer of metal between bearing surfaces.

During the scoring process, increased friction and temperatures can cause aluminum or other metal to become molten. If the bearing surfaces continue in motion with sufficient force, the molten bearing material is wiped (transferred) between them. The wiping effect is similar to a paint roller moving excess paint in front of it as the paint is applied to a flat surface.

Most scoring occurs in the presence of some lubricating oil. The discoloration surrounding an affected bearing surface can be used as evidence that some amount of oil was present when the scoring occurred. The discoloration is oil that has failed and burned (carbonized) on the surface of the component. See Figure 11-12. Discoloration often appears on a failed connecting rod at the bearing that has experienced insufficient lubrication. Discoloration rarely occurs on cylinder walls because of the constant wiping of the piston rings. If the cylinder wall is deeply scored, carbonized oil is evident deep in the scratches.

Burnt oil adhering to a scored engine component or near the overheated bearing surface indicates heat from friction with oil present. However, this evidence does not indicate the cause of the scoring. Significant scoring on a bearing surface without surrounding discoloration from burnt oil indicates that no oil was present or that the oil delivery method was malfunctioning. This condition commonly occurs on engines started without adding oil. Although this is not conclusive evidence, it can be used to determine the cause of engine failure.

A single bearing surface with scoring is usually caused by an improper clearance problem between the affected components. This assumption is made because with sufficient oil present and the lubrication method operational, lubrication should occur at all bearing surfaces. An exception is overloading of a bearing such as an excessive radial load (belt load) or an excessive axial thrust load. If more than one bearing surface contains scoring, the failure is caused by insufficient lubrication. The extent of scoring is dictated by such causes as oil level, oil quality, oil viscosity, method of lubrication, engine load, and general wear in the engine.

Cast iron and steel have higher melting points and may not transfer as much metal as aluminum, resulting in galling on piston ring running surfaces. *Galling* is a deep cut in an engine component running surface caused by metal removed by metal-to-metal contact, force, and motion. See Figure 11-13. During metal-to-metal contact, a portion of the metal may be ripped, torn, and/or wiped (transferred in a molten state). As the engine component moves, galling causes damage to the mating surfaces, such as scoring corresponding to the length of piston ring travel in a cylinder.

Figure 11-13. Galling is a deep cut formed as a portion of the metal is ripped, torn, and/or wiped from an engine component running surface.

Several areas of the engine function under conditions of boundary lubrication. *Boundary lubrication* is lubrication where a full oil film is not always present to separate engine components. For example, the compression ring when the piston is at TDC has an oil supply consisting of oil distributed on the cylinder walls by the piston rings themselves. Supply of oil in this area must be carefully controlled for proper engine operation. See Figure 11-14. An excessive amount of oil supplied results in increased exhaust emissions, oil consumption, and the risk of carbon deposit buildup on the piston ring. Any piston ring deposit buildup may cause a decrease in the sealing ability against the cylinder walls. An inadequate amount of oil supplied may allow the piston ring to excessively contact the cylinder wall.

Frequent contact between the piston rings and cylinder bore is prevented by antiwear and extreme pressure additives blended into the base oil. These additives contain chemicals that bond to engine component surfaces to form a protective chemical layer. All additives blended into the oil are consumed or deteriorate over time.

 Depending on the viscosity, up to 30% of the contents of a container of oil consists of oil additives.

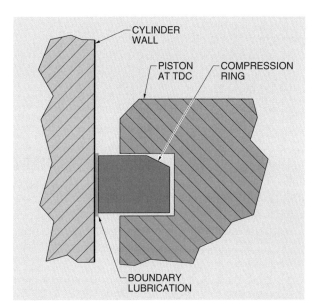

Figure 11-14. Boundary lubrication in the cylinder bore occurs as oil is distributed on the cylinder walls by the piston rings when the piston is at TDC.

Low oil and engine failure analysis

A PTO bearing with an excessive radial load and a low oil supply may indicate more pronounced evidence of failure than a connecting rod bearing. In general, other bearing surfaces in the engine show evidence of insufficient lubrication failures such as scoring and localized discoloration. It is difficult to predict which bearing will fail first. This requires the small engine service technician to thoroughly inspect all bearing surfaces before reaching any conclusion.

No-Oil Conditions. An engine operated with no oil exhibits many of the same effects as an engine operated with low oil. All evidence of friction caused by metal-to-metal contact is apparent with engine failure from no oil. The main difference is that there is little or no evidence of localized discoloration. Discoloration is not present because there is no oil in the crankcase and nothing to carbonize on or near the bearing surfaces. However, some discoloration may result from a minute amount of residual assembly oil left in the engine from the manufacturing process. When investigating a no-oil condition, the engine components responsible for the delivery of oil should be checked for proper operation. In addition, all other possible causes for engine failure should be investigated before assuming the cause was a no-oil condition.

Excessive-Oil Conditions. An *excessive-oil condition* is an undesirable condition when the amount of oil in the engine exceeds the specified amount required by the engine design. The main problem caused with an excessive-oil condition is the probability of air in the crankcase mixing with engine oil. This reduces oil lubrication effectiveness and can result in scored or damaged bearing surfaces. An excessive-oil condition is usually caused by overfilling of the crankcase. A common misconception of an excessive-oil condition is that it causes leaking or damaged oil seals and gaskets.

Engine failures caused by an excessive-oil condition are similar to those caused by a low-oil condition. Although there is a more than adequate amount of oil available for bearing surfaces, air in the crankcase becomes entrained in the lubrication system, causing air bubbles. As the crankshaft and connecting rod rotate during engine operation, the connecting rod and crankshaft dip into the overfilled crankcase oil reservoir. This creates excessive turbulence and splashing of the oil. Crankcase air mixes into the oil. Over a relatively short period of time, the crankcase becomes full of an oil-based foam.

Crankcase oil level and governor gear operation

In engines equipped with a governor gear and the proper crankcase oil level, approximately ⅓ of the circumference of the governor gear is located above the oil surface. This location allows surface tension of the oil to be broken by paddles on the rotating governor gear. This also allows oil to be consistently splashed throughout the crankcase. The submerged portion of the governor gear displaces oil with each rotation. The paddling action of the governor gear in a properly-filled crankcase is similar to the action of a swimmer in water. As the swimmer moves across the surface of the water, each stroke displaces water and splashes water into the surrounding atmosphere. As the swimmer swims under water, water is displaced with each stroke but is not splashed above the surface. Likewise, an overfilled crankcase does not allow oil to be sufficiently splashed to engine components.

In the form of foam, the amount of oil available to provide an oil film for lubricating engine components is compromised. Air bubbles in oil foam are pumped or splashed into the bearing surfaces, allowing metal-to-metal contact. As the air bubbles in the oil film compress under the load of the bearing, metal-to-metal contact causes an increase in friction, resulting in an increase in heat at the bearing surface. This can cause oil failure, severe scoring, and oil burnt onto the overheated bearing surface and/or related engine components.

A marginal excessive-oil condition (25% – 50% over specifications in a vertical shaft engine) can inhibit the ability of the governor gear to splash the required amount of oil throughout the crankcase. This may result in some long-term effects from reduced lubrication. Although the long-term effects of excessive oil may result in some oil leakage at the seals and gaskets, the principal concern is the compressibility of oil from the presence of air bubbles.

The Toro Company
The 8 HP single-cylinder engine used to power this rear engine riding mower features Magnetron® electronic ignition.

 Evidence of engine overheating is almost always an effect, not a cause.

Oil Failure. *Oil failure* is the deterioration of oil and its viscosity through oxidation, heat, and the accumulation of solids over time. Oxidation causing oil failure occurs from normal agitation with air in the presence of heat in the crankcase. Heat causing oil failure is from high combustion chamber temperatures that can result in heavy hydrocarbon residues. These residues are instrumental in the formation of sludge in the crankcase. *Sludge* is the accumulation of a semisolid, highly viscous oil material found in the crankcase of some internal combustion engines.

Solids accumulate in oil over time from unburned fuel, soot, dirt, and other combustion residues. See Figure 11-15. Water from the combustion process is always present. Greater quantities of water are produced during cold engine warm-up, adding to the deterioration of the oil and sludge formation. In addition, a slow, predictable deterioration of the properties of the oil occurs as a result of the crankcase environment and oil age.

Figure 11-15. Solids in the oil from unburned fuel, soot, dirt, and other combustion residues accumulate in the crankcase.

Overheating

Overheating is a cause of engine failure from an engine component material that has distorted beyond a specific yield point. The *yield point* is the limit of a material at which it can be exposed to heat or mechanical stress and still return to its original size and chemical composition. Overheating occurs from causes such as a lean air-fuel mixture, use of improper fuels, dirt and/or debris buildup on cooling fins, damaged cooling system components, reduction in cooling air, overspeeding, or an improper engine enclosure.

Head Gasket. The most common effect of overheating is head gasket failure (blown head gasket). A *blown head gasket* is an undesirable degradation of the head gasket which allows leakage to and from areas of the combustion chamber. A blown head gasket usually is a good indicator that an engine has experienced some degree of overheating. See Figure 11-16. A blown head gasket is a common symptom but rarely the principal cause of overheating.

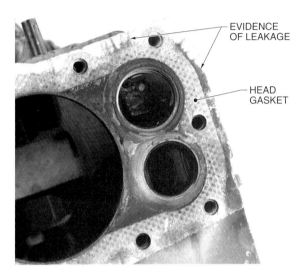

Figure 11-16. A blown head gasket is caused by engine overheating and allows leakage to and from areas of the combustion chamber.

One of the first engine components affected by overheating is the cylinder head. Intense heat produced by combustion can easily cause an aluminum cylinder head to expand, reach its yield point, and distort. For example, on an L-head engine, the aluminum cylinder head experiences thermal expansion primarily at the areas of greatest mass. The greatest mass is usually located near the cylinder head bolt bosses and sealing surface. Steel cylinder head bolts expand or contract at a different rate than the aluminum cylinder head. As an engine experiences an overheating condition, the cylinder head bolts and aluminum cylinder head expand, which increases the clamping force between the two.

When the engine is shut off, it eventually cools to ambient temperature. The aluminum cylinder head returns to its normal size, but the cylinder head bolts may not. In extreme overheating cases, this can result in a loose cylinder head. With a reduction in the clamping force caused by the stretched head bolts, the ability of the cylinder head gasket to retain a good seal is compromised. If the cylinder head gasket begins to leak during operation, maverick air can enter the combustion chamber, causing a lean air-fuel mixture. This increases combustion gas temperatures, causing a further increase in overall engine temperature. The overheating process continues until the engine no longer continues to operate or be started again.

The heating and cooling cycles of overheating have the greatest effect on the cylinder head and cylinder bolts. In extreme cases, the engine can become hot enough to distort the cylinder block. Distortion of the cylinder block can cause cylinder bore distortion or the loosening or dislodging of an exhaust valve seat insert.

Cylinder Bore Distortion. *Cylinder bore distortion* is an engine condition caused by excessive temperature variations in the combustion chamber and the inability to adequately cool a portion of the cylinder bore. The most common area affected by cylinder bore distortion is the area nearest the valves. This area is typically the most difficult to cool, and the large cast aluminum or cast iron mass results in a great amount of thermal expansion.

With higher than normal engine temperatures, the cylinder bore expands and distorts near the valves. This compromises the sealing ability of the piston rings. When piston rings no longer provide a proper seal, combustion gases have a clear path to the crankcase. A loss of the seal provided by the piston rings in the cylinder bore allows combustion gases to rapidly vaporize and burn any oil on the cylinder bore surface. This causes hot spots on the cylinder bore.

Hot Spot. A *hot spot* is a discoloration of the cylinder bore surface caused by an improper piston ring seal in a distorted cylinder block. See Figure 11-17. Hot spots are commonly associated with engine overheating. Discoloration of the cylinder bore is not caused by the metal changing chemical composition or color. Discoloration from overheating is caused by carbon stains left by the vaporization of oil on the cylinder bore. In most cases, the cylinder bore returns to its original dimensions once the cause of overheating is corrected. Hot spots in the cylinder bore usually do not cause any permanent damage.

Any abrasive particle greater in size than the thickness of the oil film can cause damage to engine components.

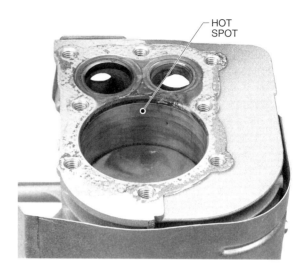
Figure 11-17. A hot spot is a discoloration in the cylinder bore caused by an improper piston ring seal and the resulting vaporization of oil.

Exhaust Valve Seat Insert. Depending on the severity of overheating, cylinder bore distortion may affect the exhaust valve seat insert. All exhaust valve seat inserts in Briggs & Stratton engines are installed with a press fit into a machined hole. The machined hole can experience thermal expansion with extreme overheating and will expand at a different rate than the exhaust valve seat insert. In extreme cases, this can cause a loosening of the exhaust valve seat insert. See Figure 11-18.

Figure 11-18. In a severe engine overheating condition, different thermal expansion rates cause the exhaust valve seat insert to become loose in the cylinder block.

The loosening of an exhaust valve seat is usually an effect rather than a cause of overheating. During normal engine operation, a loose or dislodged exhaust valve seat could partially impede exhaust gases leaving the combustion chamber. This may cause an increase in combustion chamber and overall engine temperatures.

An intake valve seat insert rarely loosens without any sign of overheating or without a loose exhaust valve seat. Intake valve seat inserts are cooled by the air-fuel mixture entering the combustion chamber and are subjected to significantly lower temperatures than an exhaust valve seat insert. A loosened intake valve seat insert without evidence of overheating can be attributed to factory manufacturing problems.

Valve Guide Failure. Excessive heat can cause valve guide failures from a decrease in working clearance between the moving valve and the stationary guide. As the valve stem expands from excessive heat, the working clearance is decreased. The valve stem diameter increases and squeezes out the oil film. With a decrease in oil film thickness, an increase in friction occurs. The increase in friction raises the temperature in the valve guide and causes galling or seizure of the valve stem and valve guide running surface.

Overheating from maverick air

An overheating engine failure typically exhibits effects such as a blown head gasket, burnt or corroded exhaust valves, and/or hot spots on the cylinder bore. With severe overheating, the exhaust valve seat insert may become loose. Although overheating appears to be the principal cause, careful analysis indicates that the engine failure started with maverick air entering the combustion chamber through a leaking exhaust valve. As the piston moves toward BDC during the intake event, a leaking exhaust valve allows unfiltered maverick air in through the exhaust system. The maverick air entering the combustion chamber contained no fuel as it bypassed the carburetor. This additional air made the air-fuel mixture much leaner than normal and increased the combustion chamber temperature and overall engine temperature. The engine failure had all the usual effects of overheating, but the actual cause was a leaking exhaust valve.

Overspeeding

Overspeeding is a cause of engine failure from damage resulting from excessive engine rpm. Overspeeding is less common than abrasive ingestion or insufficient lubrication, but can be more catastrophic. Small engines are commonly designed to operate safely at a maximum speed of less than 4000 rpm. Under 4000 rpm, internal loads on engine components are within acceptable ranges. In an overspeeding engine, these loads increase dramatically. For example, each 500 rpm increase above the recommended maximum engine speed increases the maximum force on the ends of a connecting rod by 44%. A broken connecting rod with the piston found at TDC is a common result of an overspeeding engine.

A connecting rod commonly breaks at the thinnest part of the beam, approximately 1″ from the piston pin on most connecting rods. Breakage occurs from momentum and mass of engine components rather than increased bearing and component loads. See Figure 11-19. In an overspeeding engine, the mass of the piston is instantaneously accelerated. As the piston moves up the cylinder bore, it gains momentum. With the dramatic increase in speed and direction change, the momentum of the piston applies an increasing load on the connecting rod, alternately stretching and compressing the connecting rod. This causes connecting rod fatigue and eventual fracture at the point of maximum deflection.

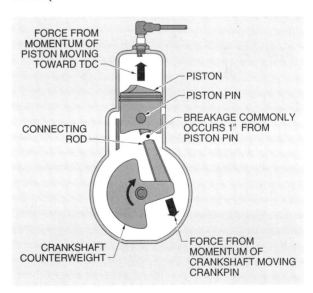

Figure 11-19. Overspeeding causes a dramatic increase in speed and direction change and alternately stretches and compresses the connecting rod until it eventually fractures.

In some instances, engine overspeeding causes a connecting rod to fracture in several places. The initial fracture cause is determined by comparing the impact points on the connecting rod pieces with corresponding impact points on other affected engine components. For example, on a Briggs & Stratton Model 13 5 HP L-head engine, overspeeding may result in a shattered connecting rod with the camshaft or cylinder block exhibiting one or more areas of impact.

The shattered connecting rod may be difficult to reassemble, but the information gained by the process justifies the time. In most cases, impact areas of the engine can be directly related to the fractures in the connecting rod. See Figure 11-20. If a fractured connecting rod is caused by overspeeding, the small engine service technician can account for all of the fractures and impacts except for one. The remaining fracture is almost always caused by the initial fracture.

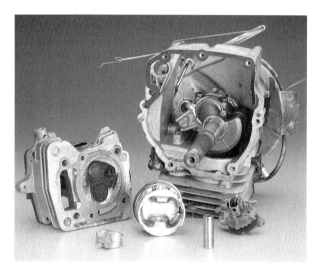

Figure 11-20. Analysis of engine parts and impact areas, and reassembly of parts, can provide valuable information pertaining to the engine failure cause.

Overspeeding-caused fractures in multiple-cylinder engines can be more difficult to determine. The additional connecting rod and crankshaft journal can cause fractures having a similar appearance. The probability of overspeeding in a multiple-cylinder engine is best determined by using all available evidence. The condition of the governor linkage, governor gear, and flyweights are strong indicators of overspeeding.

Abrasive particles and size

Abrasive ingestion engine failures are often equated with large abrasive particles similar to beach sand. Beach sand ranges in size from 150µ – 250µ. However, silica abrasive particles can be as small as 1µ, yet still exhibit a sharp-edged crystalline structure. Abrasive particles that cause excessive wear commonly average 25µ or larger. In comparison, a 25µ abrasive particle is approximately 1/20 the size of a typical pilot jet orifice.

Breakage

Breakage is a cause of engine failure from fracturing or failure of an engine component or components. Most breakage engine failures are directly caused by or intensified by vibration. Other causes of breakage include external forces, such as dropping an engine or the impact of an operating engine into a stationary object. The crankcase and crankcase cover provide the main support for the crankshaft. The crankcase cover is attached to the crankcase with mounting fasteners to control any radial or axial movement by the crankshaft and to provide support to the cylinder block against combustion pressures and vibrations.

If properly assembled and torqued, the crankcase and crankcase cover assembly distribute these loads throughout the engine similar to the strength integrity of an eggshell. If the engine assembly is compromised by a loss of torque on one or more mounting fasteners, forces from the loads may concentrate into a specific area of the engine mounting surface. For example, in an opposed two-cylinder engine, combustion loads applied to the piston are partially shared by the crankcase cover. If a crankcase cover bolt torque is lost, a crack may form at the crankcase cover gasket surface and extend to the base of the number two cylinder. In severe cases, the cylinder block may separate from the engine block entirely. See Figure 11-21.

 In an overspeeding engine, a connecting rod almost always breaks when changing direction from TDC to BDC.

Figure 11-21. In a severe case of breakage, vibration-induced forces and loads can cause the cylinder to completely separate from the engine block.

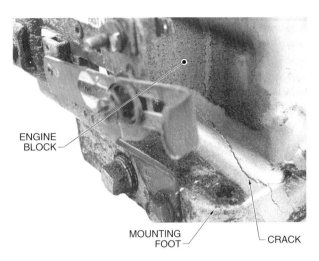

Figure 11-22. Loose mounting bolts allow engine vibration and loads to be concentrated, resulting in a crack in the engine block mounting foot area.

The base of the cylinder assembly on horizontal crankshaft engines and the sump of vertical crankshaft engines are rigid components serving as one mounting surface. Engine vibration and other forces are transferred to the mounting surface and attached equipment. If not tightly secured, the cylinder assembly is isolated and does not transfer vibrations and absorbs the load alone. Loose mounting bolts may allow undesirable movements and result in a vertical or diagonal crack in the engine block emanating from the engine mounting foot area on a horizontal crankshaft engine or a broken mounting pad on a vertical crankshaft engine. See Figure 11-22. Loose mounting bolts may also show signs of elongated bolt holes with the bolt thread impression, a polished engine mounting surface, and/or a polished equipment mounting surface.

In addition to the engine, the flatness of the equipment mounting surface must be checked before attempting engine repairs. A distorted mounting surface that deviates more than .005″ from a uniform flatness over the span of the mounting surface is unacceptable. A new engine installed on a distorted equipment mounting surface can lead to another breakage engine failure.

 A liquid-cooled engine produces less overall noise than an air-cooled engine as the engine coolant and coolant passages act to deaden sound from combustion and internal engine components.

Combination Engine Failure

A *combination engine failure* is an engine failure caused by more than one cause. Combination engine failures are very common and generally follow a pattern with other causes. For example, an equipment operator who neglects to check the engine oil level is not likely to properly maintain the air filter. In many cases, the outward appearance of the engine may be the first indication of the quality of maintenance performed. Signs of neglect on the outside of an engine usually indicate the same neglect of routine maintenance procedures.

The Briggs & Stratton Motor Wheel had a 2.5″ cylinder bore, and its most popular application was a power source for bicycles.

PREMATURE WEAR

Premature wear is commonly classified as abrasive wear or adhesive wear. *Abrasive wear* is premature wear caused by abrasive particles which remove material from engine components through friction. *Adhesive wear* is premature wear from insufficient lubrication caused by an inadequate oil film to prevent undesirable wear from metal-to-metal contact. An abrasive wear failure is easy to confuse with an adhesive wear failure from insufficient lubrication. If possible, other evidence must be analyzed and interpreted to determine the actual cause.

Abrasive particles entering an engine may not cause visible and/or measurable damage. An abrasive particle smaller than the thickness of oil film separating engine components is suspended in the oil. This causes little or no engine damage, and the abrasive particle is removed from the engine during the next oil change. An abrasive particle greater than the thickness of the oil film causes a scratch, cut, or gouge at the engine component bearing surface during movement.

Adhesive wear results in damage from metal-to-metal contact. Although adhesive wear is more evident in aluminum-to-steel bearing surfaces, scoring can occur on any surface from adhesive wear. During metal-to-metal contact, a piece of metal may be pulled from an engine component such as a cylinder wall, main bearing, or connecting rod bearing surface. The metal may then scrape across a mating surface and cause galling.

Metal from inside the galling is torn and rolled out similar to a furrow in dirt after a field is plowed. The edges of the furrow are raised above the bearing surface and create a large group of asperities on the bearing surface. The process repeats itself as metal-to-metal contact and friction create new furrows. With sustained friction and heat, larger portions of the bearing surfaces may be torn away.

Ransomes America Corporation

This power rake, powered by a 7 HP engine, has a 16.79 cu in. displacement and a 2.75 pt oil capacity.

Coleman Powermate, Inc.

This electric generator is powered by a 5.5 HP engine with a splash lubrication system.

ENGINE APPLICATION AND SELECTION

12 CHAPTER

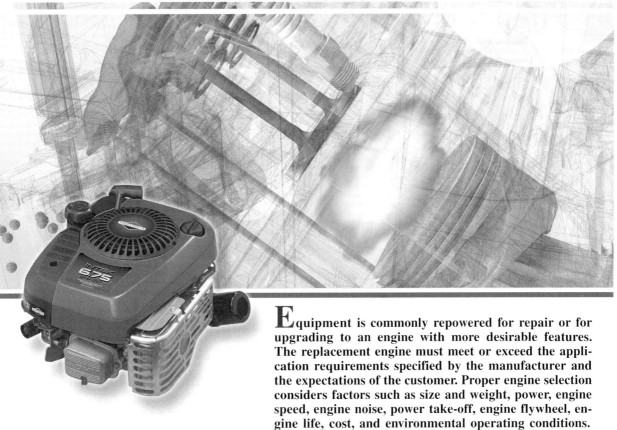

Equipment is commonly repowered for repair or for upgrading to an engine with more desirable features. The replacement engine must meet or exceed the application requirements specified by the manufacturer and the expectations of the customer. Proper engine selection considers factors such as size and weight, power, engine speed, engine noise, power take-off, engine flywheel, engine life, cost, and environmental operating conditions.

REPOWERING

Repowering is the process of replacing the original engine on a piece of equipment with another engine. Engine-driven equipment is commonly repowered due to engine failure, the need for more horsepower, or for upgrading to an engine with more desirable features. Sometimes, repowering is less costly than the parts and labor required to repair an engine. Repowering considers startability, parasitic load, serviceability, operating temperatures, noise, vibration, angle of operation, fuel system (tank and fuel line type and location), and mounting. A *parasitic load* is any load applied to an engine that is over and above the frictional load of an engine. A small engine service technician must select an engine which meets or exceeds both application requirements specified by the manufacturer and expectations of the customer.

Repowering commonly requires modifications to the engine and/or application components. Without careful consideration, this can result in a compromise of quality and/or safety. Quality and safety must never be compromised for any reason. The replacement engine installation must meet the standards and guidelines established by the American National Standards Institute (ANSI), the Outdoor Power Equipment Institute (OPEI), and the Consumer Product Safety Commission (CPSC).

All guards and other safety devices on the original application and engine must be transferred when the repowering installation is complete. Engine controls such as the rewind starter, throttle, choke, kill switch, and blade brake system must be functional. The replacement engine must be installed on the application in an acceptable location and orientation. See Figure 12-1.

250 SMALL ENGINES

Figure 12-1. Safety devices on the original application and engine must be transferred and functional when the repowering installation is complete.

When selecting an engine for repowering, the general features of the original engine are compared with those of the replacement engine. For example, horsepower, torque, weight, operating speed, and overall size should be similar. If the reason for repowering is premature engine failure, the specific cause should be determined. If the cause of the failure was wear, insufficient lubrication, overspeeding, or overheating, the features of the replacement engine should help to reduce the possibility of a similar problem recurring. For example, if the original engine was equipped with a splash lubrication system and failed due to insufficient lubrication, the replacement engine should be equipped with a low oil shutdown. A pressure lubricated engine should be equipped with a low oil pressure warning device.

Technical data is often available from manufacturers for repowering equipment. For example, a Snapper mower originally powered by a Briggs & Stratton 12 HP engine can be repowered with a Briggs & Stratton Vanguard™ 14 HP engine. See Figure 12-2. Benefits of the 14 HP replacement engine include longer engine life from engine features such as a cast iron cylinder sleeve, overhead valve design, a spin-on oil filter, improved fuel economy, and lower operating temperatures. Installation requirements, such as fasteners, drive system requirements, and other installation parts may be provided in repowering data from the equipment and engine manufacturer.

SPECIFICATIONS

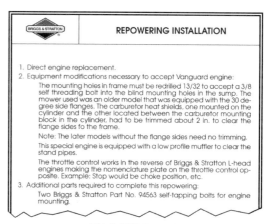

PROCEDURES

Figure 12-2. Technical data from the manufacturer lists parts, equipment modifications, and procedures required for repowering with a specific engine.

ENGINE SELECTION

Engine selection for an application is usually determined by engineers at the original equipment manufacturer (OEM). However, the small engine service technician can become more proficient in engine selection for repowering by knowing application requirements and the design features of the OEM engine used. This includes original engine size, top no-load speed, and anticipated engine life to ensure proper performance and customer satisfaction.

Original engines may be replaced by a service replacement engine. A *service replacement engine* is a replacement engine that has the operating characteristics and required features of the original engine but is not supplied by the OEM. Service replacement engines are manufactured in large quantities for inventory and may have slightly different or more features than the original engine they replace. Proper engine selection requires consideration of engine size and weight, power, speed, noise, power take-off, flywheel, engine life, and environmental operating conditions.

Engine Size and Weight

Engine size and weight requirements are determined by the specific application. Some applications do not have engine size and weight restrictions, allowing more flexibility in engine selection. Other application requirements dictate engine size and weight and/or crankshaft and cylinder orientation. For example, a rototiller, rotary lawn mower, or sidewalk edger must be light enough to maneuver without operator fatigue, but powerful enough to perform the task. The application design must allow for adequate space for proper mounting to match the engine footprint. An *engine footprint* is the area covered by the mounting base of an engine.

Other space and clearance requirements of the application may require a specific engine envelope. An *engine envelope* is the vertical, horizontal, and peripheral clearance space required for engine installation. An engine may require a large amount of horizontal space and little vertical space, or just the opposite. The engine envelope affects the type of engine selected, such as single- or multi-cylinder or air- or liquid-cooled engines. A footprint cube from the manufacturer provides a representation of an engine at actual size to aid in engine selection and installation when repowering. See Figure 12-3.

Engine Power

Engine power required for an application is determined by variables that are difficult to replicate in a small engine service facility. Generally, engine power requirements for a specific application are best determined by the output of the original OEM engine supplied with the application. If the application operated properly when new, the original engine size and power should provide a starting point for selecting a replacement engine. The replacement engine must also provide enough power to drive any additional attachments or external loads while maintaining an acceptable engine life. This includes enough reserve power to handle variations in load changes and changes in environmental operating conditions such as higher altitudes and temperature variation.

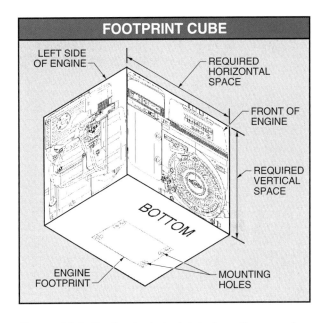

Figure 12-3. Engine technical data from the manufacturer lists engine features and installation information for repowering.

Horsepower required for some applications can be accurately calculated because output requirements are measurable. This includes applications such as electric generators, welders, air compressors, and pumps. A measurable output quantity simplifies the selection of the replacement engine. For example, if a 4000 W (4 kW) generator required a replacement engine, the replacement engine output must be in excess of 4 kW.

Output above the required 4 kW allows for inherent inefficiencies in power production and energy conversion. For example, a 6 HP engine produces 4.476 kW. The efficiency conversion factor for a generator is commonly 70%. At an efficiency conversion factor of 70%, a 6 HP engine produces 3.133 kW (4476 W × 70% = 3133 W) and is inadequate to power a 4000 W generator. The minimum horsepower required is 7.7 HP (5.744 kW). The 30% efficiency loss is from heat produced by the induction process at the armature, friction, and other measurable energy losses in the engine and/or application. In some applications, losses from power production and energy conversion can be as high as 60%. Efficiency factors are assigned based on the application. When selecting motors, round up to the next highest .5 HP. See Figure 12-4.

For example, engine horsepower required for a generator can be determined by kilowatts output, a horsepower to kilowatt conversion constant, and generator efficiency. Horsepower required for an electric generator is found by applying the formula:

$$HP = kW \times \frac{1.34}{E}$$

where
HP = horsepower
kW = kilowatts
1.34 = horsepower to kilowatt conversion constant
$(\frac{1000\,W}{746\,W} = 1.34)$
E = generator efficiency (70% if unknown)

What is the horsepower required to drive a 6 kW generator?

$$HP = kW \times \frac{1.34}{E}$$

$$HP = 6 \times \frac{1.34}{.70}$$

$$HP = \frac{8.04}{.70}$$

$$HP = \mathbf{11.49\ HP}\ (11.5\ HP\ motor)$$

Homelite, Inc.

This ventilating blower uses a 3.5 HP engine and variable speed control to produce air output from 500 to 1428 cubic feet per minute (CFM).

DETERMINING HORSEPOWER REQUIREMENTS

What is the horsepower required to drive a 5 kW generator?

$$HP = kW \times \frac{1.34}{E}$$

$$HP = 5 \times \frac{1.34}{.70}$$

$$HP = \frac{6.70}{.70}$$

$$HP = \mathbf{9.57\ HP}\ (10\ HP\ motor)$$

APPLICATION EFFICIENCY CONVERSION FACTORS

Application	Efficiency Conversion Factor*
Hydraulic pump	80
Generator	70
Welder	70
Water pump	50
Trash pump	40

* in %

Figure 12-4. Efficiency factors are assigned based on loss from heat, friction, elevation, ambient air temperature, and other measurable energy losses in the engine and/or application.

 The first time that 1,000,000 engines were produced in a single year by Briggs & Stratton was 1952.

Power Curve. A *power curve* is a graphic representation of horsepower and torque output of a specific engine in a test laboratory. See Figure 12-5. Power curves are used by equipment designers to determine engine performance capabilities. The power curve includes maximum brake horsepower (BHP), recommended maximum operating BHP, and torque produced by the engine. A *maximum BHP curve* is a graphic representation that indicates engine power developed with the throttle in WOT position across the usable engine rpm range in a test laboratory. The engine is operated under load on a dynamometer to maintain a set speed (usually 3600 rpm) at WOT position. The load is then increased, resulting in a decrease in rpm. The speed decrease and power output are plotted on a graph.

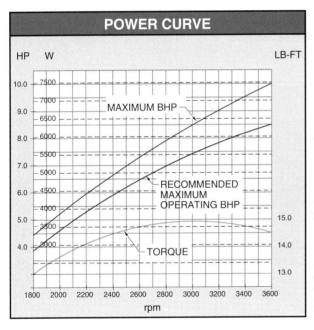

Figure 12-5. A power curve provides a graphic representation of engine performance.

A *recommended maximum operating BHP curve* is a graphic representation that indicates engine power at 85% of maximum BHP. Production engines shipped from the manufacturer should not develop less than 85% of the recommended maximum operating BHP. The recommended maximum operating BHP curve is represented on all power curves published by Briggs & Stratton. A *torque curve* is a graphic representation that indicates maximum engine torque produced at a specific rpm. It is not possible to produce engine power in excess of the maximum BHP curve. An engine can produce maximum rated power only at WOT position at the rated speed. A power curve represents the absolute maximum power that a particular engine can develop at the rated speed. Questions related to applications requiring loads and speeds not listed on the power curve should be referred to the application manufacturer.

Governor System. Available engine power is affected by the governor system. The governor system is usually associated with controlling the engine speed. However, maximum engine power available at any given speed occurs when the throttle is in WOT position. A governed engine develops no measurable torque or horsepower at its maximum top no-load speed. As the load increases, rpm decreases. The rpm decrease causes the governor to open the throttle plate in proportion to the amount of applied load in an attempt to maintain the engine speed. The opening of the throttle plate results in an increase of power and torque.

Governor droop is the amount of rpm decrease between top no-load governed speed and rpm where the power is delivered. For proper operation, the amount of acceptable governor droop should be approximated when selecting an engine for a specific application. A *governor droop curve* is a combination of rpm decrease and maximum brake horsepower (BHP) curves. A governor droop curve provides information regarding specific engine speed requirements and the need to set the top no-load speed of an engine above the rpm that delivers the power required by the load. The shape of the governor droop curve is determined by factors such as the governor spring rate, governor sensitivity, and governor geometry.

The governor droop for a walk-behind lawn mower is greater than the governor droop for an electric generator to allow for a wider range of engine rpm variation from load. See Figure 12-6. For example, a walk-behind lawn mower has a preset top no-load speed of 3200 rpm. When moved from short grass to tall grass, there is a decrease in engine speed. This decrease, or governor droop, occurs before the engine has compensated for the increased load at maximum BHP occurring at at 2600 rpm. If the governor droop is too great, the engine may not be able to recover from an increased load, possibly causing poor application performance. The governor used depends on the response required by the application. The electronic governor is the most responsive, with mechanical and pneumatic governors respectively less responsive.

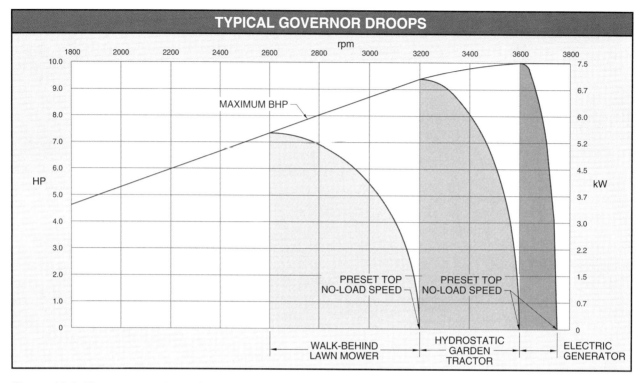

Figure 12-6. The governor droop for a walk-behind lawn mower is greater than the governor droop for an electric generator to allow for a wider range of engine rpm variation from load.

Engine Speed

The engine speed selected for a specific application must provide desirable operating performance with adequate power to operate under load at lower rpm and resist stalling during a sudden load change. The operating speed of the engine must be set higher than the rpm at which peak torque occurs. This allows torque to increase as engine rpm decreases in response to an increased load.

Many engines have the same or similar horsepower ratings, but have a different power range or operating characteristics. See Figure 12-7. For example, Briggs & Stratton Model 243400 10 HP engine has a stroke of 3.25″. Briggs & Stratton Model 222400 10 HP engine has a stroke of 2.38″. Both engines are rated at 10 HP, but the difference in stroke changes the rpm range where maximum torque occurs. The top no-load speed for Model 243400 is set at 3200 rpm. If a moderate to heavy load is applied, the rpm decrease can be as much as 600 rpm before the torque peak is reached. The top no-load speed for Model 222400 is set at 3200 rpm. If a moderate to high load is applied, the rpm decrease cannot exceed 200 rpm before the torque peak is reached.

Model 243400, with a longer stroke, produces torque, but at a lower loaded engine rpm than Model 222400, which has a shorter stroke. If the top no-load speed setting causes the engine rpm to fall below the point of maximum torque output, the engine eventually stalls. Setting the top no-load speed in excess of the speed at which the load is applied is critical to overall engine and application performance.

Engine speed and torque curve vary with the application. For example, a generator set requires an engine that can maintain a constant rpm under load with little fluctuation during load variation. When driving a generator, engine rpm is directly related to the frequency or cycles of electricity generated. If the frequency varies, damage can occur to electrical equipment connected to the generator. Generator applications require close control of the amount of governor droop to maintain 3600 rpm under load. Deviation from the desired speed must be minimal and should not exceed 180 rpm. Generator sets in North America and many countries commonly generate electricity at 60 cycles per second.

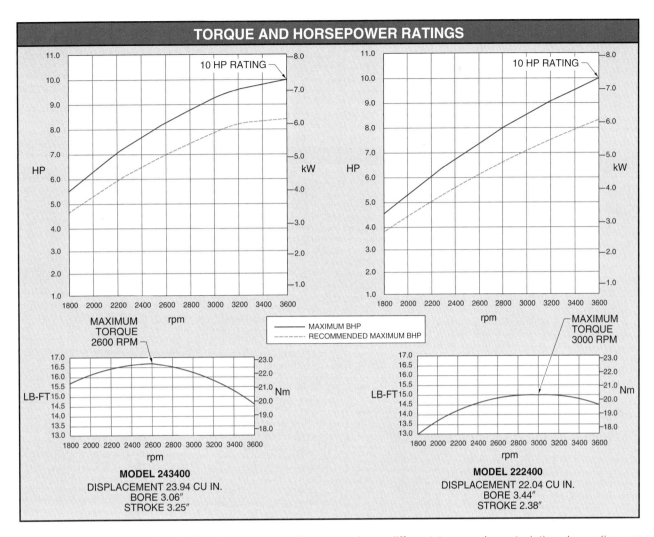

Figure 12-7. Engines with similar horsepower ratings may have different torque characteristics depending on engine stroke.

Speed Control. Speed controls for small engines vary with the application. An engine designed to operate at a single specified top no-load speed uses an adjustable fixed speed control. An *adjustable fixed speed control* is a control mounted on the engine that uses a speed adjustment nut to maintain an engine at a specific top no-load speed. An adjustable fixed speed control helps to more accurately set the speed of the engine on mechanical or pneumatic governor systems. Adjustable fixed speed controls are commonly used on generators, pumps, and air compressors. See Figure 12-8.

A *remote speed control* is a control that allows the operator to vary engine speed from a remote location. For example, a rotary lawn mower has a remote manual friction control mounted on the mower handle. A *manual friction speed control* is a control mounted on the engine that uses a friction lever to maintain an engine at a specific top no-load speed. These controls are commonly used on farm and outdoor power equipment to adjust engine speed from idle to maximum top no-load speed.

Caution: Never exceed the maximum safe engine speed. Maximum safe engine speed is determined by the engine manufacturer based on engine design, material, and operating characteristics. The maximum speed setting is also influenced by the constraints of the specific application.

Briggs & Stratton recommends that equipment manufacturers thoroughly test installed engines for a minimum of 200 hours to ensure proper performance.

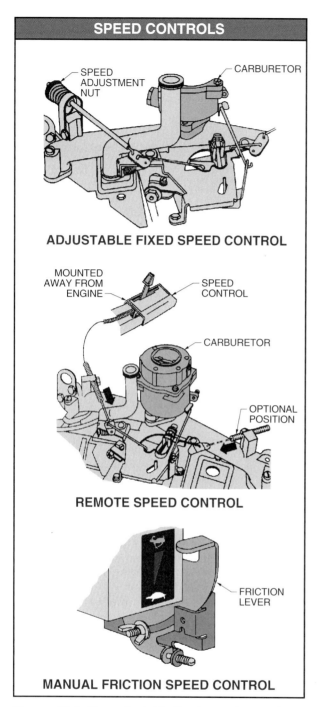

Figure 12-8. The adjustable fixed speed control and manual friction speed control are mounted on the engine. The remote speed control uses a cable to control engine speed away from the engine.

The lowest engagement speed for a centrifugal clutch should be 2200 rpm to prevent accidental clutch engagement when the engine is started.

Engine Noise

Small engines produce noise from combustion and the motion of engine components. Other sources of engine noise are the carburetor and the cooling fan. Engine noise from combustion is reduced with a muffler. A *muffler* is an engine component fitted with baffles and plates that subdues noise produced from exhaust gases exiting the combustion chamber.

A muffler is selected based on the application of the engine, the load, and the noise produced by the driven equipment. Noise is measured in decibels using a sound level meter that indicates the noise level at a particular time. A *decibel (dB)* is a unit of expressing relative intensity of sound on a scale from 0 dB (average least perceptible) to 140 dB (deafening). Noise is identified by frequency. *Frequency* is the number of vibrations per second a noise contains measured in hertz (Hz). Frequency scales commonly used when measuring noise are the A and C frequency scales. The A frequency scale is the most common and most closely resembles human hearing. Most noise requirements are specified in decibels measured in the A frequency scale (dBA). For example, a Chicago noise control ordinance specifies that equipment sound measured at 50′ must not exceed 70 dBA. Small engine sound commonly ranges from 70 dBA to 85 dBA at an average distance of 10 meters (m). The C frequency scale is used for specialized industrial noise testing.

The starting point for reducing total noise of equipment is the selection of a proper muffler to reduce exhaust noise. Exhaust noise is a significant contributor to total engine noise. Generally, the size of the muffler and the amount of restriction of exhaust gases increase with the noise reduction desired. Once exhaust noise is lower than other engine or machine noises, additional noise reduction is attained by reducing other equipment noise. Exhaust noise also is reduced by redirection of exhaust gases or the addition of a spark arrester.

Power Take-Off

The power take-off (PTO) extension is usually the end of the crankshaft that extends outside of the engine crankcase. The PTO extension is used to drive attached equipment. Torque produced by the engine is transferred through the PTO extension to the equipment. Most engines have a variety of PTO extensions available for special requirements of various applications. See Figure 12-9. PTO specifications are listed in engine manufacturer catalogs. Not all PTO extensions are an integral extension of the crankshaft. Some PTO extensions utilize the engine flywheel or a camshaft extension.

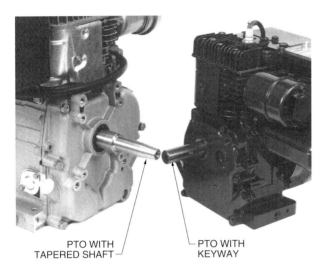

Figure 12-9. PTO design and specifications are selected for application requirements.

The connection from the application to the engine PTO must be secure. Maximum power of the engine is generated in a relatively small portion of the power stroke. During the power stroke, acceleration resulting in instantaneous crankshaft torque commonly reaches seven to eight times the rated engine torque. Adequate belts and pulleys must be used to avoid power loss from slippage when instantaneous crankshaft torque occurs. Secure drive connections and direct PTO connection components prevent undesirable motion of parts, which could result in noise, fretting, and damage to other engine and application parts.

Fretting is damage that occurs to machine parts where two contacting surfaces are subject to slippage. Fretting is usually indicated by red rust-like color on steel components accompanied by pitting of the metal. Fretting commonly occurs on drive pulleys and centrifugal clutches. A tight connection to the PTO is critical when using components that have high inertia. High inertia components absorb most of the load during power stroke acceleration, store most of the rotational energy, and supply this energy back to the machine as it coasts through to the next power stroke cycle.

Load-bearing pulleys and other drive members should be mounted as close to the crankshaft main bearings as possible. This decreases the radial loads on the crankshaft and crankshaft bearing by reducing the leverage gained from a drive member located at the outboard end of the PTO extension. A *radial load* is a load applied perpendicular to the shaft. See Figure 12-10.

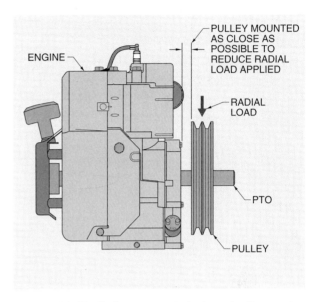

Figure 12-10. Pulleys mounted close to the engine decrease the radial loads on the PTO by reducing leverage from drive members.

Sometimes, a direct-coupled system is used to attach the engine to an application, such as a high pressure washer or a rotary lawn mower. Direct-coupled systems require that the engine and the driven equipment shafts be concentric and parallel. The coupling used should allow for minor misalignment with the driven equipment to avoid excessive loads on the engine and the equipment shafts. In some applications, a flexible coupling is used to compensate for greater misalignment. Equipment, such as pumps and generators attached to the engine crankshaft, must be mounted concentric and square to the engine. SAE standard pattern flange-mounting, piloting projections, and mounting pads designed into the engine must be used.

Applications such as centrifugal pumps, boat propellers, and some clutches must not exceed the axial load capabilities of the main bearings or thrust surfaces of the engine. An *axial load* is a load applied parallel to the shaft. An axial load is acceptable up to 50 lb, unless the engine has special thrust capabilities from ball bearings or roller bearings. See Figure 12-11.

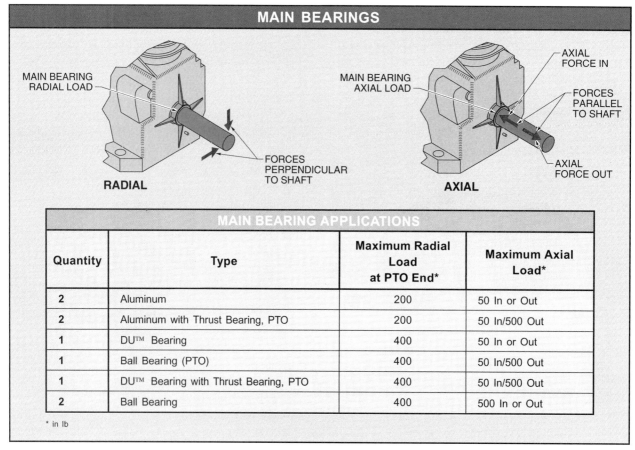

Figure 12-11. Radial and axial loads applied require a specific number and type of main bearings.

Jacobsen Division of Textron Inc.

This greens mower is powered by an 18 HP engine with an 8 gal. fuel tank and a pleated paper fuel filter.

Ball bearings can withstand more axial and radial thrust load than plain bearings. The type of bearing used determines the maximum amount of radial and axial thrust load an engine can withstand. Engines with one ball bearing control an axial thrust load based on the direction of the force that can be absorbed by the bearing.

For example, a centrifugal water pump attached to an engine may have a single ball bearing on the PTO side. If the thrust is away from the engine and toward the pump, the load is carried through the bearing. If the axial thrust is toward the engine, the aluminum bearing on the flywheel side may not be capable of carrying the thrust. The engine and application manufacturer must be consulted for bearing requirements for specific axial loads. Some applications require power to be taken from the flywheel side of the engine either as the primary power source or in conjunction with power from the PTO side of the engine.

Larger engines may utilize a booster fan. A *booster fan* is an engine component mounted on the flywheel side of the engine that consists of a hub with bosses and tapped holes for connecting driven equipment components. The booster fan may incorporate a pulley used to drive a belt. Power taken from the booster fan generally should not exceed 50% of rated horsepower. See Figure 12-12.

Figure 12-12. Power taken from the booster fan should not exceed 50% of rated horsepower.

In some cases, the booster fan provides threaded holes for an extension shaft to mount a pulley or other components. In this case, the 50% rated horsepower limit should also be followed. If the connection from the booster fan is directly coupled to the application via a drive shaft or other means, 100% of the engine power can be taken from the booster fan side. This percentage applies to pure torsional power only. Any radial load applied to the shaft reduces the amount of power that can be taken from the shaft to 50%. If the application requires more than 50% of the rated horsepower from an extension shaft and belt pulley, consult the engine and application manufacturer.

If a belt drive pulley is mounted to a stub shaft attached to the booster fan, a minimum clearance between the pulley and the rotating screen on the booster fan must be maintained. To permit an adequate flow of cooling air, a 1.25″ minimum clearance is generally adequate for pulleys up to 6″ in diameter.

The National Safety Council estimates that 1,700,000 workers in the U.S. between ages 50 and 59 years of age have compensable noise-induced hearing loss.

Engine Flywheel

The engine flywheel supplies a portion of the inertia required for startup without kickback and proper engine operation. On some engines, the flywheel provides the majority of inertia required. Inertia is required for proper engine operation. As the flywheel or other inertia-exhibiting component turns, momentum causes the engine to continue through non-combustion strokes of the engine at a relatively uniform speed.

Small vertical shaft engines commonly use a low-inertia flywheel. A *low-inertia flywheel* is a flywheel that uses an external load connected to the engine to provide the minimum engine inertia required. Engines with low-inertia flywheels are used extensively on rotary lawn mowers where the blade is connected directly to the PTO. The mower blade supplies most of the required inertia for proper operation. Without a mower blade, a vertical shaft engine must be supplied with a high-inertia flywheel. A *high-inertia flywheel* is a flywheel that exhibits greater mass than a low-inertia flywheel and is designed to provide the minimum engine inertia required independent of the external parasitic load. The minimum required inertia for adequate starting is dependent on engine displacement, optional compression release, and external inertial load.

Many engines gain additional inertia from flywheels and/or other components with a significant amount of mass from attached equipment. It is not uncommon for this additional inertia to be several times greater than typical engine flywheel inertia. The additional inertia necessary for proper equipment operation is usually helpful for engine starting and smooth operation. The additional inertia also helps to absorb and store accelerative forces caused by the engine. In some cases, an application with a high-inertia load is not connected to the engine with a direct coupling. For example, a large chipper/shredder may use a centrifugal clutch to allow easier starting with the high-inertial load of the cutter blade and the chipper flywheel.

Inadequate inertia causes engine operation problems. Excessive inertia can also cause problems. Excessive inertia connected directly (or indirectly, with a clutch system) to the crankshaft can cause:

- Hard starting because of difficulty with the cranking effort to adequately accelerate the engine.
- Slow or sluggish acceleration when started.

- Difficulty stopping the engine.
- Difficulty stopping engine within the mandatory three second period for CPSC compliance mowers.
- Hot restart problems resulting from flooding from long coast-down time.
- Engine damage due to marginal lubrication from longer coast-down cycles, inducing combustion chamber and oil reservoir flooding.

Single-cylinder engines have unique characteristics that must be considered in power output, power transmission, and vibration. See Figure 12-13. The horizontal line is the mean or average engine speed. Engine speed is sporadic during normal operation, with the slowest engine speed occurring just before ignition or TDC. After ignition, rpm increases significantly to maximum rpm, then gradually slows before repeating the cycle. An engine and driven equipment that have low inertia exhibit the most radical engine speed fluctuation. The addition of greater inertia decreases the radical engine speed fluctuations significantly. Speed fluctuation that occurs during normal operation of all engines makes it necessary to use belts and pulleys of the proper size to avoid slippage and power loss. Tight connections on the crankshaft PTO are also required to avoid rocking, fretting, and/or damage leading to failure.

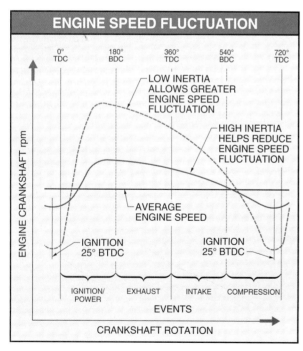

Figure 12-13. Speed fluctuation within the operating cycle of a single-cylinder engine is reduced by inertia from the engine flywheel and driven equipment.

Generators and rpm control

A typical portable generator operating at 3600 rpm generates electricity at 60 cycles per second. The rpm required to produce electricity at this frequency is 60 rpm/sec. (3600 rpm divided by 60 seconds in a minute). With a standard generator governor droop tolerance of ±3% of the 3600 rpm operating speed (± 108 rpm), the engine must operate within the rpm range of 3492 rpm to 3708 rpm to maintain the proper frequency.

Engine Life

Engine life is variable, depending on engine features, maintenance, and use. Engine life can be extended with premium features in the engine design, such as Cobalite™ valves and seats, cast iron cylinder sleeves, oil filtration, heavy-duty air cleaners and debris guards, and other accessories. Engine life can also be extended by selecting an oversized engine. An *oversized engine* is an engine that is typically capable of producing 20% – 30% more power than required by the application at full load.

An oversized engine has an abundance of reserve power to maintain engine performance at an acceptable level for a longer period of time. This reserve power compensates for the normal power loss from decreased volumetric efficiency caused by the accumulation of combustion chamber deposits and other results of long hours of use. An oversized engine is usually an excellent choice if the additional cost can be justified. However, other factors such as additional weight or the potential for producing greater or differing vibration characteristics must be considered.

Engine life is reduced by using an undersized engine for an application. An *undersized engine* is an engine which consistently works close to maximum capacity with little or no reserve power. A standard test for identifying power output required for the application is the throttle plate while the engine is under load. The throttle plate position should not exceed 50% of WOT at full load. For example, a properly-powered water pump operating at 3600 rpm should have a throttle plate position of up to 50% of WOT. With this throttle plate position, the engine is producing about 75% of the power it is capable of producing to rotate the impeller. If the engine were replaced with an engine with less horsepower, the throttle plate position would be open wider proportionally (approximately 90% – 100% of WOT) at the same (3600) rpm.

The water pump operates at the same (3600) rpm with both engines. However, the engine with less horsepower is more susceptible to factors that may tend to decrease engine output. For example, elevation, air filter maintenance, spark plug condition, and/or deviations in valve adjustment can reduce power enough to cause a decrease in engine rpm and horsepower. Less normal wear should occur on an engine that exhibits a smaller throttle plate opening under load over the same number of hours of operation.

Engineers select engines for specific applications based on test data using measurements such as test hours or horsepower hours. However, most users in the field are likely to judge engine performance on perceived power in use and engine life in months or years of service. An oversized engine should provide better performance and increased engine life.

Environmental Operating Conditions

Environmental operating conditions are conditions present which affect engine performance. These conditions are variable with altitude, climate, atmospheric conditions, dust, moisture, and chemicals. For example, environmental operating conditions are different in Denver, Colorado (elevation 5280′) than in Baltimore, Maryland (elevation 155′). Dry, hilly conditions are different from humid, flat conditions. Environmental operating conditions commonly include air density, angle of operation, and rough terrain.

Air Density. Air density is directly related to ambient temperature, altitude, humidity, and barometric pressure. The denser the air, the greater the amount of oxygen present in the charge. With more oxygen, more fuel can be introduced into the combustion chamber, resulting in a power increase. Ambient temperature and altitude have the greatest effect on air density. Hot air is less dense than cool air. Engine horsepower decreases by 1% for each 10°F above the standard SAE testing temperature of 60°F. See Figure 12-14.

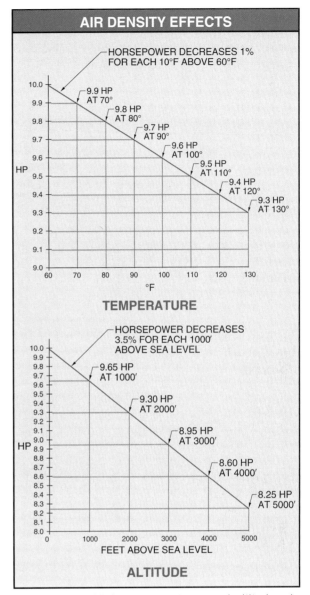

Figure 12-14. Higher temperatures and altitudes decrease air density and reduce engine performance.

As altitude increases, air density decreases, resulting in a reduction of engine horsepower. Engine horsepower decreases by 3.5% for each 1000′ above sea level. The power decrease is in direct relation to the oxygen content of the atmosphere at a set altitude. As the air density decreases, the number of oxygen molecules in the atmosphere decreases.

Humidity also affects engine output as air with a high moisture content displaces more atmospheric oxygen than air with a low moisture content. Barometric pressure, although a factor in engine operation, is of less measurable significance than temperature or altitude.

Since ambient temperature, altitude, humidity, and barometric pressure can be variable for any application, an engine seldom operates in ideal conditions. Engine selection for a specific application should not require more than 85% of the maximum BHP. Most horsepower ratings are established in accordance with ANSI/SAE J607-AUG88, *Small Spark Ignition Engine Test Code*. Power curves are corrected to standard conditions of sea level, barometer of 29.92″ mercury (Hg), and 60°F. Engine performance specifications are derived from laboratory test engines equipped with a standard air cleaner and muffler.

Engines built during and after 1987 are rated in accordance with ANSI/SAE J1349-JAN90, *Engine Power Test Code – Spark Ignition and Diesel*. This standard permits both gross and net power ratings. Test conditions are at a higher elevation, 328′, with a lower barometer reading of 29.31″ Hg, and a higher temperature of 77°F. The most recent test standard (ANSI/SAE J1349-JAN90) results in a lower correction factor for power ratings, and the corrected BHP is approximately 3.5% less than the older test standard (ANSI/SAE J607-AUG88).

Angle of Operation. The angle of operation must be considered when selecting and installing an engine. If the engine is permanently installed at an angle, oil fill and oil level checking provisions must be designed to maintain the required oil capacity in the crankcase. The standard angle of operation limit for most small engines is a constant 15° angle under normal operation. See Figure 12-15.

If the angle of operation causes the fuel tank to be lower than the carburetor on a gravity feed system, fuel may not be properly supplied to the carburetor under load. In addition, the fuel level in the carburetor may become a problem as fuel seeks its own level in the fuel bowl. If the angle of operation exceeds 15°, the air bleed system and the fuel level in the emulsion tube may cause an extremely lean or rich fuel mixture. In most instances, carburetor function dictates the maximum angle of operation of the engine. Most engines operate satisfactorily up to an angle of 30° intermittently.

Rough Terrain. Rough terrain creates special vibration problems from radical, intermittent acceleration rates of component parts and applied loads. Engines subjected to rough terrain applications require the engine-to-frame mounting to be resistant to fatigue failures. Rough terrain installations may also require extra brackets for engine accessory installations such as fuel tank, air cleaner, and carburetor mountings.

The fuel system must contain fuel in the tank while providing adequate venting without the loss of fuel. This may require a special labyrinth system. A *labyrinth system* is a set of baffles and/or a foam insert used to reduce sudden movement of fuel in the fuel tank resulting from motion of the application. See Figure 12-16.

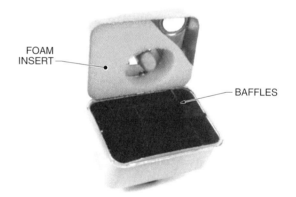

Figure 12-16. Rough terrain installations require special fuel tanks to contain the fuel, provide adequate venting, and reduce sudden movement of fuel.

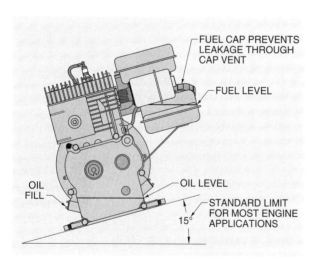

Figure 12-15. The oil fill and level checking provisions are affected by the angle of operation.

A similar labyrinth system is used in some carburetor bodies to reduce the possibility of splashing liquid gasoline into an internal vent channel. Under extremely rough conditions, the splashing of gasoline in the fuel bowl can enter the internal vent channel and be ingested in the engine. This can cause rough operation on rough terrain.

Operating conditions similar to those caused by rough terrain can be created by conditions of the applications. For example, a compactor operating on very hard soil bounces severely, causing similar conditions to a vehicle driven over rough terrain.

> ### Engine life testing
>
> All Briggs & Stratton engines are life tested under full load for a specific length of time based on the intended use of the engine. Maintenance procedures prescribed in the owner/operator manual are followed. Consumer engines are tested for 500 hours at full load and full speed. Industrial/commercial engines are tested for 1000 hours at full load and full speed. Vanguard™ engines are tested for 2000 hours at full load and full speed. This amount is equivalent to driving around the circumference of the earth six times. At full load, the distance traveled is similar to continually traveling up, but never down, a steep mountain grade.

FUEL SYSTEMS

The fuel system of a small gas engine consists of a fuel tank, fuel lines, fuel filter, and a carburetor. There are four types of carburetors commonly used on small engines today. They are Vacu-Jet, Pulsa-Jet, Flo-Jet, and Flo-Jet with integral fuel pump. Careful consideration is required when choosing the carburetor, fuel tank, and fuel lines to ensure maximum safety, operating efficiency, and customer satisfaction.

Fuel Tank

The fuel tank is mounted directly to the Vacu-Jet and Pulsa-Jet carburetors. Special consideration for fuel tank and fuel line installation is not applicable. The integrated fuel tank is mounted below carburetor level, minimizing the potential for fuel leakage. Some Pulsa-Jet fuel systems provide optional fuel tank sizes. When selecting an optional fuel tank size, adequate room must be allowed for the tank to fit in the space provided. Fuel filtration is provided by a small mesh screen on the end of the fuel pick-up tube.

Flo-Jet carburetors are usually supplied without a fuel pump. This requires the fuel tank to be located in such a position as to provide gravity flow to the carburetor. A variety of fuel tank sizes can be used with Flo-Jet carburetors. The size of the tank is determined primarily by the fuel consumption rate of the engine and the space available. Fuel filtration is provided by using an in-line fuel filter between the fuel tank and the carburetor. The Flo-Jet carburetor installation provides flexibility of fuel tank size and location, providing there is sufficient elevation for gravity flow of the fuel from the tank to the carburetor. The fuel tank should not be located higher than 18″ above the inlet fitting of the carburetor when using a gravity feed system. Excessive pressure from the weight of the fuel can cause severe carburetor flooding. The fuel tank outlet for a gravity feed system must be at least 1″ above the fuel inlet fitting of the carburetor to ensure proper and consistent fuel delivery.

Flo-Jet carburetors with integral fuel pumps allow greater flexibility when choosing the size and location of the fuel tank. The fuel tank can be located without the considerations of height for gravity feed as the fuel pump provides fuel flow. The maximum lift distance for a Briggs & Stratton fuel pump is 9″ for engines with a rewind starter and 18″ for engines with an electric starter.

Fuel Tank Construction. Fuel tanks should be constructed of material that is resistant to corrosion. Common materials used for fuel tanks include coated steel, aluminum, or plastic. Fabricated steel tanks that are not coated to resist corrosion should be avoided. During normal use, foreign materials such as water, alcohol, and salt (in colder climates) may enter the fuel system. These impurities, in conjunction with gasoline that becomes stale with prolonged storage, can be very corrosive.

Fuel tanks should be mounted rigidly to avoid excessive vibration. They should also be located in an area that is not subject to heat from the exhaust gas stream or near the muffler. Heating of the fuel, in combination with vibration, accelerates the release of gasoline vapors and tends to create pressure in the tank. The buildup of pressure, in conjunction with any vibration, makes it more difficult to contain the fuel. The fuel tank must be vented so that positive and negative pressures are equalized within the tank. The most common method for venting a fuel tank is a vented fuel cap. See Figure 12-17.

264 SMALL ENGINES

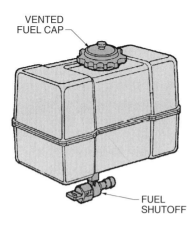

Figure 12-17. The fuel shutoff valve eliminates inadvertent leakage and pressure applied on the inlet needle and seat from the weight of fuel in the tank during transportation.

On fuel tanks that use a gravity feed system, a fuel shutoff valve is recommended to prevent inadvertent leakage of the fuel when the engine is transported. The fuel shutoff valve should be located where it is easily accessible. The fuel shutoff valve also eliminates extra pressure applied to the inlet needle and seat from the weight of fuel in the fuel tank during transportation. The consistent use of the fuel shutoff valve reduces the chances of fuel leakage both externally and internally.

Generac Corporation

This portable generator is equipped with an Oil Gard® to shut down the engine if a low oil level occurs.

Fuel Filter

Fuel filter selection is based on anticipated environmental operating conditions and method of fuel delivery. Extremely dusty and dirty conditions require better filtering capabilities. A 75 micron (μ) or finer fuel filter should be used on engines equipped with a fuel pump. A $75\mu - 150\mu$ fuel filter can be used on engines equipped with a gravity-feed fuel system.

Fuel Lines

Fuel lines must be routed using gradual curves to minimize possible formation of low or high pockets. Low pockets in the fuel line can accumulate any water in the fuel which precipitates out of solution. In cold climates, accumulated water can freeze, causing blockage of the fuel line. High pockets in the fuel line can allow fuel vapors to accumulate, forming a vapor bubble which restricts the flow of liquid fuel (vapor lock). See Figure 12-18.

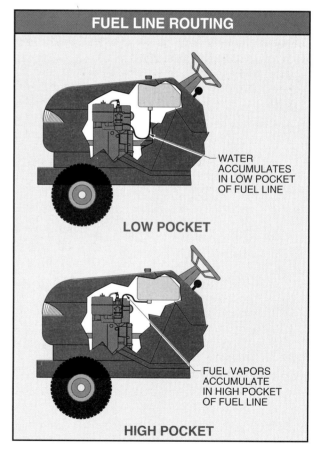

Figure 12-18. Fuel lines are routed using gradual curves to minimize the formation of low or high pockets.

Fuel lines must be located to permit free flow of fuel and to avoid contact with sharp or hot components that can sever or possibly melt the fuel line. Fuel line material is selected based on the type of fuel used and the operating environment. Fuel lines must be resistant to deterioration from fuel, heat, or ultraviolet light. Fuel line protectors may be required in installations where the fuel line must be routed in less than optimum locations.

Fuel lines currently manufactured have a strong resistance to fuel additives such as alcohol and various ethers. Some older fuel lines were susceptible to damage from gasoline additives. Strong concentrations of alcohol additives and/or other additives can cause fuel lines to become brittle, resulting in eventual fracture. This is caused by the propensity of alcohol to seek and absorb water from contacted materials. Some older fuel lines contained substantial quantities of water in the rubber compound. Removal of this water by alcohol additives caused the rubber fuel line to become very brittle.

Other fuel line materials were affected by additives causing excessive elasticity and softening. This stretching effect of the fuel line was caused by the interaction of the rubber and polymer molecules with the active fuel additives that have been used over the last 20 years.

Fuel line hoses, connectors, and mounts are required to maintain proper location to prevent snagging or accidental disconnection. Safety specifications commonly require fuel-wetted hose to withstand a minimum pull-off force of 10 lb at the fuel line connections on the carburetor and fuel tank outlet fitting.

ELECTRICAL SYSTEMS

The electrical system of an engine installation includes the electrical components required for starting and operating the engine and equipment accessories. Battery size or capacity should be large enough to supply sufficient electrical power to the starter motor, enabling it to develop enough torque to crank the engine above the minimum speed for starting. The starter motor should be capable of cranking the engine a minimum speed of 350 rpm with any parasitic load attached. Most lawn and garden equipment and some industrial commercial applications use a 30 to 50 amp-hour (A/hr) battery.

If the alternator on the replacement engine is a regulated/rectified system, a 30 A/hr to 50 A/hr battery size is sufficient. If the alternator on the replacement engine is not a regulated system, a battery of 50 A/hr or greater is recommended. Battery cable length and size must be appropriate for the starter motor and battery used. See Figure 12-19.

BATTERY CABLE SPECIFICATIONS		
Total Cable Length*	Models 8 to 13** (SAE Gauge No.)	Models 16 to 42** (SAE Gauge No.)
2.5	10	6
3.0	8	6
4.0	6	6
5.0	6	4
6.0	4	4

* in feet
** model number based on engine cu in. displacement

Figure 12-19. The battery cable diameter required increases with total cable length.

The alternator selected for the application must supply enough current to sustain the operation of any accessory items which may be drawing current from the charging system. Accessory items such as lights and electric lifting systems for front or rear mounted implements may require extra output from the charging system. Insufficient output from the charging system can eventually completely discharge the battery.

Most service replacement engines are equipped with the highest output alternator system available for a specific engine model. This output ensures the required output for all applications even if a lower output alternator system was supplied on the original engine. Engines commonly manufactured and sold to OEMs that include a specific electrical system, such as the Briggs & Stratton Dual Circuit alternator systems, are also available as service replacement engines. Alternator specifications for replacement engines can be used when repowering to meet OEM engine requirements. See Figure 12-20.

 Briggs & Stratton has over 32,000 authorized service facilities in 102 countries.

BRIGGS & STRATTON ALTERNATOR SPECIFICATIONS

Alternator	Output at 12 V, 3600 rpm	Model Number*
System 3 & 4	1.0 A at 6 V unregulated to battery or .5 A at 12 V unregulated to battery	9, 11, 12
DC only	1.2 A DC+ unregulated to battery	13, 1047
AC only	60 W – 100 W for lighting	17, 19, 22, 25, 28, 32, 40, 42
DC only	3 A DC+ unregulated	17, 19, 22, 25, 28, 32, 40, 42
Dual circuit	60 W – 100 W for lighting 3 A DC+ unregulated	16, 17, 19, 22, 25, 26, 28, 29, 30, 32, 40, 42
Tri-circuit with harness	60 W – 100 W for lighting and 5 A DC+ unregulated	16, 17, 19, 22, 25, 26, 28, 29, 30, 32, 40, 42
5 A	5 A DC+ regulated	16, 17, 19, 22, 25, 26, 28, 29, 30, 40, 42
9 A	9 A DC+ regulated	22, 25, 26, 28, 29, 30, 40, 42
10 A	10 A DC+ regulated	16, 17, 19, 22, 25, 26, 28, 29, 30, 40, 42
16 A	16 A DC+ regulated	22, 25, 28, 29, 30, 32, 40, 42

* model number based on engine cu in. displacement

Figure 12-20. Replacement engine alternator specifications must meet or exceed requirements of the OEM engine.

With the exception of the starter cable, all battery powered wiring should be protected with fuses or circuit breakers located as close as possible to the battery or alternator. Wiring should also be protected from contact with moving parts which could cause a break, a short, and/or a spark, creating a potential fire hazard. Generally, No. 14 wire is used for battery charging and headlight circuits, and No. 16 and/or No. 18 wire is used for all other wiring.

Murray Inc.

This lawn tractor features a partial enclosure engine installation to facilitate engine cooling and provide access to engine components.

VIBRATION

Vibration of equipment can cause operator discomfort and fatigue, noise, and failure of components. Vibration should be kept to a minimum. Single-cylinder engines are only partially balanced with respect to forces required to stop and start the reciprocating piston. Most single-cylinder engines use 50% counterweighting in which rotating counterweights on the crankshaft balance approximately one-half of piston accelerative forces. In multiple-cylinder engines, piston vibration forces are offset to reduce overall engine vibration.

One of the primary sources of vibration in equipment is the engine itself. As the engine operates, piston, connecting rod, and crankshaft motion create accelerative forces in all directions. Vibration is measured in the horizontal, vertical, and perpendicular planes as the engine operates as a free body with no load on a springboard in a laboratory. See Figure 12-21.

The small engine service technician does not normally have the equipment required for vibration measurements. However, some simple guidelines have proven to be accurate in predicting the probability of vibration-induced problems.

When an engine is installed in the equipment, vibration should be less than when the engine is free-standing. This reduction is due to the additional mass provided by the attached equipment. Once installed, vibration of the equipment is evaluated by operating the engine through the normal speed and load range. Operating characteristics are checked using touch, sight, and sound. If all three characteristics are satisfactory, the installation should be satisfactory.

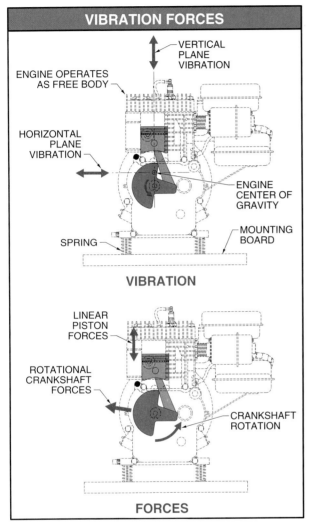

Figure 12-21. Engine vibration is created by forces resulting from piston, connecting rod, and crankshaft motion inside the engine.

Resonance

An engine and an application are comprised of several components which are affected differently by vibration. Some components, such as rubber mounts, function as dampers and absorb energy from vibration. Other components, such as springs, hoses, and tires, are mass and do not normally absorb energy from vibration. Each engine and application component has a natural frequency at which vibration occurs if impacted by a force. If the frequency of the vibration affecting a component is much lower or higher than the natural frequency, there is little net effect. Combining of dissimilar frequencies of vibration commonly yields little increase in force or intensity as each vibration wave tends to retain its own signature, frequency, and intensity. If the vibration waves were diametrically opposed in intensity and frequency, the waves would cancel each other. See Figure 12-22.

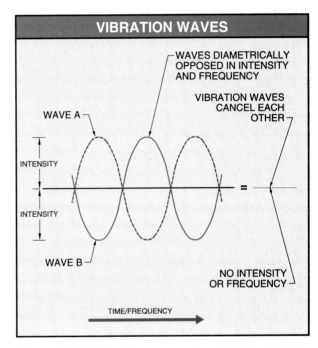

Figure 12-22. Vibration waves diametrically opposed in intensity and frequency cancel each other.

If any vibration wave passing through the application is close to the natural frequency of the application component, the system is in resonance. *Resonance* is the state of the vibration wave frequency being equal to the natural vibration wave frequency of the component. When an object is in resonance, the energy or force of vibration waves generated is multiplied. When the vibration wave is approximately equal to the natural vibrating frequency of the application, the wave form is the product of all similar vibration waves. The level or intensity of the resonant vibration is determined by the dampening effect of application components and the frequency and strength of the vibration wave.

High levels of vibration can result in structural fatigue and eventual failure of engine and application components. The source of the vibration is not always the most common location for vibration-induced structural failure. Resonance most commonly causes the greatest amount of deflection or motion at the farthest point from the source of the energy. For example, on a Briggs & Stratton engine, the greatest amount of visible motion (deflection) from vibration is on air filter system parts, fuel tank and straps, or the carburetor. The component farthest from the source of the vibration is the one that typically exhibits the maximum deflection.

Vibration caused by resonance is usually eliminated by one of two methods. The first and easiest method is to identify the speed or speeds at which the resonant vibration occurs. If possible, the installation is modified to eliminate the need to operate at the undesirable speed or speeds. If this is impossible, the resonant component or components are isolated. The natural frequency of the component(s) can be raised above the normal operating speed by changing the stiffness of the component.

For example, mounting hardware such as straps and brackets can be used to connect an engine directly to another component or frame of the application. Mounting hardware is commonly used on larger walk-behind blowers and air compressors to increase the natural frequency of the component(s) susceptible to problems from resonant vibration. This mounting hardware is installed at the factory to maintain the overall natural frequency of the application and must be reinstalled when repowering with a similar engine size and style and/or after any service procedure.

Ransomes America Corporation
This greens mower is powered by an 18 HP engine that develops 30 lb-ft of torque at 2400 rpm.

Vibration frequency

Soldiers often break the marching cadence when crossing a suspension bridge. The soldiers marching in perfect time across the suspended structure could produce a powerful vibration wave. If the frequency and magnitude of the vibration wave matched the natural vibration frequency of the suspension bridge, the resulting resonance could cause exaggerated motion, deterioration in structural integrity, and possible structural failure. Resonance in small engines at specific frequency and magnitude can result in fractured air cleaner assemblies, muffler components, and/or engine mounting supports.

COOLING SYSTEM

Adequate cooling for an engine is required for maximum safety and efficiency. Approximately two-thirds of the heat energy generated during combustion is lost in the form of waste heat. Approximately 30% is rejected to cooling air. For example, a typical 18 HP engine consumes fuel at 10,000 Btu/HP/hr (180,000 Btu of heat input). Of the heat input, 30% or 54,000 Btu/hr is rejected to cooling air. This amount is equivalent to a furnace that can heat an average-sized house located in a colder climate.

Removal of the waste heat is necessary to ensure the proper operation of the engine. All internal combustion engines are directly and indirectly cooled by air. Air-cooled engines transfer heat both directly to the air and indirectly through lubricating oil. An air-cooled engine is only partially cooled by ambient air. A significant portion of the heat generated by the engine is carried away from internal components by lubricating oil. As the lubricating oil is splashed around the inside of the crankcase, it transfers accumulated heat to the cylinder block walls where, through conduction, it moves to the cooling fins. Cooling fins increase the outside surface area of the engine and complete the cooling process by transferring heat to moving air from the flywheel. Thus, an air-cooled engine is actually cooled by lubricating oil and air. Liquid-cooled engines transfer heat to a liquid coolant, and then to the air using a radiator or other heat exchanger. Lubricating oil also provides some cooling in a liquid-cooled engine.

Different applications require open, partially enclosed, or totally enclosed engine installations. See Figure 12-23. An *open installation* is an engine installation that does not have any barriers which impede access of air to and away from the engine. The installation allows the maximum amount of air to enter and exit the engine.

The main consideration for an open installation is the path of the rejected heated air coming from the engine. The heated air must always be properly routed to prevent possible heat-induced problems caused by the heated air. Engines in an open installation rarely have cooling problems.

A *partial enclosure installation* is an engine installation that allows partial access to engine components. A *total enclosure installation* is an engine installation that does not allow a person to touch the engine by hand nor see the engine by normal line-of-sight vision. Partial or total enclosure installations must provide adequate:

- Ventilation of the enclosure when the engine is shut down to avoid overheating the fuel system and prevent accumulation of hazardous vapors. This is accomplished by using a series of louvers or holes strategically cut and placed on the engine enclosure.
- Control of the waste heat to the outside air. This is accomplished by providing an unimpeded path for the waste heat to completely exit the engine compartment.
- Inward flow of suitable cooling air. This is accomplished through the use of a duct called a plenum to ensure that there is adequate ambient temperature air available to cool the engine.
- Supply of cool carburetor intake air. This is accomplished by utilizing a common plenum used for engine cooling air or by locating the actual air intake outside of the engine enclosure using a snorkel-type device.

Total enclosure installations are usually used for noise reduction or environmental protection. This is accomplished by using tight-fitting enclosure components that do not allow leaks. Openings at the bottom of a total enclosure installation are permissible.

MAINTENANCE AND SERVICE

Proper engine selection and installation allows routine maintenance and adjustments without special tools or equipment. All systems requiring periodic maintenance must be readily accessible without extensive removal of equipment parts. Maintenance and service work tends to be ignored if access is difficult for the individual performing the service. Routine service tasks commonly include checking the oil, changing the oil and oil filter, changing the air filter, and removing debris. Accessibility to other engine components for periodic service is also required. See Figure 12-24.

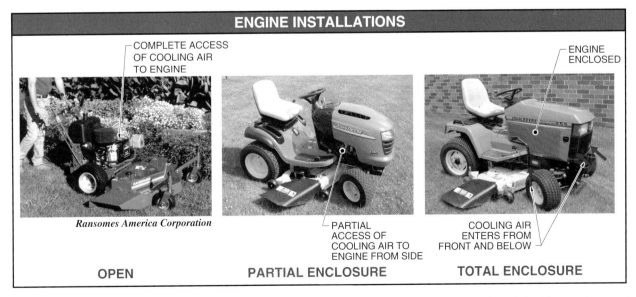

Figure 12-23. The engine installation must provide proper evacuation of heat produced by the engine and equipment components.

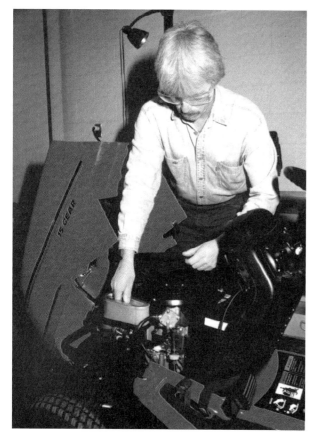

Figure 12-24. Access to engine components is required for performing routine service tasks.

Access to other engine components such as the carburetor, fuel filter, governor, controls, and spark plugs should also be considered. Service procedures other than periodic maintenance tasks can be simplified by the engine installation. For example, after many hours of operation, engines may gradually lose power as a result of combustion chamber deposits. Combustion chamber deposits require the removal of the cylinder head. Easy access to the engine for the removal of the cylinder head reduces service time required.

 Fuel filters are designed to restrict particles larger than the smallest orifice in the fuel system.

SAFETY CONSIDERATIONS

Safety is the most important consideration when selecting and installing an engine for a specific application. For example, the rewind starter rope must not contact any sharp edges when it is extended or retracted. A CPSC requirement on walk-behind lawn mowers states that the rewind starter rope grip must be positioned a miniumum distance of 24″ from the engine. See Figure 12-25. If possible, the rope grip should be positioned so that it can be used by the right or left hand without hitting any obstructions when the starter rope is pulled.

The oil level check should be located so that the oil level can be easily checked on a regular basis. An extended oil dipstick and fill tube improves accessibility and minimizes the possibility of dirt entering the crankcase through the oil fill tube. The oil drain should be located to provide easy draining and capture of waste oil. It may be necessary to use an oil drain extension tube to meet this requirement. Easy access must also be provided for the oil filter (if equipped) and air filter. Equipment operated in dusty environments requires that the air cleaner be checked on a regular basis.

During normal engine operation, debris may pass through the rotating screen and become lodged between the cooling fins on an air-cooled engine. Debris lodged in the cooling fins reduces the cooling capacity of the engine, and eventually results in overheating. Cooling fins should be cleaned after every 100 hours of engine operation. For easy cleaning of the cooling fins, access to blower housing screws should be considered when repowering.

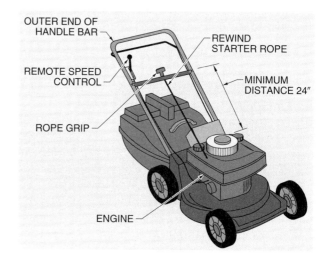

Figure 12-25. The rewind starter rope grip must be a minimum distance of 24″ from the engine and must be free of any obstructions when the rope is pulled.

Engine Exhaust

All small gasoline engines produce toxic gases and must always be operated in a well-ventilated area. Engine installations should discharge exhaust gases away from the application and operator safely and efficiently. Improper exhaust system modification can result in a safety hazard. The minimum engine exhaust pipe diameter should be the same size as the exhaust port diameter. The exhaust system should be secured and guarded to prevent injury and damage. Airborne combustibles should not be allowed to collect on or near mufflers and/or hot engine components.

Repowering may require reworking the exhaust system if the exhaust exit on the replacement engine is not in the same location as on the original engine. Most service replacement engines come with multi-position exhaust deflectors for use during repowering. An exhaust deflector should be installed so that exhaust gases are routed away from the operator, fuel system components, and direct paths close to the ground. See Figure 12-26. An exhaust gas deflector should never be installed with the opening pointed upward, because this allows rainwater, dirt, and/or debris to collect. This could possibly result in a fire hazard and/or muffler and engine damage.

Local, state, and/or federal law may require the use of a spark arrester to prevent ignition of dry grass, wood, or other combustibles from sparks in exhaust gases. A *spark arrester* is a component in the exhaust system that redirects the flow of exhaust gases through a screen to trap sparks discharged from the engine. Spark arrester designs vary with engine type and size, and all must be U.S. Forestry Service approved.

Fuel System Safety

Engine installations must consider the possibility of fire hazard and explosive danger of gasoline. Fuel tanks and fuel lines should be located away from any heat or spark source. See Figure 12-27. In addition, the fuel tank should be located so that if a spill occurs during filling, it does not create a potential hazard. Spilled fuel should not collect on the equipment or come in contact with the high-tension ignition cable. It should flow away from the muffler outlet. Fuel tanks and lines must be securely fastened, connected, and protected from any sharp edges that may puncture and cause a fuel leak. Burned out, damaged, and/or modified mufflers can be an ignition source for unconfined gasoline.

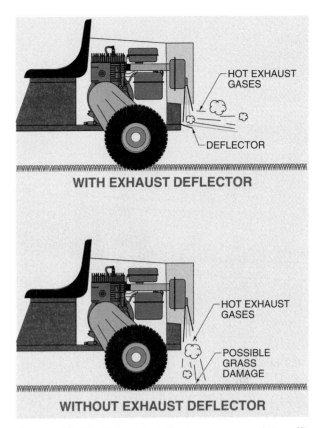

Figure 12-26. Exhaust deflectors are used to efficiently redirect hot exhaust gases to prevent possible injury or damage.

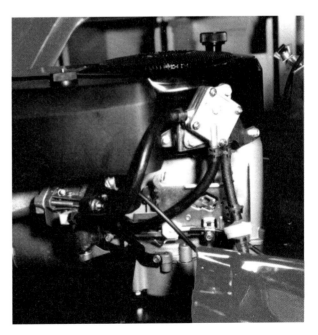

Figure 12-27. Fuel tanks and fuel lines must be securely fastened and located away from any heat or spark source.

272 SMALL ENGINES

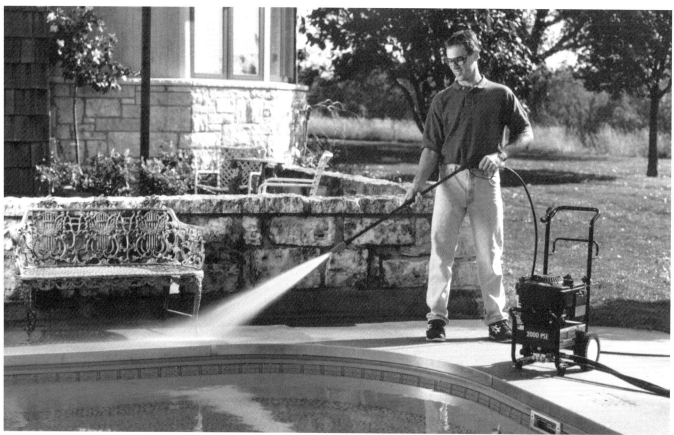
Mi-T-M Corporation

This pressure washer has a vertical direct-drive engine installation and pumps 2 gal./min.

Ransomes America Corporation

The 14 HP engine used to power this utility vehicle produces 24 lb-ft of torque at 2200 rpm.

APPENDIX

Engine Classification 274	Metric System 286
Industry and Standards Organizations . . 275	Metric Prefixes 287
Safety Color Coding 276	Metric Conversions 287
Potential Effects of Carbon Monoxide Exposure 276	Inch – Millimeter Equivalents 288
Hazardous Material Container Labeling 277	Twist Drill Fractional, Number, and Letter Sizes 289
Hazardous Materials 278	Metric Screw Threads 289
Air Density Effects 279	Drilled Hole Tolerances 289
Math Symbols 279	Standard Series Threads – Graded Pitches 290
Coefficient of Friction 279	Metric Drill Sizes 290
Units of Power 279	Pounds Per Square Inch – Kilopascals Conversion Table 291
Reading Micrometers 280	Ambient Pressure and Engine Coolant* Boiling Point 291
Fuel/Oil Mix 280	
Alphabet of Lines 281	Ohm's Law 291
Welding Symbol 282	Measures 292
Welding Symbol – Arrow 282	Warranty Claim 292
Welding Symbol – Tail 282	Failure Analysis 293
Welding Symbol – Reference Line 283	Governor System Troubleshooting 296
Welding Symbol – Weld Symbol 283	Vanguard™ Three-Cylinder Gasoline Engine Maintenance Schedule 298
Welding Symbol – Dimensions 283	
Weld Joints and Types 284	Briggs & Stratton Master Service Technician Program 299
English System 285	Industrial Electrical Symbols 300

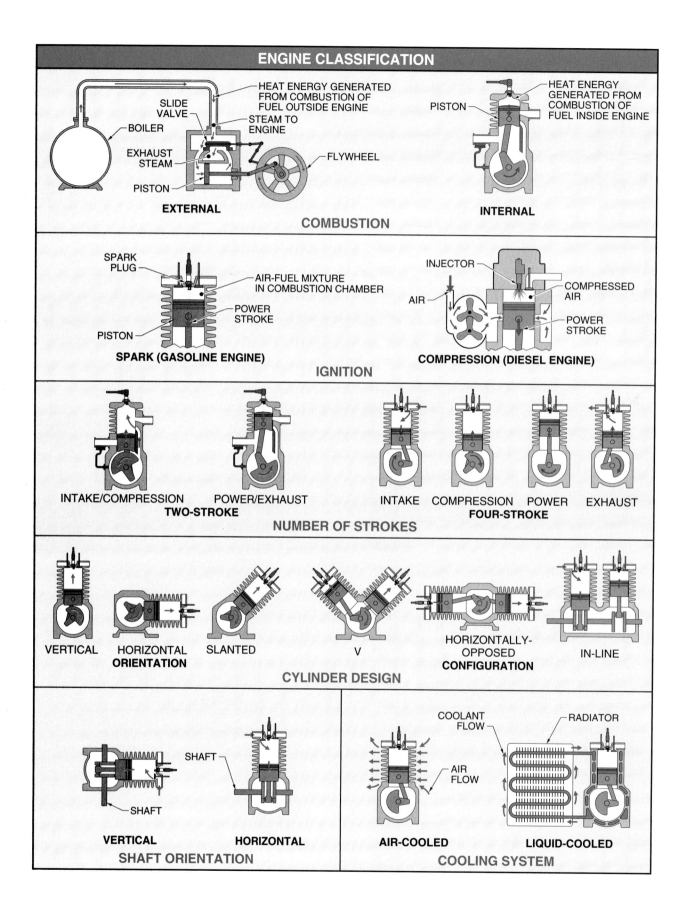

INDUSTRY AND STANDARDS ORGANIZATIONS

CPSC
Consumer Products Safety Commission
4330 East West Hwy
Bethesda, MD 20814

EPA
Environmental Protection Agency
401 M Street SW
Washington, DC 20460

DOD
Department of Defense
Defense Printing Service
700 Robbins Avenue
Philadelphia, PA 19111-5094

DOT
Department of Transportation
400 7th Street SW
Washington, DC 20590

NIOSH
National Institute for Occupational
Safety and Health
4676 Columbia Parkway
Cincinnati, OH 45226

OSHA
Occupation Safety and
Health Administration
230 South Dearborn Street
Chicago, IL 60604

GOVERNMENT AGENCIES

ANSI
American National Standards Institute
11 West 42nd Street
New York, NY 10036

CSA
Canadian Standards Association
178 Rexdale Blvd
Rexdale, ON M9W 1R

ISO
International Organization for Standardization
Case Postale 56 CH - 1211
Geneve 20 Switzerland

STANDARDS ORGANIZATIONS

SAE
Society of Automotive Engineers
400 Commonwealth Dr.
Warrendale, PA 15096

ASAE
American Society of Agricultural Engineers
2950 Niles Rd.
St. Joseph, MI 49085

ASTM International
American Society for Testing and Materials
1916 Race St.
Philadelphia, PA 19103

TECHNICAL SOCIETIES

NFPA
National Fire Protection Association
Batterymarch Park
Quincy, MA 02269

UL
Underwriters Laboratories Inc.
333 Pfingsten Rd.
Northbrook, IL 60062

TRADE ORGANIZATIONS

API
American Petroleum Institute
1220 L St. NW
Washington, DC 20005

OPEI
Outdoor Power Equipment Institute
341 S. Patrick St.
Old Town Alexandria, VA 22314

PRIVATE ORGANIZATIONS

EETC
Equipment & Engine Training Council
P.O. Box 648
W307 N5480 Anderson Road
Hartland, WI, 53029

OPEESA
Outdoor Power Equipment and
Engine Service Association
210 Allen Drive
Exton, PA 19341

AED
AED Foundation
Associated Equipment Distributors, Inc.
615 W. 22nd Street
Oak Brook, IL 60523

TRAINING-RELATED ORGANIZATIONS

SkillsUSA
SkillsUSA–VICA
P.O. Box 3000
Leesburg, VA 20177

FFA
National FFA Center
P.O. Box 68960, 6060 FFA Drive
Indianapolis, IN 46268

STUDENT-RELATED ORGANIZATIONS

SAFETY COLOR CODING

Color		Use	Applications
Red	■	Identify: 1. Fire protection equipment and apparatus 2. Danger 3. Stop	1. Fire exit signs, fire alarm boxes, fire extinguishers, fire hose locations, fire hydrants 2. Safety cans with a flash point of 100°F or less, danger signs 3. Emergency stop box on hazardous machines and stop buttons used for emergency stopping of machinery
Orange	■	Designate dangerous parts of machines or energized equipment which may cut, crush, shock, or otherwise cause injury and to emphasize such hazards when doors are open or guards are removed	Inside mowing guards; safety starting buttons; inside transmission guards for gears, pulleys, chains, etc.; exposed parts of pulleys, gears, rollers, etc.
Yellow	■	Designate caution and mark physical hazards such as striking against, stumbling, falling, and tripping. Solid yellow, yellow and black stripes, or yellow and black checks may be used in any combination to attract the most attention	Construction equipment such as bulldozers, tractors, carryalls, etc.; coverings or guards for guy wires; exposed and unguarded edges of platforms, pits, and walls; handrails, guardrails, on top and bottom treads of stairways where caution is needed; markings for projections, doorways, etc.; pillars, posts, or columns
Purple	■	Designate radiation hazards. Yellow is used in combination with purple for markers such as tags, labels, signs, etc.	Rooms and areas where radioactive materials are stored or handled, burial grounds, disposable cans for contaminated materials, containers of radioactive materials, etc.
Green	■	Designate safety and location of first aid equipment	Safety bulletin boards, gas masks, first aid kits, stretchers, etc.
Black, White, or B/W	▦	Designate: 1. Traffic 2. Housekeeping areas	1. Dead ends of aisles or passageways; location and width of aisleways, stairways, and direction signs 2. Location of refuse cans, food dispensing equipment, etc.

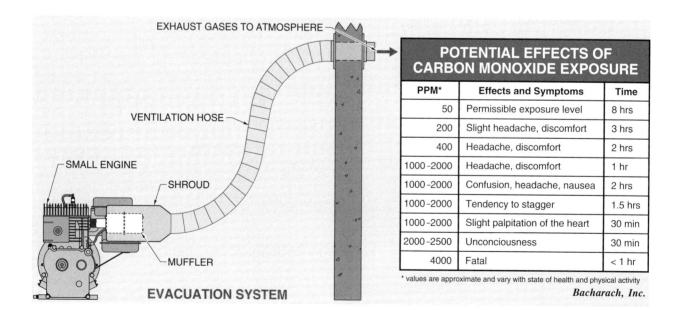

EVACUATION SYSTEM

POTENTIAL EFFECTS OF CARBON MONOXIDE EXPOSURE

PPM*	Effects and Symptoms	Time
50	Permissible exposure level	8 hrs
200	Slight headache, discomfort	3 hrs
400	Headache, discomfort	2 hrs
1000-2000	Headache, discomfort	1 hr
1000-2000	Confusion, headache, nausea	2 hrs
1000-2000	Tendency to stagger	1.5 hrs
1000-2000	Slight palpitation of the heart	30 min
2000-2500	Unconsciousness	30 min
4000	Fatal	< 1 hr

* values are approximate and vary with state of health and physical activity

Bacharach, Inc.

INDUSTRY AND STANDARDS ORGANIZATIONS

CPSC
Consumer Products Safety Commission
4330 East West Hwy
Bethesda, MD 20814

EPA
Environmental Protection Agency
401 M Street SW
Washington, DC 20460

DOD
Department of Defense
Defense Printing Service
700 Robbins Avenue
Philadelphia, PA 19111-5094

DOT
Department of Transportation
400 7th Street SW
Washington, DC 20590

NIOSH
National Institute for Occupational Safety and Health
4676 Columbia Parkway
Cincinnati, OH 45226

OSHA
Occupation Safety and Health Administration
230 South Dearborn Street
Chicago, IL 60604

GOVERNMENT AGENCIES

ANSI
American National Standards Institute
11 West 42nd Street
New York, NY 10036

CSA
Canadian Standards Association
178 Rexdale Blvd
Rexdale, ON M9W 1R

ISO
International Organization for Standardization
Case Postale 56 CH - 1211
Geneve 20 Switzerland

STANDARDS ORGANIZATIONS

SAE
Society of Automotive Engineers
400 Commonwealth Dr.
Warrendale, PA 15096

ASAE
American Society of Agricultural Engineers
2950 Niles Rd.
St. Joseph, MI 49085

ASTM International
American Society for Testing and Materials
1916 Race St.
Philadelphia, PA 19103

TECHNICAL SOCIETIES

NFPA
National Fire Protection Association
Batterymarch Park
Quincy, MA 02269

UL
Underwriters Laboratories Inc.
333 Pfingsten Rd.
Northbrook, IL 60062

TRADE ORGANIZATIONS

API
American Petroleum Institute
1220 L St. NW
Washington, DC 20005

OPEI
Outdoor Power Equipment Institute
341 S. Patrick St.
Old Town Alexandria, VA 22314

PRIVATE ORGANIZATIONS

EETC
Equipment & Engine Training Council
P.O. Box 648
W307 N5480 Anderson Road
Hartland, WI, 53029

OPEESA
Outdoor Power Equipment and Engine Service Association
210 Allen Drive
Exton, PA 19341

AED
AED Foundation
Associated Equipment Distributors, Inc.
615 W. 22nd Street
Oak Brook, IL 60523

TRAINING-RELATED ORGANIZATIONS

SkillsUSA
SkillsUSA–VICA
P.O. Box 3000
Leesburg, VA 20177

FFA
National FFA Center
P.O. Box 68960, 6060 FFA Drive
Indianapolis, IN 46268

STUDENT-RELATED ORGANIZATIONS

SAFETY COLOR CODING

Color	Use	Applications
Red	Identify: 1. Fire protection equipment and apparatus 2. Danger 3. Stop	1. Fire exit signs, fire alarm boxes, fire extinguishers, fire hose locations, fire hydrants 2. Safety cans with a flash point of 100°F or less, danger signs 3. Emergency stop box on hazardous machines and stop buttons used for emergency stopping of machinery
Orange	Designate dangerous parts of machines or energized equipment which may cut, crush, shock, or otherwise cause injury and to emphasize such hazards when doors are open or guards are removed	Inside mowing guards; safety starting buttons; inside transmission guards for gears, pulleys, chains, etc.; exposed parts of pulleys, gears, rollers, etc.
Yellow	Designate caution and mark physical hazards such as striking against, stumbling, falling, and tripping. Solid yellow, yellow and black stripes, or yellow and black checks may be used in any combination to attract the most attention	Construction equipment such as bulldozers, tractors, carryalls, etc.; coverings or guards for guy wires; exposed and unguarded edges of platforms, pits, and walls; handrails, guardrails, on top and bottom treads of stairways where caution is needed; markings for projections, doorways, etc.; pillars, posts, or columns
Purple	Designate radiation hazards. Yellow is used in combination with purple for markers such as tags, labels, signs, etc.	Rooms and areas where radioactive materials are stored or handled, burial grounds, disposable cans for contaminated materials, containers of radioactive materials, etc.
Green	Designate safety and location of first aid equipment	Safety bulletin boards, gas masks, first aid kits, stretchers, etc.
Black, White, or B/W	Designate: 1. Traffic 2. Housekeeping areas	1. Dead ends of aisles or passageways; location and width of aisleways, stairways, and direction signs 2. Location of refuse cans, food dispensing equipment, etc.

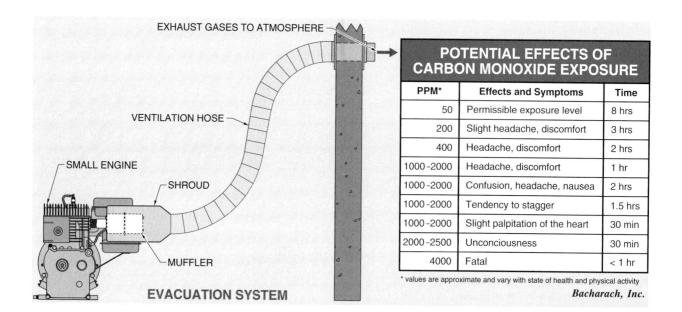

EVACUATION SYSTEM

POTENTIAL EFFECTS OF CARBON MONOXIDE EXPOSURE

PPM*	Effects and Symptoms	Time
50	Permissible exposure level	8 hrs
200	Slight headache, discomfort	3 hrs
400	Headache, discomfort	2 hrs
1000–2000	Headache, discomfort	1 hr
1000–2000	Confusion, headache, nausea	2 hrs
1000–2000	Tendency to stagger	1.5 hrs
1000–2000	Slight palpitation of the heart	30 min
2000–2500	Unconsciousness	30 min
4000	Fatal	< 1 hr

* values are approximate and vary with state of health and physical activity

Bacharach, Inc.

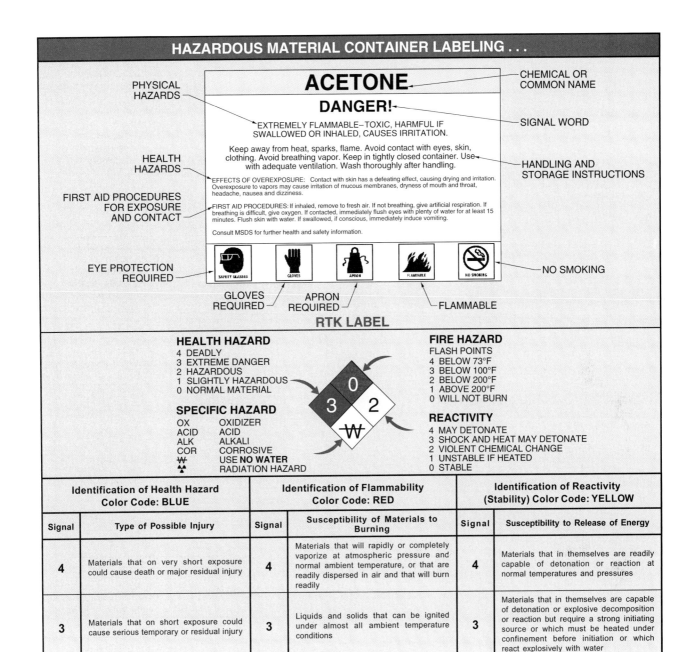

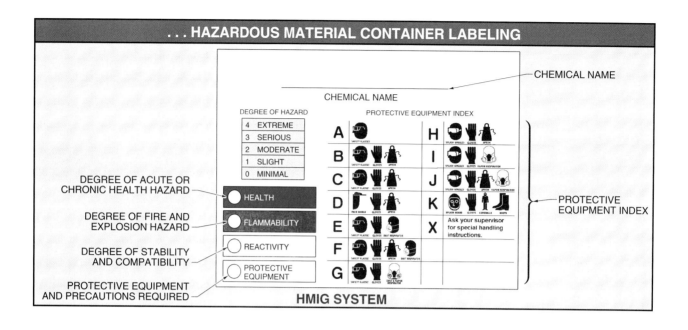

HAZARDOUS MATERIALS

Chemical	Hazard Rating				Chemical Abstract Service Number	Chemical	Hazard Rating				Chemical Abstract Service Number
	H	F	R	S/H			H	F	R	S/H	
Acetic acid	2	2	0	—	64-19-7	Isopropyl ether	2	3	1	—	108-20-3
Acetone	1	3	0	—	67-64-1	Methanol	1	3	0	—	67-56-1
Acetonitrile	3	3	0	—	75-05-8	Methyl acetate	1	3	0	—	79-20-9
Acrolein	3	3	3	—	107-02-8	Methyl bromide	3	1	0	—	74-83-9
Allyl alcohol	3	3	1	—	107-16-6	Methyl isobutyl ketone	2	3	0	—	108-10-1
Ammonia anhydrous	3	1	0	—	7664-41-7	Methylamine	3	4	0	—	74-89-5
Aniline	3	2	0	—	65-53-3	Morpholine	2	3	0	—	110-91-8
Bromine	3	0	0	OX	7726-95-6	Naphtha	1	4	0	—	8030-30-6
1 – 3 Butadiene	2	4	2	—	106-99-0	Naphthalene	2	2	0	—	91-20-3
Butyl acetate	1	3	0	—	123-86-4	Nitric acid	3	0	0	OX	7697-37-2
tert-Butyl alcohol	1	3	0	—	75-65-0	Nitrobenzene	3	2	1	—	98-95-3
Caustic soda	3	0	1	—	1310-73-2	p-Nitrochlorobenzene	2	1	3	—	100-00-5
Chlorine	3	0	0	OX	7782-50-5	Octane	0	3	0	—	111-65-9
Chloroform	2	0	0	—	67-66-3	Oxalic acid	2	1	0	—	144-62-7
o-Cresol	3	2	0	—	1319-77-3	Pentane	1	4	0	—	109-66-0
Cumene	2	3	1	—	98-82-8	Petroleum distillates	1	4	0	—	8002-05-9
Cyclohexane	1	3	0	—	110-82-7	Phenol	3	2	0	—	108-95-2
Cyclohexanol	1	2	0	—	108-93-0	Propane gas	1	4	0	—	74-98-6
Cyclohexanone	1	2	0	—	108-94-1	1-Propanol	1	3	0	—	71-23-8
Diborane	3	4	3	W	19287-45-7	Propyl acetate	1	3	0	—	109-60-7
Dimethylamine	3	4	0	—	124-40-3	n-Propyl alcohol	1	3	0	—	71-23-8
p-Dioxane	2	3	1	—	123-91-1	Propylene oxide	4	2	2	—	75-56-9
Ethyl acetate	1	3	0	—	141-78-6	Pyridine	2	3	0	—	110-86-1
Ethyl ether	2	4	1	—	60-29-7	Sodium cyanide	3	0	0	—	143-33-9
Formic acid	3	2	0	—	64-18-6	Sodium hydroxide	3	0	1	—	1310-73-2
n-Heptane	1	3	0	—	142-82-5	Stoddard solvent	0	2	0	—	8052-41-3
n-Hexane	1	3	0	—	110-54-3	Sulfur dioxide	3	0	0	—	7446-09-5
Hydrazine	3	3	3	—	302-01-2	Sulfuric acid	3	0	2	W	7664-93-9
Hydrochloric acid	3	0	0	—	7647-01-0	Tetrahydrofuran	2	3	1	—	109-99-9
Hydrogen peroxide	2	0	1	OX	7722-84-1	1-1-1 Trichloroethane	2	1	0	—	71-55-6
Iodine	—	—	—	—	7553-56-2	Triethylamine	2	3	0	—	121-44-8
Isobutyl alcohol	1	3	0	—	78-83-1	Xylene	2	3	0	—	1330-20-7
Isopropyl alcohol	1	3	0	—	67-63-0						

Lab Safety Supply, Inc.

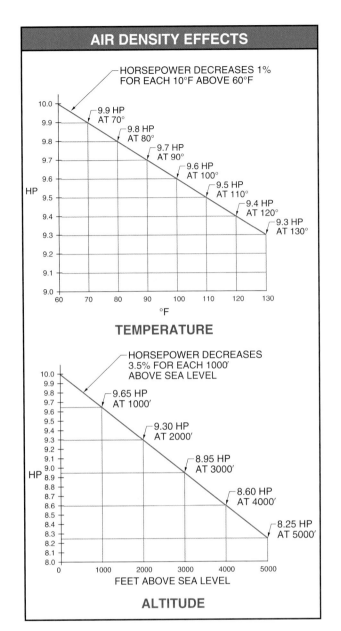

MATH SYMBOLS	
Symbol	Meaning
+	Plus; add
−	Minus; subtract
×	Times; multiply
÷	Divided by; divide
Σ	Sum of
=	Equals; is equal to
≈	Approximately equals; approximately equal to
>	Is greater than
<	Is less than
√	Square root
$\sqrt[3]{}$	Cube root
E^x	Power of

COEFFICIENT OF FRICTION				
Materials	Dry (μ_s)*	Dry (μ_k)**	Wet (μ_s)	Wet (μ_k)
Steel on steel	0.20	0.12	0.15	0.09
Steel on ice	0.03	0.01	—	—
Hemp rope on wood	0.50	0.40	—	—
Leather on oak	0.50	0.30	0.40	0.25
Wrought iron on cast iron or brass	0.31	0.18	0.16	0.10
Tire rubber on concrete	1.00	0.70	0.70	0.50

* static
** kinetic

UNITS OF POWER				
Power	W	ft-lb/sec	HP	kW
1 W	1	0.7376	1.341×10^{-3}	0.001
1 ft-lb/sec	1.356	1	1.818×10^{-3}	1.356×10^{-3}
1 HP	745.7	550	1	0.7457
1 kW	1000	736.6	1.341	1

SMALL ENGINES

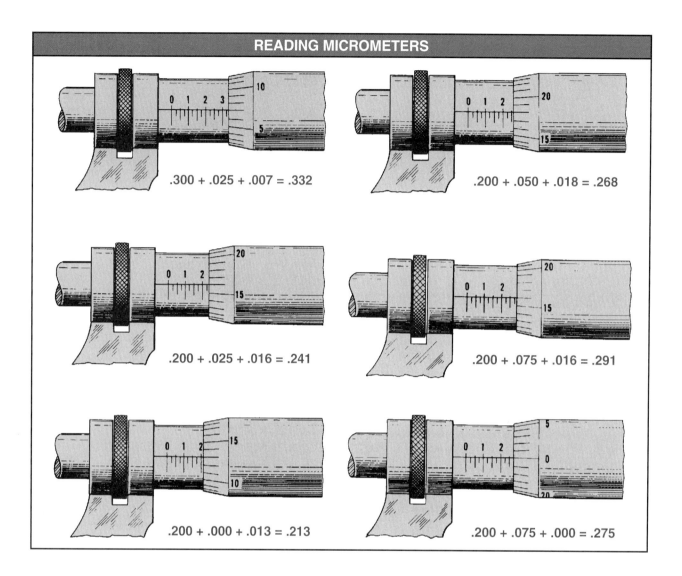

FUEL/OIL MIX

	US Gallons		Imperial Gallons		Metric			US Gallons		Imperial Gallons		Metric	
16 : 1	Fuel*	Oil**	Fuel*	Oil**	Petrol***	Oil***	32 : 1	Fuel*	Oil**	Fuel*	Oil**	Petrol***	Oil***
	1	8	1	10	4	.250		1	4	1	5	4	.125
	3	24	3	30	12	.750		3	12	3	15	12	.375
	5	40	5	50	20	1.250		5	20	5	25	20	.625
	6	48	6	60	24	1.500		6	24	6	30	24	.750
24 : 1	1	5.33	1	6.4	4	.160	50 : 1	1	2.5	1	3	4	.080
	3	16	3	19.2	12	.470		3	8.0	3	9	12	.240
	5	26.66	5	32.0	20	.790		5	13.0	5	15	20	.400
	6	32	6	38.4	24	.940		6	15.5	6	18.5	24	.480

* in gal.
** in oz
*** in l

ALPHABET OF LINES

NAME AND USE	CONVENTIONAL REPRESENTATION	EXAMPLE	
OBJECT LINE — Define shape. Outline and detail objects.	THICK	OBJECT LINE	
HIDDEN LINE — Show hidden features.	$\frac{1}{8}''$ (3 mm), $\frac{1}{32}''$ (0.75 mm) THIN	HIDDEN LINE	
CENTER LINE — Locate centerpoints of arcs and circles.	$\frac{1}{16}''$ (1.5 mm), $\frac{1}{8}''$ (3 mm), $\frac{3}{4}''$ (18 mm) TO $1\frac{1}{2}''$ (36 mm) THIN	CENTER LINE, CENTERPOINT	
DIMENSION LINE — Show size or location. **EXTENSION LINE** — Define size or location.	DIMENSION LINE, 2'-6", EXTENSION LINE THIN	DIMENSION LINE, $1\frac{3}{4}$, EXTENSION LINE	
LEADER — Call out specific features.	OPEN ARROWHEAD, CLOSED ARROWHEAD, X, 3X THIN	$1\frac{1}{2}$ DRILL, LEADER	
CUTTING PLANE — Show internal features.	$\frac{1}{8}''$ (3 mm), $\frac{1}{16}''$ (1.5 mm), $\frac{3}{4}''$ (18 mm) TO $1\frac{1}{2}''$ (36 mm), A—A THICK	LETTER IDENTIFIES SECTION VIEW, CUTTING PLANE LINE	
SECTION LINE — Identify internal features.	$\frac{1}{16}''$ (1.5 mm) THIN	SECTION LINES	
BREAK LINE — Show long breaks. **BREAK LINE** — Show short breaks.	$\frac{3}{4}''$ (18 mm) TO $1\frac{1}{2}''$ (36 mm) THIN; FREEHAND THICK	LONG BREAK LINE, SHORT BREAK LINE	

282 SMALL ENGINES

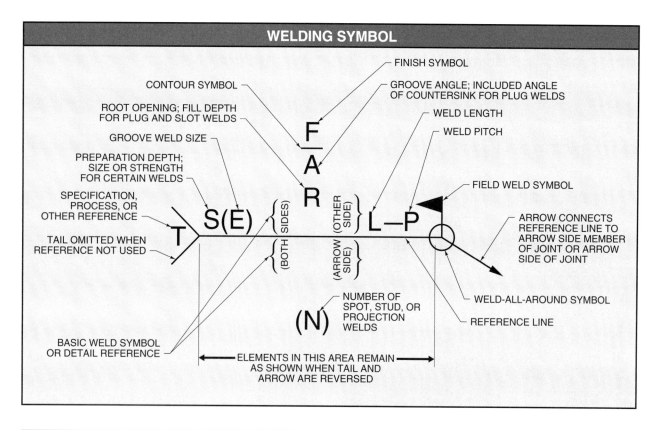

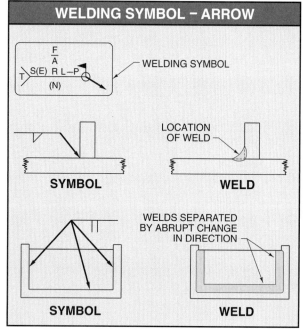

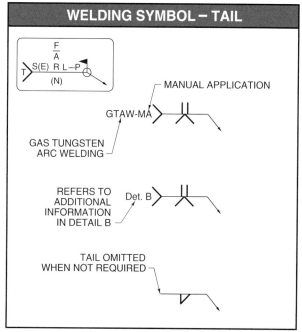

Appendix **283**

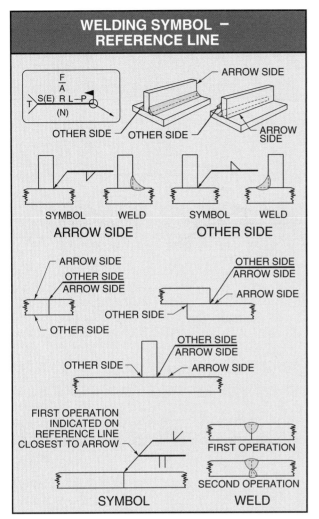

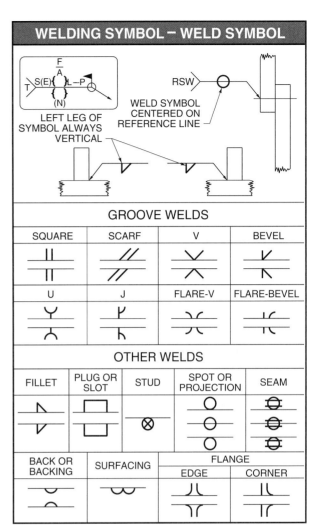

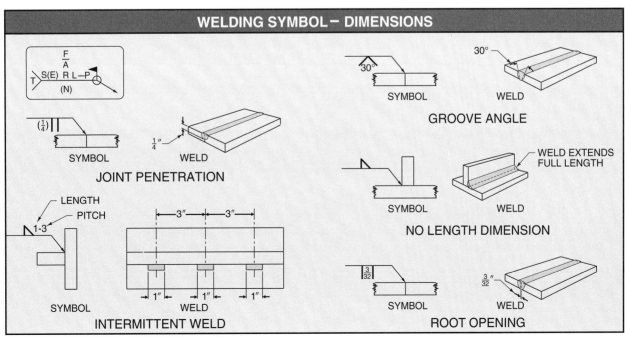

WELD JOINTS AND TYPES

APPLICABLE WELDS	WELD SYMBOL	BUTT	LAP	T	EDGE	CORNER
SQUARE-GROOVE	‖	✓	—	✓	✓	✓
BEVEL-GROOVE	⌐	✓	✓	✓	✓	✓
V-GROOVE	∧	✓	—	—	✓	✓
U-GROOVE	⌒	✓	—	—	✓	✓
J-GROOVE	⌐	✓	✓	✓	✓	✓
FLARE-BEVEL-GROOVE	⌐	✓	✓	✓	✓	✓
FLARE-V-GROOVE	⌒	✓	—	—	✓	✓
FILLET	▽	—	✓	✓	—	✓
PLUG	⊔	—	✓	✓	—	✓
SLOT	⊔	—	✓	✓	—	✓
EDGE-FLANGE	⌐	✓	—	—	✓	—
CORNER-FLANGE	⌐	—	—	—	✓	✓
SPOT	○	—	✓	✓	—	✓
PROJECTION	○	—	✓	✓	—	✓
SEAM	⊖	—	✓	✓	✓	✓
BRAZE	◁BRAZE	✓	✓	✓	—	✓

ENGLISH SYSTEM

		Unit	Abbreviation	Equivalents
LENGTH		mile	mi	5280', 320 rd, 1760 yd
		rod	rd	5.50 yd, 16.5'
		yard	yd	3', 36"
		foot	ft or '	12", .333 yd
		inch	in. or "	.083', .028 yd
AREA		square mile	sq mi or mi^2	640 a, 102,400 sq rd
		acre	A	4840 sq yd, 43,560 sq ft
$A = l \times w$		square rod	sq rd or rd^2	30.25 sq yd, .00625 A
		square yard	sq yd or yd^2	1296 sq in., 9 sq ft
		square foot	sq ft or ft^2	144 sq in., .111 sq yd
		square inch	sq in. or in^2	.0069 sq ft, .00077 sq yd
VOLUME		cubic yard	cu yd or yd^3	27 cu ft, 46,656 cu in.
$V = l \times w \times t$		cubic foot	cu ft or ft^3	1728 cu in., .0370 cu yd
		cubic inch	cu in. or in^3	.00058 cu ft, .000021 cu yd
CAPACITY	*U.S. liquid measure*	gallon	gal.	4 qt (231 cu in.)
		quart	qt	2 pt (57.75 cu in.)
WATER, FUEL, ETC.		pint	pt	4 gi (28.875 cu in.)
		gill	gi	4 fl oz (7.219 cu in.)
		fluidounce	fl oz	8 fl dr (1.805 cu in.)
		fluidram	fl dr	60 min (.226 cu in.)
		minim	min	1/60 fl dr (.003760 cu in.)
	U.S. dry measure	bushel	bu	4 pk (2150.42 cu in.)
VEGETABLES, GRAIN, ETC.		peck	pk	8 qt (537.605 cu in.)
		quart	qt	2 pt (67.201 cu in.)
		pint	pt	½ qt (33.600 cu in.)
	British imperial liquid and dry measure	bushel	bu	4 pk (2219.36 cu in.)
		peck	pk	2 gal. (554.84 cu in.)
		gallon	gal.	4 qt (277.420 cu in.)
		quart	qt	2 pt (69.355 cu in.)
DRUGS		pint	pt	4 gi (34.678 cu in.)
		gill	gi	5 fl oz (8.669 cu in.)
		fluidounce	fl oz	8 fl dr (1.7339 cu in.)
		fluidram	fl dr	60 min (.216734 cu in.)
		minim	min	1/60 fl dr (.003612 cu in.)
MASS AND WEIGHT	*avoirdupois*	ton		2000 lb
		short ton		2000 lb
COAL, GRAIN, ETC.		long ton		2240 lb
		pound	lb or #	16 oz, 7000 gr
		ounce	oz	16 dr, 437.5 gr
		dram	dr	27.344 gr, .0625 oz
		grain	gr	.037 dr, .002286 oz
GOLD, SILVER, ETC.	*troy*	pound	lb	12 oz, 240 dwt, 5760 gr
		ounce	oz	20 dwt, 480 gr
		pennyweight	dwt or pwt	24 gr, .05 oz
		grain	gr	.042 dwt, .002083 oz
	apothecaries'	pound	lb ap	12 oz, 5760 gr
		ounce	oz ap	8 dr ap, 480 gr
DRUGS		dram	dr ap	3 s ap, 60 gr
		scruple	s ap	20 gr, .333 dr ap
		grain	gr	.05 s, .002083 oz, .0166 dr ap

METRIC SYSTEM

LENGTH	Unit	Abbreviation	Number of Base Units
	kilometer	km	1000
	hectometer	hm	100
	dekameter	dam	10
	***meter**	m	1
	decimeter	dm	.1
	centimeter	cm	.01
	millimeter	mm	.001
AREA $A = l \times w$	square kilometer	sq km or km^2	1,000,000
	hectare	ha	10,000
	are	a	100
	square centimeter	sq cm or cm^2	.0001
VOLUME $V = l \times w \times t$	cubic centimeter	cu cm, cm^3, or cc	.000001
	cubic decimeter	dm^3	.001
	***cubic meter**	m^3	1
CAPACITY WATER, FUEL, ETC.	kiloliter	kl	1000
	hectoliter	hl	100
	dekaliter	dal	10
	***liter**	l	1
	cubic decimeter	dm^3	1
	deciliter	dl	.10
	centiliter	cl	.01
	milliliter	ml	.001
MASS AND WEIGHT COAL, GRAIN, ETC. GOLD, SILVER, ETC.	metric ton	t	1,000,000
	kilogram	kg	1000
	hectogram	hg	100
	dekagram	dag	10
	***gram**	g	1
	decigram	dg	.10
	centigram	cg	.01
	milligram	mg	.001

* base units

METRIC PREFIXES

Multiples and Submultiples	Prefixes	Symbols	Meaning
$1{,}000{,}000{,}000{,}000 = 10^{12}$	tera	T	trillion
$1{,}000{,}000{,}000 = 10^{9}$	giga	G	billion
$1{,}000{,}000 = 10^{6}$	mega	M	million
$1000 = 10^{3}$	kilo	k	thousand
$100 = 10^{2}$	hecto	h	hundred
$10 = 10^{1}$	deka	d	ten
Unit $1 = 10^{0}$			
$.1 = 10^{-1}$	deci	d	tenth
$.01 = 10^{-2}$	centi	c	hundredth
$.001 = 10^{-3}$	milli	m	thousandth
$.000001 = 10^{-6}$	micro	μ	millionth
$.000000001 = 10^{-9}$	nano	n	billionth
$.000000000001 = 10^{-12}$	pico	p	trillionth

METRIC CONVERSIONS

Initial Units	Final Units										
	giga	mega	kilo	hecto	deka	base unit	deci	centi	milli	micro	nano	pico
giga		3R	6R	7R	8R	9R	10R	11R	12R	15R	18R	21R
mega	3L		3R	4R	5R	6R	7R	8R	9R	12R	15R	18R
kilo	6L	3L		1R	2R	3R	4R	5R	6R	9R	12R	15R
hecto	7L	4L	1L		1R	2R	3R	4R	5R	8R	11R	14R
deka	8L	5L	2L	1L		1R	2R	3R	4R	7R	10R	13R
base unit	9L	6L	3L	2L	1L		1R	2R	3R	6R	9R	12R
deci	10L	7L	4L	3L	2L	1L		1R	2R	5R	8R	11R
centi	11L	8L	5L	4L	3L	2L	1L		1R	4R	7R	10R
milli	12L	9L	6L	5L	4L	3L	2L	1L		3R	6R	9R
micro	15L	12L	9L	8L	7L	6L	5L	4L	3L		3R	6R
nano	18L	15L	12L	11L	10L	9L	8L	7L	6L	3L		3R
pico	21L	18L	15L	14L	13L	12L	11L	10L	9L	6L	3L	

INCH – MILLIMETER EQUIVALENTS

Inches		mm	Inches		mm	Inches		mm
	.00004	.001		.11811	3		.550	13.970
	.00039	.01	1/8	.1250	3.175		.55118	14
	.00079	.02		.13780	3.5	9/16	.56250	14.2875
	.001	.025	9/64	.14063	3.5719		.57087	14.5
	.00118	.03		.150	3.810	37/64	.57813	14.6844
	.00157	.04	5/32	.15625	3.9688		.59055	15
	.00197	.05		.15748	4	19/32	.59375	15.0812
	.002	.051	11/64	.17188	4.3656		.600	15.24
	.00236	.06		.1750	4.445	39/64	.60938	15.4781
	.00276	.07		.17717	4.5		.61024	15.5
	.003	.0762	3/16	.18750	4.7625	5/8	.6250	15.875
	.00315	.08		.19685	5		.62992	16
	.00354	.09		.20	5.08	41/64	.64063	16.2719
	.00394	.1	13/64	.20313	5.1594		.64961	16.5
	.004	.1016		.21654	5.5		.650	16.51
	.005	.1270	7/32	.21875	5.5562	21/32	.65625	16.6688
	.006	.1524		.2250	5.715		.66929	17
	.007	.1778	15/64	.23438	5.9531	43/64	.67188	17.0656
	.00787	.2		.23622	6	11/16	.68750	17.4625
	.008	.2032	1/4	.250	6.35		.68898	17.5
	.009	.2286					.700	17.78
	.00984	.25		.25591	6.5	45/64	.70313	17.8594
	.01	.254	17/64	.26563	6.7469		.70866	18
	.01181	.3		.275	6.985	23/32	.71875	18.2562
1/64	.01563	.3969		.27559	7		.72835	18.5
	.01575	.4	9/32	.28125	7.1438	47/64	.73438	18.6531
	.01969	.5		.29528	7.5		.74803	19
	.02	.508	19/64	.29688	7.5406	3/4	.750	19.050
	.02362	.6		.30	7.62			
	.025	.635	5/16	.3125	7.9375	49/64	.76563	19.4469
	.02756	.7		.31496	8		.76772	19.5
	.0295	.75	21/64	.32813	8.3344	25/32	.78125	19.8438
	.03	.762		.33465	8.5		.78740	20
1/32	.03125	.7938	11/32	.34375	8.7375	51/64	.79688	20.2406
	.0315	.8		.350	8.89		.800	20.320
	.03543	.9		.35433	9		.80709	20.5
	.03937	1	23/64	.35938	9.1281	13/16	.81250	20.6375
	.04	1.016		.37402	9.5		.82677	21
3/64	.04687	1.191	3/8	.375	9.525	53/64	.82813	21.0344
	.04724	1.2	25/64	.39063	9.9219	27/32	.84375	21.4312
	.05	1.27		.39370	10		.84646	21.5
	.05512	1.4		.400	10.16		.850	21.590
	.05906	1.5	13/32	.40625	10.3188	55/64	.85938	21.8281
	.06	1.524		.41339	10.5		.86614	22
1/16	.06250	1.5875	27/64	.42188	10.7156	7/8	.875	22.225
	.06299	1.6		.43307	11		.88583	22.5
	.06693	1.7	7/16	.43750	11.1125	57/64	.89063	22.6219
	.07	1.778		.450	11.430		.900	22.860
	.07087	1.8		.45276	11.5		.90551	23
	.075	1.905	29/64	.45313	11.5094	29/32	.90625	23.0188
5/64	.07813	1.9844	15/32	.46875	11.9062	59/64	.92188	23.4156
	.07874	2		.47244	12		.92520	23.5
	.08	2.032	31/64	.48438	12.3031	15/16	.93750	23.8125
	.08661	2.2		.49213	12.5		.94488	24
	.09	2.286	1/2	.50	12.7		.950	24.130
	.09055	2.3				61/64	.95313	24.2094
3/32	.09375	2.3812		.51181	13		.96457	24.5
	0.9843	2.5	33/64	.51563	13.0969	31/32	.96875	24.6062
	.1	2.54	17/32	.53125	13.4938		.98425	25
	.10236	2.6		.53150	13.5	63/64	.98438	25.0031
7/64	.10937	2.7781	35/64	.54688	13.8906	1	1.0000	25.4

TWIST DRILL FRACTIONAL, NUMBER, AND LETTER SIZES

Drill No.	Frac	Deci	Drill No.	Frac	Deci	Drill No.	Frac	Deci	Drill No.	Frac	Deci
80	—	.0135	42	—	.0935	7	—	.201	X	—	.397
79	—	.0145	—	3/32	.0938	—	13/64	.203	Y	—	.404
—	1/64	.0156				6	—	.204			
78	—	.0160	41	—	.0960	5	—	.206	—	13/32	.406
77	—	.0180	40	—	.0980	4	—	.209	Z	—	.413
			39	—	.0995				—	27/64	.422
76	—	.0200	38	—	.1015	3	—	.213	—	7/16	.438
75	—	.0210	37	—	.1040	—	7/32	.219	—	29/64	.453
74	—	.0225				2	—	.221			
73	—	.0240	36	—	.1065	1	—	.228	—	15/32	.469
72	—	.0250	—	7/64	.1094	A	—	.234	—	31/64	.484
			35	—	.1100				—	1/2	.500
71	—	.0260	34	—	.1110	—	15/64	.234	—	33/64	.516
70	—	.0280	33	—	.1130	B	—	.238	—	17/32	.531
69	—	.0292				C	—	.242			
68	—	.0310	32	—	.116	D	—	.246	—	35/64	.547
—	1/32	.0313	31	—	.120	—	1/4	.250	—	9/16	.562
			—	1/8	.125				—	37/64	.578
67	—	.0320	30	—	.129	E	—	.250	—	19/32	.594
66	—	.0330	29	—	.136	F	—	.257	—	39/64	.609
65	—	.0350				G	—	.261			
64	—	.0360	—	9/64	.140	—	17/64	.266	—	5/8	.625
63	—	.0370	28	—	.141	H	—	.266	—	41/64	.641
			27	—	.144				—	21/32	.656
62	—	.0380	26	—	.147	I	—	.272	—	43/64	.672
61	—	.0390	25	—	.150	J	—	.277	—	11/16	.688
60	—	.0400				—	9/32	.281			
59	—	.0410	24	—	.152	K	—	.281	—	45/64	.703
58	—	.0420	23	—	.154	L	—	.290	—	23/32	.719
			—	5/32	.156				—	47/64	.734
57	—	.0430	22	—	.157	M	—	.295	—	3/4	.750
56	—	.0465	21	—	.159	—	19/64	.2297	—	49/64	.766
—	3/64	.0469				N	—	.302			
55	—	.0520	20	—	.161	—	5/16	.313	—	25/32	.781
54	—	.0550	19	—	.166	O	—	.316	—	51/64	.797
			18	—	.170				—	13/16	.813
53	—	.0595	—	11/64	.172	P	—	.323	—	53/64	.828
—	1/16	.0625	17	—	.173	—	21/64	.328	—	27/32	.844
52	—	.0635				Q	—	.332			
51	—	.0670				R	—	.339			
50	—	.0700	16	—	.177	—	11/32	.344	—	55/64	.859
			15	—	.180				—	7/8	.875
49	—	.0730	14	—	.182	S	—	.348	—	57/64	.891
48	—	.0760	13	—	.185	T	—	.358	—	29/32	.906
—	5/64	.0781	—	3/16	.188	—	23/64	.359	—	59/64	.922
47	—	.0785				U	—	.368			
46	—	.0810	12	—	.189	—	3/8	.375	—	15/16	.938
			11	—	.191				—	61/64	.953
45	—	.0820	10	—	.194	V	—	.377	—	31/32	.969
44	—	.0860	9	—	.196	W	—	.386	—	63/64	.984
43	—	.0890	8	—	.199	—	25/64	.391	—	1	1.000

METRIC SCREW THREADS

Coarse (general purpose)		Fine	
Nom Size & Thd Pitch	Tap Drill Dia (mm)	Nom Size & Thd Pitch	Tap Drill Dia (mm)
M1.6 × 0.35	1.25	—	—
M1.8 × 0.35	1.45	—	—
M2 × 0.4	1.6	—	—
M2.2 × 0.45	1.75	—	—
M2.5 × 0.45	2.05	—	—
M3 × 0.5	2.50	—	—
M3.5 × 0.6	2.90	—	—
M4 × 0.7	3.30	—	—
M4.5 × 0.75	3.75	—	—
M5 × ..8	4.20	—	—
M6.3 × 1	5.30	—	—
M7 × 1	6.00	—	—
M8 × 1.25	6.80	M8 ö 1	7.00
M9 × 1.25	7.75		
M10 × 1.5	8.50	M10 ö 1.25	8.75
M11 × 1.5	9.50		
M12 × 1.75	10.30	M12 ö 1.25	10.50
M14 × 2	12.00	M14 ö 1.5	12.50
M16 × 2	14.00	M16 ö 1.5	14.50
M18 × 2.5	15.50	M18 ö 1.5	16.50
M20 × 2.5	17.50	M20 ö 1.5	18.50
M22 × 2.5	19.50	M22 ö 1.5	20.50
M24 × 3	21.00	M24 ö 2	22.00
M27 × 3	24.00	M27 ö 2	25.00
M30 × 3.5	26.50	M30 ö 2	28.00
M33 × 3.5	29.50	M30 ö 2	31.00
M36 × 4	32.00	M36 ö 3	33.00
M39 × 4	35.00	M39 ö 3	36.00
M42 × 4.5	37.50	M42 ö 3	39.00
M45 × 4.5	40.50	M45 ö 3	42.00
M48 × 5	43.00	M48 ö 3	45.00
M52 × 5	47.00	M52 ö 3	49.00
M56 × 5.5	50.50	M56 ö 4	52.00
M60 × 5.5	54.50	M60 ö 4	56.00
M64 × 6	58.00	M64 ö 4	60.00
M68 × 6	62.00	M68 ö 4	64.00
M72 × 6	66.00	—	—
M80 × 6	74.00	—	—
M90 × 6	84.00	—	—
M100 × 6	94.00	—	—

DRILLED HOLE TOLERANCES

Drill Size	Tolerance*	
	Plus	Minus
.0135 (No. 80) – .185 (No. 13)	.003	.002
.1875 – .246 (D)	.004	.002
.250 (E) – .750	.005	.002
.756 – 1.000	.007	.003
1.0156 – 2.000	.010	.004
2.0312 – 3.500	.015	.005

* generally accepted tolerances for good practice

STANDARD SERIES THREADS – GRADED PITCHES

Nominal Diameter	UNC		UNF		UNEF	
	TPI	Tap Drill	TPI	Tap Drill	TPI	Tap Drill
0 (.0600)			80	3/64		
1 (.0730)	64	No. 53	72	No. 53		
2 (.0860)	56	No. 50	64	No. 50		
3 (.0990)	48	No. 47	56	No. 45		
4 (.1120)	40	No. 43	48	No. 42		
5 (.1250)	40	No. 38	44	No. 37		
6 (.1380)	32	No. 36	40	No. 33		
8 (.1640)	32	No. 29	36	No. 29		
10 (.1900)	24	No. 25	32	No. 21		
12 (.2160)	24	No. 16	28	No. 14	32	No. 13
1/4 (.2500)	20	No. 7	28	No. 3	32	7/32
5/16 (.3125)	18	F	24	I	32	9/32
3/8 (.3750)	16	5/16	24	Q	32	11/32
7/16 (.4375)	14	U	20	25/64	28	13/32
1/2 (.5000)	13	27/64	20	29/64	28	15/32
9/16 (.5625)	12	31/64	18	33/64	24	33/64
5/8 (.6250)	11	17/32	18	37/64	24	37/64
11/16 (.6875)					24	41/64
3/4 (.7500)	10	21/32	16	11/16	20	45/64
13/16 (.8125)					20	49/64
7/8 (.8750)	9	49/64	14	13/16	20	53/64
15/16 (.9375)					20	57/64
1 (1.000)	8	7/8	12	59/64	20	61/64

METRIC DRILL SIZES

Drill Diameter						Drill Diameter					
mm	in.	mm	in.	mm	in.	mm	in.	mm	in.	mm	in.
.40	.0157	1.95	.0768	4.70	.1850	8.00	.3150	13.20	.5197	25.50	1.0039
.42	.0165	2.00	.0787	4.80	.1890	8.10	.3189	13.50	.5315	26.00	1.0236
.45	.0177	2.05	.0807	4.90	.1929	8.20	.3228	13.80	.5433	26.50	1.0433
.48	.0189	2.10	.0827	5.00	.1969	8.30	.3268	14.00	.5512	27.00	1.0630
.50	.0197	2.15	.0846	5.10	.2008	8.40	.3307	14.25	.5610	27.50	1.0827
.55	.0217	2.20	.0866	5.20	.2047	8.50	.3346	14.50	.5709	28.00	1.1024
.60	.0236	2.25	.0886	5.30	.2087	8.60	.3386	14.75	.5807	28.50	1.1220
.65	.0256	2.30	.0906	5.40	.2126	8.70	.3425	15.00	.5906	29.00	1.1417
.70	.0276	2.35	.0925	5.50	.2165	8.80	.3465	15.25	.6004	29.50	1.1614
.75	.0295	2.40	.0945	5.60	.2205	8.90	.3504	15.50	.6102	30.00	1.1811
.80	.0315	2.45	.0965	5.70	.2244	9.00	.3543	15.75	.6201	30.50	1.2008
.85	.0335	2.50	.0984	5.80	.2283	9.10	.3583	16.00	.6299	31.00	1.2205
.90	.0354	2.60	.1024	5.90	.2323	9.20	.3622	16.25	.6398	31.50	1.2402
.95	.0374	2.70	.1063	6.00	.2362	9.30	.3661	16.50	.6496	32.00	1.2598
1.00	.0394	2.80	.1102	6.10	.2402	9.40	.3701	16.75	.6594	32.50	1.2795
1.05	.0413	2.90	.1142	6.20	.2441	9.50	.3740	17.00	.6693	33.00	1.2992
1.10	.0433	3.00	.1181	6.30	.2480	9.60	.3780	17.25	.6791	33.50	1.3189
1.15	.0453	3.10	.1220	6.40	.2520	9.70	.3819	17.50	.6890	34.00	1.3386
1.20	.0472	3.20	.1260	6.50	.2559	9.80	.3858	18.00	.7087	34.50	1.3583
1.25	.0492	3.30	.1299	6.60	.2598	9.90	.3898	18.50	.7283	35.00	1.3780
1.30	.0512	3.40	.1339	6.70	.2638	10.00	.3937	19.00	.7480	35.50	1.3976
1.35	.0531	3.50	.1378	6.80	.2677	10.20	.4016	19.50	.7677	36.00	1.4173
1.40	.0551	3.60	.1417	6.90	.2717	10.50	.4134	20.00	.7874	36.50	1.4370
1.45	.0571	3.70	.1457	7.00	.2756	10.80	.4252	20.50	.8071	37.00	1.4567
1.50	.0591	3.80	.1496	7.10	.2795	11.00	.4331	21.00	.8268	37.50	1.4764
1.55	.0610	3.90	.1535	7.20	.2835	11.20	.4409	21.50	.8465	38.00	1.4961
1.60	.0630	4.00	.1575	7.30	.2874	11.50	.4528	22.00	.8661	40.00	1.5748
1.65	.0650	4.10	.1614	7.40	.2913	11.80	.4646	22.50	.8858	42.00	1.6535
1.70	.0669	4.20	.1654	7.50	.2953	12.00	.4724	23.00	.9055	44.00	1.7323
1.75	.0689	4.30	.1693	7.60	.2992	12.20	.4803	23.50	.9252	46.00	1.8110
1.80	.0709	4.40	.1732	7.70	.3031	12.50	.4921	24.00	.9449	48.00	1.8898
1.85	.0728	4.50	.1772	7.80	.3071	12.80	.5039	24.50	.9646	50.00	1.9685
1.90	.0748	4.60	.1811	7.90	.3110	13.00	.5118	25.00	.9843		

POUNDS PER SQUARE INCH – KILOPASCALS CONVERSION TABLE*

Pounds Per Square Inch to Kilopascals (1 lb/in.^{2} = 6.894757 kPa)

lb/in.^{2}	0	1	2	3	4	5	6	7	8	9
0	—	6.895	13.790	20.864	27.579	34.474	41.369	48.263	55.158	62.053
10	68.948	75.842	82.737	89.632	96.527	103.421	110.316	117.211	124.106	131.000
20	137.895	144.790	151.685	158.579	165.474	172.369	179.264	186.158	193.053	199.948
30	206.843	213.737	220.632	227.527	234.422	241.316	248.211	255.106	262.001	268.896
40	275.790	282.685	289.580	296.475	303.369	310.264	317.159	324.054	330.948	337.843
50	344.738	351.633	358.527	365.422	372.317	379.212	386.106	393.001	399.896	406.791
60	413.685	420.580	427.475	434.370	441.264	448.159	455.054	461.949	468.843	475.738
70	482.633	489.528	496.423	503.317	510.212	517.107	524.002	530.896	537.791	544.686
80	551.581	558.475	565.370	572.265	579.160	586.054	592.949	599.844	606.739	613.633
90	620.528	627.423	634.318	641.212	648.107	655.002	661.897	668.791	675.686	682.581
100	689.476	696.370	703.265	710.160	717.055	723.949	730.844	737.739	744.634	751.529

Kilopascals to Pounds Per Square Inch (1 kPa = 0.1450377 lb/in.^{2})

kPa	0	1	2	3	4	5	6	7	8	9
0	—	0.145	0.290	0.435	0.580	0.725	0.870	1.015	1.160	1.305
10	1.450	1.595	1.740	1.885	2.031	2.176	2.321	2.466	2.611	2.756
20	2.901	3.046	3.191	3.336	3.481	3.626	3.771	3.916	4.061	4.206
30	4.351	4.496	4.641	4.786	4.931	5.076	5.221	5.366	5.511	5.656
40	5.802	5.947	6.092	6.237	6.382	6.527	6.672	6.817	6.962	7.107
50	7.252	7.397	7.542	7.687	7.832	7.977	8.122	8.267	8.412	8.557
60	8.702	8.847	8.992	9.137	9.282	9.427	9.572	9.718	9.863	10.008
70	10.153	10.298	10.443	10.588	10.733	10.878	11.023	11.168	11.313	11.458
80	11.603	11.748	11.893	12.038	12.183	12.328	12.473	12.618	12.763	12.908
90	13.053	13.198	13.343	13.489	13.634	13.779	13.924	14.069	14.214	14.359
100	14.504	14.649	14.794	14.939	15.084	15.229	15.374	15.519	15.664	15.809

* 1 kPa = 1 kilonewton/m^2

AMBIENT PRESSURE AND ENGINE COOLANT* BOILING POINT

PSI Above Atmospheric Pressure (14.7) PSI	Boiling Point**
1	229
2	232
3	235
4	238
5	241
6	244
7	247
8	249
9	251
10	253
11	255
12	257
13	259
14	261
15	264
16	266
17	268
18	271
19	274
20	278

* 50/50 mixture of antifreeze and water
** in °F

OHM'S LAW

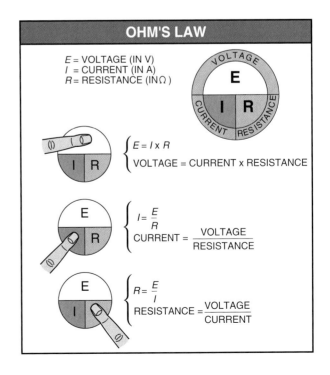

E = VOLTAGE (IN V)
I = CURRENT (IN A)
R = RESISTANCE (IN Ω)

$E = I \times R$
VOLTAGE = CURRENT × RESISTANCE

$I = \dfrac{E}{R}$
CURRENT = $\dfrac{\text{VOLTAGE}}{\text{RESISTANCE}}$

$R = \dfrac{E}{I}$
RESISTANCE = $\dfrac{\text{VOLTAGE}}{\text{CURRENT}}$

MEASURES

Linear
12 inches = 1 foot
3 feet = 1 yard
36 inches = 1 yard
5.5 yards = 1 rod
16.5 feet = 1 rod
40 rods = 1 furlong
660 feet = 1 furlong
8 furlongs = 1 mile
320 rods = 1 mile
1760 yards = 1 mile
5280 feet = 1 mile

Angular
60 seconds = 1 minute
60 minutes = 1 degree
57.3 degrees = 1 radian
180 degrees = π radians
360 degrees = 2π radians

Time
60 seconds = 1 minute
60 minutes = 1 hour
24 hours = 1 day
7 days = 1 week
52 weeks = 1 year
365.26 days = 1 year

Avoirdupois
437.5 grains = 1 ounce
16 ounces = 1 pound
100 pounds = 1 hundredweight
1000 pounds = 1 kip
2 kips = 1 ton
2000 pounds = 1 ton
2240 pounds = 1 long ton

Dry
2 pints = 1 quart
4 quarts = 1 gallon
2 gallons = 1 peck
8 quarts = 1 peck
4 pecks = 1 bushel

Liquid
4 gills = 1 pint
2 pints = 1 quart
57.75 cubic inches = 1 quart
4 quarts = 1 gallon
231 cubic inches = 1 gallon
31.5 gallons = 1 barrel

Square
144 square inches = 1 square foot
9 square feet = 1 square yard
1296 square inches = 1 square yard
30.25 square yards = 1 square rod
160 square rods = 1 acre
4840 square yards = 1 acre
43,560 square feet = 1 acre
640 acres = 1 square mile

Cubic
7.48 gallons = 1 cubic foot
1728 cubic inches = 1 cubic foot
27 cubic feet = 1 cubic yard
202 gallons = 1 cubic yard
128 cubic feet = 1 cord

Surveyor's Linear
7.92 inches = 1 link
16.5 feet = 1 rod
25 links = 1 rod
4 rods = 1 chain
66 feet = 1 chain
100 links = 1 chain
80 chains = 1 mile

Surveyor's Square
625 square links = 1 square rod
16 square rods = 1 square chain
10 square chains = 1 acre
640 acres = 1 square mile
1 square mile = 1 section
36 square miles = 1 township
36 sections = 1 township

WARRANTY CLAIM

BRIGGS & STRATTON WARRANTY CLAIM — A2121231

Please TYPE or PRINT CLEARLY

1 Type of Claim (check 1 only)
- Warranty Repair
- New Defective Service Parts
- Questionable/Disputed
- Policy Adjustment*

*Requires Authorization Number

Purchase Date Mo./Yr.
Failure Date Mo./Yr.
Hours Used
Repair Date Mo./Yr.

2 Owner Information (check 1 only): Consumer / Commercial / Store / Equipment Manufacturer
First Name
Last Name or Store Name
Address
City / State
Zip Code / Phone
Customer Signature

3 Warranty Performed by:
Briggs & Stratton Warranty ID No.
Dealer Name
Address
City / State
Phone / Zip Code
Dealer Signature (Certifies Accuracy of All Statements Herein)

4 Model Number — Engine or Short Block / Type Number Engine / Code Number Engine / Equipment Manufacturer / Type of Equipment (See reverse side for Code Numbers) / Unit Originally Sold by (Store Name)

5 Defective Part Reference Number (1 number only—see Illustrated Parts List)

DEFECT CODES (In Box 6 Enter Defect Code That Best Describes Condition Found)
- AW—Assembled Wrong
- BC—Broken/Cracked
- BL—Blown
- BT—Burnt
- BW—Bent/Warped
- CD—Casting Deficiency/Porous
- CL—Came Loose/Off
- CP—Corroded/Pitted
- EF—Electrical Failure
- FM—Foreign Material
- LK—Leaked
- LM—Loose Magnets
- MI—Missing
- NS—Not Seating
- OA—Out of Adjustment
- PA—Paint
- PM—Part Made/Machined Incorrectly
- SG—Scored/Galled
- SS—Stuck/Seized
- ST—Stripped
- UO—Unknown/Other
- VC—Valve Clearance

6 Defect Code (1 defect code only)

7 Condition Found and Probable Cause of Defect: (Word "Defective" Not Sufficient)

8 Warranty Work Performed (Itemize Any Miscellaneous Charges, If Applicable)

9 Part Number / Qty. / Description — Do Not Enter Engine or Short Block Part Numbers Here (use Box 12)

10 Labor Required — Hrs. / Mins.
- Repair
- Remove & Replace Engine
- Total

11 Miscellaneous Charges — Dollars / Cents
Total Miscellaneous Charges, Itemized in Box 8

12 Short Block Part Number (Requires Authorization Number) / Qty.

Service Replacement Engine Part Number (Requires Authorization Number)
Model Number / Type Number / Qty.

13 Authorization Number

14 For Factory Use Only / F/C

ORIGINAL: RETURN TO FACTORY BLUE: DEALER COPY

Retain and Tag All Material Until Warranty Reimbursement Is Received

FAILURE ANALYSIS...

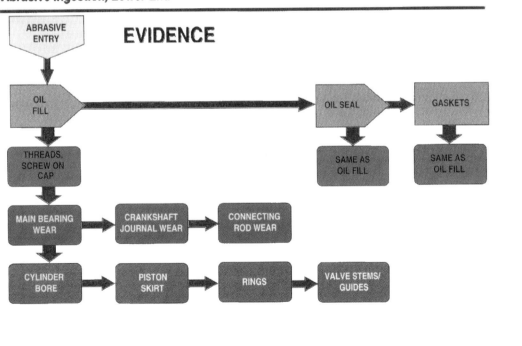

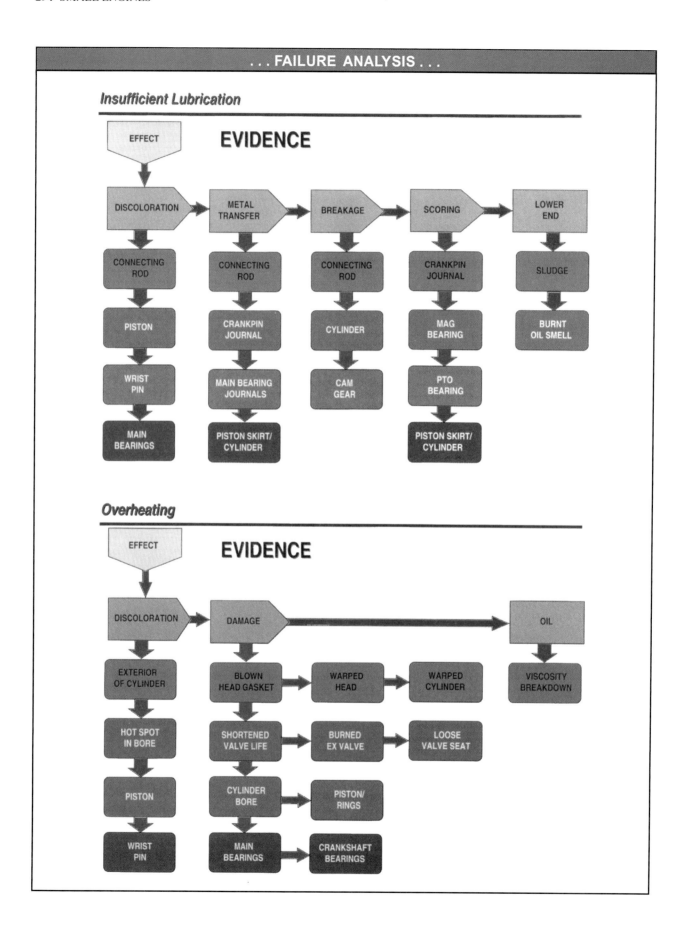

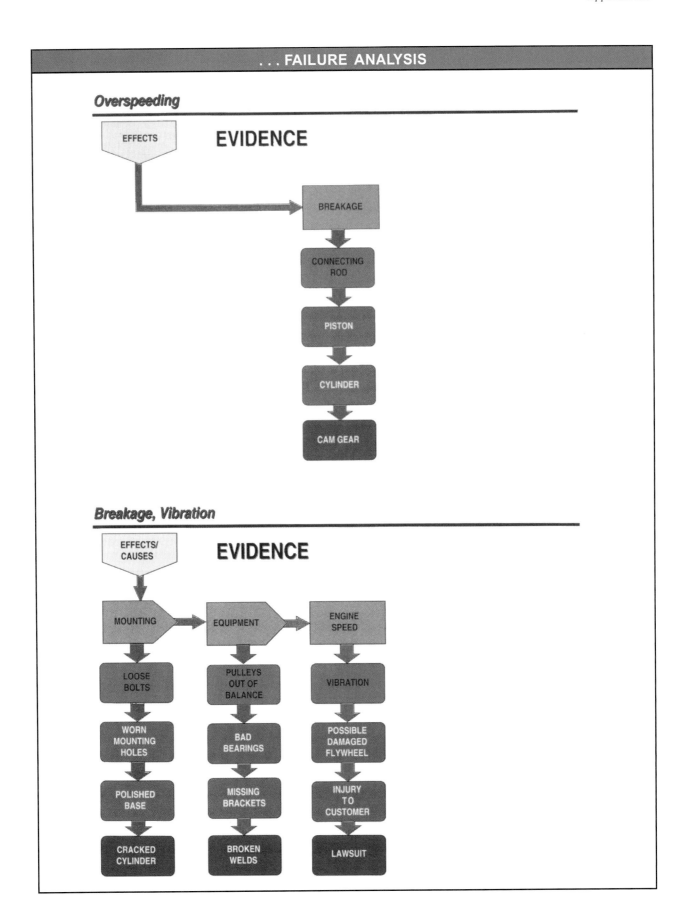

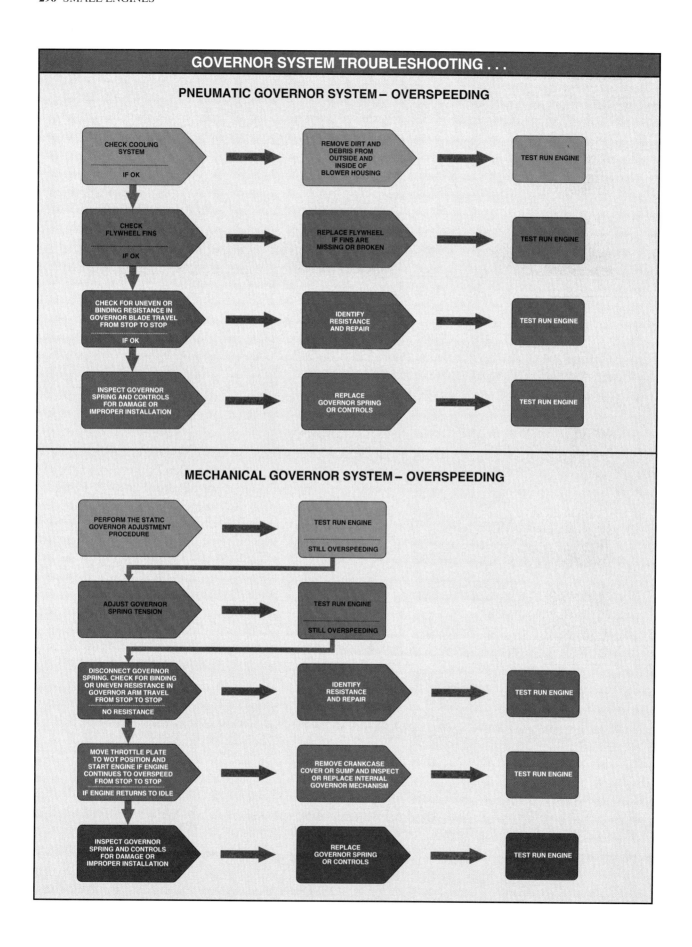

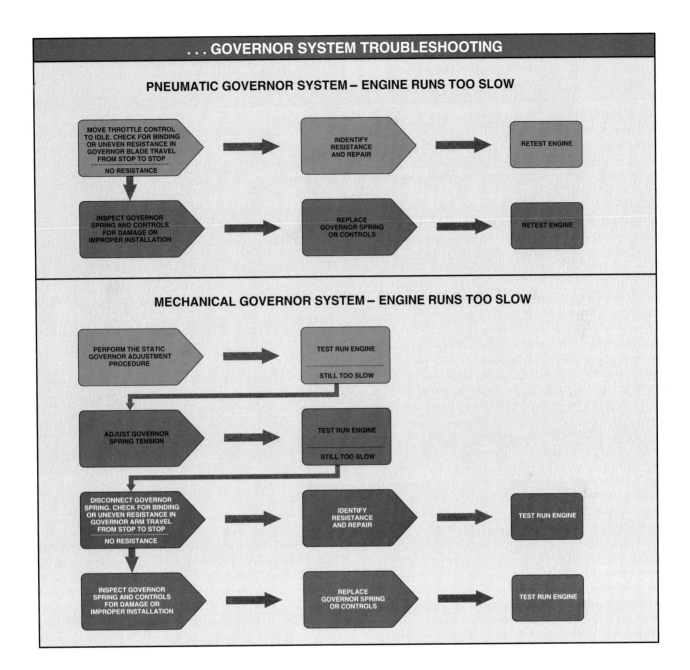

VANGUARD™ THREE-CYLINDER GASOLINE ENGINE MAINTENANCE SCHEDULE

System	Maintenance Operation	Daily	Every 50 Hours	Every 100 Hours	Every 200 Hours	Every 600 Hours	Yearly
Lubrication System	Check oil level	✓					
	Check for oil leaks	✓					
	Change oil		✓[1]		✓[2,4]		
	Change oil filter		✓[1]		✓[2,4]		
Cooling System	Check coolant	✓					
	Change coolant						✓
	Check fan belt			✓			
Engine	Service air cleaner		✓[1]	✓[2,4]		✓[3,4]	
	Check cylinder head bolt torque					✓	
	Check valve clearance					✓	
Electrical System	Check battery electrolyte				✓		
	Change spark plugs						✓
Fuel System	Change fuel filter					✓	

[1] Perform first maintenance operation after 50 hours
[2] Then perform maintenance operation at this interval
[3] Replace after every 600 hours of operation
[4] Service more often when operating under heavy load or in high temperatures

BRIGGS & STRATTON MASTER SERVICE TECHNICIAN PROGRAM

Becoming A Master Service Technician

There are many reasons for becoming a Master Service Technician (MST). Most important from a business standpoint is the customers' perception of greater value. Imagine yourself faced with making a decision on who to trust for service on your powered equipment – a dealership advertising a Briggs & Stratton Master Service Technician or one that cannot make that claim. Other service industries have repeatedly proven that customers looking for reliable, efficient, cost-effective service will invariably choose establishments that have proven themselves through some method of certification.

Another equally important reason to achieve MST status relates to the new dealer category of Master Sales and Service Dealer which requires an MST on the service staff. This creates an ongoing need for the MST. And from a personal view, there is the gratification that comes from knowing that you are among an elite group of service technicians that have PROVEN they are the "Best of the Best."

Master Service Technician Exam

The MST exam is a comprehensive test of your Briggs & Stratton product knowledge. This means that it is important to study and understand not just how to solve problems systematically but also troubleshooting, parts identification, 4 cycle theory, warranty policy and procedures.

The complete exam takes about 4.5 hours. It will be given by a representative of the regional Central Sales and Service Distributor (CSSD). The MST exam may be given at the CSSD facility or in the field at the direction of the CSSD. Contact your CSSD for details. All materials needed for the exam will be supplied to you at the test site.

There are 311 questions in the exam. Your grade will be based on the amount of correct answers divided by the 311 possible. Therefore, time management is at a premium. No single question should use up an excessive amount of time. If you have difficulty answering a question, continue through the rest of the section and complete those that are less difficult for you. Once you have completed the less difficult questions, return to those that gave you difficulty and finish what you can. A guess at any answer is better than leaving it blank. **Any unanswered questions count as an incorrect answer.**

Again, as in all of the sections of the exam, it is important to budget your time an not spend a great deal of time on any one question. Remember that your score is based on the amount of correct responses divided by the total possible responses. Answer all of the less difficult questions first, for the best chance at a passing score.

Exam Preparation

It is recommended by the Customer Education Department at Briggs & Stratton that the first step before attempting the exam is to have attended the four day, Authorized Field Service School (within the past four years), available at your Central Sales and Service Distributor. This is an excellent educational experience in preparation for the MST exam.

Familiarity with the current Single Cylinder "L" Head Repair Manual (part #270962) and Opposed Twin Cylinder Repair Manual (part #271172) is also important. This portion of the MST exam is mostly "open book" and the test administrator will supply the necessary repair manuals for use during the exam. What you need to study in the current Single Cylinder "L" Head Repair Manual (part #270962) and Opposed Twin Cylinder Repair Manual (part #271172) is the format, layout, and location of the various sections. **Do not however, spend endless hours reading and committing to memory, the specifications and procedures found in these manuals.** Because the exam is timed, it is important to be able to access information quickly and accurately.

It is important to know your way around the Briggs & Stratton microfiche card set, and the use of microfiche readers. You may be asked to find part numbers for a specific engine as well as service replacement engines and/or short block part numbers. All of the microfiche cards could be used during the exam including the less used cards containing service bulletins, decal kits, crankshaft conversion data, etc. **All materials and microfiche readers will be supplied by the test administrator.**

Other excellent study resources would be the annual Technical Update Seminar notebooks from the past few years. Subjects covered in these notebooks and the information regarding service procedures and theory will be very helpful during the exam. It will not be permissible to bring the Update Seminar notebooks or this Information and Study Guide with you on exam day.

Other areas and/or literature with which you should be familiar are, Engine Failure Analysis wall poster (MS-8310) and Alternator Specification and Identification wall poster (MS-2288). It is also important to familiarize yourself with the Warranty Policy and Procedure sheet (MS-2860 or MS-4259) and the Warranty Claim Form (MS-2900-1/93).

Additional Information

Additional information regarding the MST exam can be obtained by contacting a Briggs & Stratton CSSD representative or the Briggs & Stratton Customer Education Department, P.O. Box 702, Milwaukee, WI 53201 – 0702.

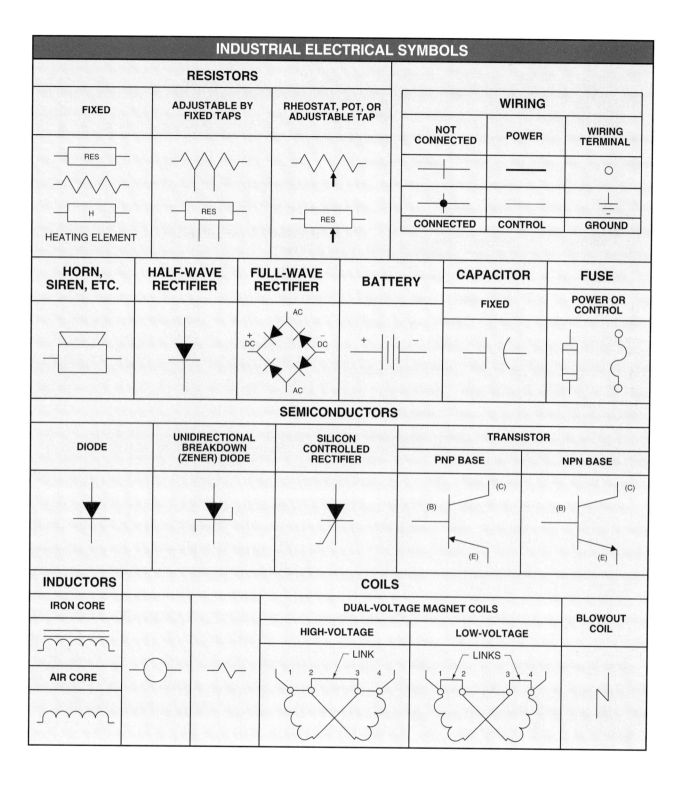

GLOSSARY

The terms in this glossary are defined as they relate to small engines.

abrasive ingestion: A cause of engine failure through the undesirable introduction of abrasive particles into a small engine.

abrasive particle: A particle with enough hardness to cause the grinding or wearing away of material through friction.

abrasive wear: Premature wear caused by abrasive particles which remove material from engine components through friction.

accident report: A document that details facts about an accident in the facility.

AC sine wave: A symmetrical waveform that contains 360°.

AC voltage test: A test that uses a DMM to indicate the voltage potential of the alternator stator.

adhesive wear: Premature wear from insufficient lubrication caused by an inadequate oil film to prevent undesirable wear from metal-to-metal contact.

adiabatic process: A process in which heat is derived from the process itself.

adjustable fixed speed control: A control mounted on the engine that uses a speed adjustment nut to maintain an engine at a specific top no-load speed.

adjustable orifice jet: An assembly used to regulate passage of a fluid through an opening.

afterfire: An engine condition that occurs when the engine continues to operate after the ignition switch is shut OFF.

air bleed: A passage in the carburetor that directs air and atmospheric pressure into the main and idle circuits to facilitate the mixture of air and fuel.

air-cooled engine: An engine that circulates air around the cylinder block and cylinder head to maintain the desired engine temperature.

air density: The mass of air per unit volume.

air guide: A sheet metal component used to direct cooling air from the blower housing to the cylinder cooling fins.

alcohol: A fuel, or fuel additive, used to enhance the octane rating of gasoline.

alternating current (AC): The flow of electrons that reverses direction at regular intervals.

alternator: A charging system device that produces AC voltage and amperage.

aluminum: A nonferrous metal commonly alloyed with zinc or copper.

aluminum valve guide: An integrally machined part of an aluminum cylinder block and is not a separate engine component.

American Petroleum Institute (API): A trade association of the United States petroleum industry.

American Society for Testing and Materials (ASTM): The largest organization in the world devoted to developing and publishing voluntary, full-consensus standards.

American Society of Agricultural Engineers (ASAE): A professional and technical organization of members interested in engineering knowledge and technology in food and agriculture, associated industries, and related resources.

amperes: The measurement of the number of electrons flowing through a conductor per unit of time.

ANSI International: A national organization that helps identify industrial and public needs for national standards.

anti-afterfire solenoid: A device that shuts OFF the fuel at the carburetor to prevent the engine from receiving fuel after the ignition switch is turned OFF.

antifreeze: An ethylene glycol chemical mixture used with water to lower the freezing point of engine coolant.

antifriction bearing: A bearing that contains moving elements to provide a low friction support surface for rotating or sliding surfaces.

antiknock index (AKI): The numerical value assigned to gasoline that indicates the ability to eliminate knocking and/or pinging in an operating engine.

applied load: A resistive force opposing engine forces.

applied pressure: Pressure applied from combustion gases to the piston ring, causing it to expand.

area: The number of unit squares equal to the surface of an object.

armature: A rotating part of a DC motor that consists of wire wound around an armature shaft.

asperities: Tiny projections from the machining process which produce surface roughness or unevenness.

asymmetrical: A shape in which one-half is not the mirror image of the other half.

atom: A small unit of a material that consists of protons, electrons, and neutrons.

austenitic steel: A heat-resistive metal alloy consisting of cobalt, tungsten, and chromium.

autoignition: The spontaneous combustion of the charge commonly caused by low octane fuel or excessive compression ratio.

automatic switch: A switch that stops the flow of current any time current limits are reached.

axial load: A load applied parallel to the shaft.

babbitt: A nonferrous metal alloy consisting of copper, lead, and tin or lead and tin.

barrel-faced compression ring: A piston ring that has a curved running surface to provide consistent lubrication of the piston ring and cylinder wall.

battery: An electrical energy storage device.

battery loading device: An electrical test tool that applies an electrical load to the battery while measuring amperage and voltage.

bearing: A component used to reduce friction and to maintain clearance between stationary and rotating components of an engine.

bearing journal: A precision ground surface on which the crankshaft rotates.

Bernoulli's principle: A principle in which air flowing through a narrowed portion of a tube increases in velocity and decreases in pressure.

blower housing: A sheet metal or composite material component that encompasses the fan to direct cooling air to the cylinder block and cylinder head.

blown head gasket: An undesirable degradation of the head gasket which allows leakage to and from areas of the combustion chamber.

body: When referring to force, anything with mass.

bonding: The use of metal-to-metal contact or a wire between two containers to prevent possible ignition from static electricity sparks.

booster fan: An engine component mounted on the PTO that consists of a hub with bosses and tapped holes for connecting driven equipment components.

bore: The diameter of the cylinder bore.

bottom dead center (BDC): The point at which the piston is farthest from the cylinder head.

boundary lubrication: Lubrication where a full oil film is not always present to separate engine components.

bowl-style carburetor: A carburetor that has a fuel reservoir (bowl) located in the carburetor.

bowl vent: A passage drilled into the carburetor connecting the fuel bowl to the atmosphere.

brake horsepower (BHP): The amount of usable power taken from an engine.

brake mean effective pressure (BMEP): The average effective pressure placed on a piston during one complete operating cycle.

brass valve guide: A separate machined valve guide insert manufactured from brass alloy.

breakage: A cause of engine failure from fracturing or failure of an engine component or components.

break-away clutch: An engine component that allows slippage to prevent damage to the pinion gear during a misfire or unexpected reverse rotation.

breaker point ignition system: An ignition system that uses a mechanical switch to control timing of ignition.

breaker points: An ignition system component that has two points (contact surfaces) that function as a mechanical switch.

break-in: The process that causes the running surfaces of piston rings and the surface of the cylinder bore to conform to each other.

break-in period: The period of time required for the running surfaces of piston rings and the surface of the cylinder bore to conform to one another after initial startup.

bridge rectifier: A device that uses four interconnected diodes to change one cycle of AC current into two DC pulses.

British thermal unit (Btu): The amount of heat energy required to raise the temperature of 1 pound (lb) of water 1°F (Fahrenheit).

brushes: Carbon components in contact with the commutator that carry battery current to operate the starter motor.

bronze: A nonferrous metal alloy that consists of brass and zinc.

calorie: The amount of heat energy required to change the temperature of one gram of water 1°C.

cam gear: The portion of the camshaft that meshes with the crankgear.

cam lobe: An egg-shaped protrusion on the camshaft that moves a tappet to open a valve.

camshaft: An engine component that includes the cam gear, cam lobes, and bearing surfaces.

Canadian Standards Association (CSA): A Canadian national organization that develops standards and provides facilities for certification testing to national and international standards.

capacitor: An electrical component that stores voltage.

carbon monoxide (CO): A toxic (poisonous) gas produced by incomplete combustion of gasoline (HC-based fuels).

carburetor: An engine component that provides the required air-fuel mixture to the combustion chamber based on engine operating speed and load.

charge: The volume of compressed air-fuel mixture trapped inside the combustion chamber ready for ignition.

charging system: A system that replenishes the electrical power drawn from the battery during starting and accessory operation.

check ball: A component that functions as a one-way valve to allow fuel to flow in one direction only.

check valve: A valve that allows the flow of material in one direction.

chemical hazard: A solid, liquid, gas, mist, dust, and/or vapor that is toxic when inhaled, absorbed, or ingested.

choke plate: A flat plate placed in the carburetor body between the throttle plate and air intake that restricts air flow to help start a cold engine.

chromium plating: A piston ring surface treatment that adds a layer of chromium to increase hardness and durability.

circuit: A complete path that controls the rate and direction of electron flow on which voltage is applied.

cleaning tank: A tank used for cleaning parts in flammable solvents with a lid that automatically closes to contain flames during a fire.

code: A regulation or minimum requirement.

coefficient of thermal expansion: The unit change in dimension of a material by changing the temperature 1°F.

cohesion: The molecular attraction by which atoms and molecules are united throughout the mass.

coil: A circular wound wire (winding) consisting of insulated conductors arranged to produce lines of magnetic flux.

cold cranking amps (CCA): The number of amps produced by the battery for 30 sec at 0°F while maintaining 1.2 V per cell.

combination engine failure: An engine failure caused by more than one cause.

combustible liquid: A liquid that has a flash point at or above 100°F.

combustion: The rapid, oxidizing chemical reaction in which a fuel chemically combines with oxygen in the atmosphere and releases energy in the form of heat.

commutator: A sectional piece of copper that is directly connected to many loops of copper wire in contact with brushes.

compression: The process of reducing or squeezing a charge from a large volume to a smaller volume in the combustion chamber.

compression event: An engine operation event in which the trapped air-fuel mixture, or just air on some engines, is compressed inside the cylinder.

compression ignition engine: An engine that ignites fuel by compression.

compression ratio: A comparison of the volume of the combustion chamber with the piston at BDC to the volume of the combustion chamber with the piston at TDC.

compression release system: A system that relieves excess pressure during the compression event by allowing a small amount of compressed gas to be released through the muffler or carburetor.

compression ring: The piston ring located in the ring groove closest to the piston head.

condenser: A capacitor used in an ignition system that stores voltage and resists any change in voltage.

conduction: Heat transfer that occurs from atom to atom when molecules come in direct contact with each other, and through vibration, when kinetic energy is passed from atom to atom.

conductor: A material that allows the free flow of electrons.

connecting rod: An engine component that transfers motion from the piston to the crankshaft and functions as a lever arm.

Consumer Product Safety Commission (CPSC): A federal commission empowered to implement consumer safety standards throughout the United States.

convection: Heat transfer that occurs when heat is transferred by currents in a fluid.

convolution: An irregularly-shaped ridge or pocket that acts as a small reservoir for lubricating oil.

cooling air plenum: A duct made from sheet metal, plastic, or similar materials that provides a specific path for the cooling air to enter the engine cooling system.

cooling fan: An engine component that supplies cooling air to the engine when rotated.

cooling fin: An integral thin cast strip designed to provide efficient air circulation and dissipation of heat away from the engine cylinder block into the air stream.

counterweight: A protruding mass integrally cast into the crankshaft which partially balances the forces of a reciprocating piston and reduces the load on crankshaft bearing journals.

crankcase: An engine component that houses and supports the crankshaft.

crankcase breather: An engine component that relieves crankcase pressure created by the reciprocating motion of the piston during engine operation.

crankcase cover: An engine component that provides access to internal parts in the crankcase and supports the crankshaft.

crankgear: A gear located on the crankshaft that is used to drive other parts of an engine.

crankpin journal: A precision ground surface that provides a rotating pivot point to attach the connecting rod to the crankshaft.

crankshaft: An engine component that converts linear (reciprocating) motion of the piston into rotary motion.

current: The flow of electrons moving past a point in a circuit.

cycle: One complete wave of alternating voltage that contains 360°.

cylinder block: The engine component which consists of the cylinder bore, cooling fins on air-cooled engines, and valve train components, depending on the engine design.

cylinder bore: A hole in an engine block that aligns and directs the piston during movement.

cylinder bore distortion: An engine condition caused by excessive temperature variations in the combustion chamber and the inability to adequately cool a portion of the cylinder bore.

cylinder head: A cast aluminum alloy or cast iron engine component fastened to the end of the cylinder block farthest from the crankshaft.

cylinder leakdown test: A test that checks the sealing capability of compression components of a small engine using compressed air.

cylinder leakdown tester: A test tool designed to test the sealing capability of compression components of a small engine.

damping diaphragm: The diaphragm in a fuel pump that flexes from pressurized fuel to increase volume in the fuel section of the secondary chamber.

DC amperage test: A test that uses a DMM to indicate the current that should enter the battery if all connections to the battery are good.

decibel (dB): A unit of expressing relative intensity of sound on a scale from 0 dB (average least perceptible) to 140 dB (deafening).

Department of Defense (DOD): A federal agency responsible for developing United States Military Standards (MIL Standards).

Department of Transportation (DOT): A federal agency responsible for traffic control, enforcement of safety regulations, and aids to navigation.

depletion region: The region of a diode which separates P-type material and N-type material.

detonation: An undesirable engine condition in which there is spontaneous combustion of a significant portion of the charge before the spark-induced flame front reaches it.

diaphragm: A rubber membrane that separates chambers and flexes when a pressure differential occurs.

diesel engine: A reciprocating internal combustion engine that ignites fuel by high compression.

digital multimeter (DMM): A test tool used to measure two or more electrical values.

dinitrogen monoxide (N₂O): An oxide of nitrogen that is commonly known as laughing gas.

diode: An electrical semiconductor device that can be used to convert AC to DC.

dipper: An engine component attached to the connecting rod which directs oil from the oil reservoir to bearing surfaces.

direct current (DC): The flow of electrons in one direction only.

displacement (swept volume): The volume that a piston displaces in an engine when it travels from TDC to BDC during the same piston stroke.

distillation test: A test for determining the composition and volatility characteristics of the components of a given fuel sample.

dither effect: An effect based on the theory that less time is required to accelerate mass that is already in motion.

downdraft carburetor: A carburetor that has the air intake opening above the fuel bowl.

dry bulb primer system: A primer system consisting of a rubber bulb filled with air connected to the fuel bowl by a passageway.

DU™: A low-friction, self-lubricating bearing that can be used with or without lubrication.

duty cycle: The length of time (expressed as a percentage) that equipment can operate continuously at its rated output within a given time period.

dynamic leak: An undesirable discharge of gasoline which occurs when the engine is operating.

dynamometer: A device that applies a load to an operating engine and measures torque, load, speed, or horsepower.

ear muffs: Ear protection devices worn over the ears.

earplugs: Ear protection devices made of moldable rubber, foam, or plastic which are inserted into the ear canal.

easy-likely method: A troubleshooting method that isolates the cause of a malfunction by grouping possible causes as easy, difficult, likely, and unlikely.

eddy current: Undesirable current induced in the metal structure of an electrical device due to the rate of change in the induced magnetic field.

eddy current dynamometer: A dynamometer used to measure engine torque using load produced by a magnetic field.

efficiency: The ratio of fuel energy supplied to work produced.

elastic limit: The last point at which a material can be deformed and still return to its original physical dimensions.

electrical symbols: Graphic illustrations used in electrical system diagrams to show the function of a device or component.

electric dynamometer: A dynamometer used to measure brake horsepower by converting mechanical energy into electrical energy.

electricity: Energy created by the flow of electrons in a conductor.

electric starting system: A group of electrical components activated by the operator to rotate the crankshaft when starting an engine.

electrolyte: A mixture of water and sulfuric acid (H_2SO_4) used in a lead-acid battery.

electronic governor system: A governor system that uses a limited angle torque (LAT) motor in place of the governor spring and speed-sensing device used in a mechanical governor system.

electrons: The parts of the atom that have a negative electrical charge.

elliptical shape: An oval shape in which one-half is a mirror image of the other half.

emergency plan: A document that details the exact action to be taken in the event of an emergency.

emulsion tube: A small, hollow, cylindrical component placed in the carburetor with one opening submerged in the fuel bowl and the other opening projecting through the inner wall of the tube.

emulsion tube well: The cavity that surrounds the emulsion tube.

energy: The resource that provides the capacity to do work.

engine: A machine that converts a form of energy into mechanical force.

engine block: The main structure of an engine which supports and helps maintain alignment of internal and external components.

engine envelope: The vertical, horizontal, and peripheral clearance space required for engine installation.

engine failure: The complete stoppage of the engine caused by the failure of one or more engine components.

engine footprint: The area covered by the mounting base of an engine.

environmental operating conditions: Conditions present which affect engine performance.

Environmental Protection Agency (EPA): A federal agency established in 1970 to control and abate pollution in the areas of air, water, solid waste, pesticides, radiation, and toxic substances.

ethanol: An alcohol additive that is distilled from fermented grain and used in gasoline as an octane enhancer.

excessive-oil condition: An undesirable condition when the amount of oil in the engine exceeds the specified amount required by the engine design.

exhaust event: An engine operation event in which spent gases are removed from the combustion chamber and released to the atmosphere.

exhaust manifold: An engine component that collects and directs exhaust gases from each cylinder to the muffler.

external combustion engine: An engine that generates heat energy from the combustion of a fuel outside the engine.

external vent carburetor: A carburetor that has the fuel bowl vent located outside the air path of the carburetor.

face shield: An eye protection device that covers the entire face with a plastic shield.

failure analysis: The analysis of the engine component or components related to the cause of an engine failure (complete engine stoppage).

ferroxiding: A piston ring treatment process that changes the outer surface of the piston ring to iron oxide.

flame arrestor: A mesh or perforated metal insert within a safety can which protects its contents from external flame or ignition.

flame front: The boundary wall that separates the charge from the combustion by-products.

flammability hazard: The degree of susceptibility of materials to burning based on the form or condition of the material and its surrounding environment.

flammable liquid: A liquid that has a flash point below 100°F.

flash point: The lowest temperature at which a liquid gives off vapor sufficient to ignite when an ignition source is introduced.

float: A carburetor component that floats at a specific level to regulate the opening and closing of the needle and seat.

flow chart: A diagram that shows a logical sequence of steps for a given set of conditions.

flywheel: A cast-iron, aluminum, or zinc disk that is mounted at one end of the crankshaft to provide inertia for the engine.

flywheel ring gear: The gear attached to the engine flywheel driven by the pinion gear during engine starting.

force: Anything that changes or tends to change the state of rest or motion of a body.

force differential: The measured difference in forces acting on a single object.

fossil fuel: A fuel derived from previously living things that have been preserved in a mineralized or petrified state.

four-stroke cycle engine: An internal combustion engine that utilizes four distinct piston strokes (intake, compression, power, and exhaust) to complete one operating cycle.

free electron: An electron that is capable of jumping in or out of the outer orbit.

free length: The overall dimension of a governor spring when unloaded.

free piston ring gap: The distance between the two ends of a piston ring in an uncompressed state.

freeze plug: A concave-shaped metal plug pressed into a hole at the water jacket used to provide a release for pressure from freezing coolant in a liquid-cooled engine.

frequency: 1. The number of complete electrical cycles per second (cps). **2.** The number of vibrations per second a noise contains measured in hertz (Hz).

fretting: Damage that occurs to machine parts where two contacting surfaces are subject to slippage.

friction: The resistance to motion that occurs when two surfaces slide against each other.

friction bearing: A bearing that consists of a fixed, nonmoving bearing surface, such as machined metal or a pressed-in bushing, that provides a low-friction support surface for rotating or sliding surfaces.

friction horsepower (FHP): The amount of power required to overcome the internal friction of engine moving parts.

friction welding: A metal joining process in which heat and pressure cause fusion as one or both pieces are rotated and pressed against each other.

fuel cup: A reservoir located high inside the fuel tank.

fuel filter: A fuel system component that removes foreign particles by straining fuel from the fuel tank.

fuel pump: An engine component that pressurizes the fuel system to advance fuel from the fuel tank to the carburetor.

fuel solenoid: An electrically actuated component that controls the flow of fuel in the injection pump.

full-wave rectification: The process of rectifying AC and recovering the B– pulse of AC that the diode blocks.

fully-ducted engine: An air-cooled engine in which cooling air flow routing and rate are controlled by air guides and a sealed blower housing.

fuse: An overcurrent protection device with a thin metal strip that melts and opens the circuit when a short circuit or overcurrent condition occurs.

galling: A deep cut in an engine component running surface caused by metal removed by metal-to-metal contact, force, and motion.

gasoline: A liquid fossil fuel derivative that primarily consists of the elements hydrogen (H) and carbon (C).

generator: An electrical device that produces an AC sine wave as a wire coil is rotated in a magnetic field or as magnets are rotated inside a wire coil.

gerotor oil pump: An oil pump that consists of a multiple-lobed inner rotor meshing with an outer rotor to discharge oil under pressure.

glass thermometer: A graduated glass tube that is filled with a material such as alcohol or mercury which expands when heated.

glaze breaking: The process of using a flexible hone consisting of small Carborundum stones rotated in the cylinder bore to remove glazing and to obtain the desired surface appearance.

glow plug: A diesel engine component that preheats air inside the combustion chamber to facilitate ignition of the charge.

goggles: An eye protection device secured on the face with an elastic headband that may be used over prescription glasses.

governed idle: A governor system function that allows an engine to accept light to moderate loads at idle speed without stalling the engine.

governed idle spring: A governor system component that has a low spring rate that offsets some of the force from the speed-sensing device to improve governor sensitivity at low rpm.

governed speed: The speed obtained at the balance point between the forces of the speed-sensing device and the governor spring.

government agencies: Federal, state, and local government organizations and departments which establish rules and regulations related to safety, health, and equipment installation and operation.

governor blade: A movable metal or plastic blade which deflects air from flywheel fins to act as the speed-sensing device of a pneumatic governor system.

governor droop: The amount of rpm decrease between the top no-load governed speed and rpm where power is delivered.

governor droop curve: A combination of rpm decrease and maximum brake horsepower (BHP) curves.

governor return spring: An electronic governor system component that applies force to close the throttle plate.

governor spring: A governor system component that pulls the throttle plate toward the wide open throttle (WOT) plate position.

governor system: A system that maintains a desired engine speed regardless of the load applied to the engine.

gravity feed fuel system: A fuel system that uses the location of the tank to provide head pressure to force fuel to flow to the fuel reservoir or fuel bowl of the carburetor.

half-wave rectifier: An electronic device used in a charging system that converts AC to DC by blocking one-half of the AC sine wave to allow current to flow in only one direction.

hand tool: A tool that is powered by hand.

hardfacing: The application of material to an engine component to improve resistance wear from load, heat, and chemical corrosion.

harmonic hunting and surging: The undesirable quick and/or slow changing of engine rpm in a cyclical duration caused by excessive governor spring vibration.

hazardous material: A material capable of posing a risk to health, safety, or property.

header pipe: A separate exhaust pipe used for each cylinder.

head gasket: The filler material placed between the cylinder block and cylinder head to seal the combustion chamber.

head pressure: The force derived from the mass of a contained liquid such as fuel stored in a fuel tank.

health hazard: The likelihood of a material to cause, either directly or indirectly, temporary or permanent injury or incapacitation due to acute exposure by contact, inhalation, or ingestion.

heat: Kinetic energy caused by atoms and molecules in motion within a substance.

helix: The component or portion of the armature shaft that has helical grooves to provide axial movement of the starter pinion gear.

hemoglobin: A blood component that transports oxygen.

hertz (Hz): The international unit of frequency equal to one cycle per second.

high-inertia flywheel: A flywheel that has greater mass than a low-inertia flywheel and is designed to provide the minimum engine inertia required without considering inertia from the load.

honing: The process of using a hone with rigid Carborundum stones rotated in the cylinder bore to remove small surface irregularities and any glazing.

Hooke's law: A law that states the amount of stretch or compression (change in spring length) is directly proportional to the applied force ($F \sim \Delta x$).

horsepower (HP): A unit of power equal to 746 watts (W) or 33,000 lb-ft per minute (550 lb-ft per second).

hot soak back: The period immediately following the initial shutdown of an engine when cooling air flow has stopped and the engine enclosure temperature increases for a brief time.

hot spot: A discoloration of the cylinder bore surface caused by an improper piston ring seal in a distorted cylinder block.

hunting: The undesirable quick changing of engine rpm when set at a desired speed.

hydrocarbon (HC) molecule: A molecule held together by a loose bond between hydrogen and carbon atoms that occurs naturally in all fossil fuels.

hydrometer: An instrument used to measure the specific gravity of a liquid.

hysteresis: The undesirable motion of governor system components caused by engine vibration and governor system friction characteristics.

idle circuit: The path from the fuel supply to a small hole in the throat on the engine side of the throttle plate that provides the fuel required at idle speed.

ignition armature: A component containing two or more coils which, when acted upon by a magnetic field, induce electrical energy.

ignition (combustion) event: An engine operation event in which the charge is ignited and rapidly oxidized through a chemical reaction to release heat energy.

ignition system: A system that provides a high-voltage spark in the combustion chamber at the proper time.

indicated horsepower (IHP): The power produced inside the engine cylinder and the sum of the usable power (BHP) plus the power used to drive the engine (FHP).

induction: The production of voltage and current by the proximity and motion of a magnetic field or electric charge.

induction principle: A theory which states that with a conductor, any one of the following (current, a magnetic field, or motion) can be produced by the remaining two.

inductive field coil: A coil of wire attached to a segmented iron core that produces a magnetic field when current is passed through it.

inertia: The property of matter by which any physical body persists in its state of rest or uniform motion until acted upon by an external force.

inherent pressure: The internal spring force that expands a piston ring based on the design and properties of the material used.

injection pump: A diesel engine component that provides pressurized fuel to the cylinder at precise intervals.

injector: A diesel engine component that functions as an ON-OFF valve to introduce fuel into the cylinder.

in-line engine: An engine that has two or more parallel cylinders adjoining each other.

insufficient lubrication: A cause of engine failure from the absence, loss, or degradation of the oil film between two bearing surfaces.

insulator: A material through which current cannot flow easily.

intake event: An engine operation event in which the air-fuel mixture, or just air in diesel engines, is introduced to fill the combustion chamber.

intake manifold: An engine component that distributes the air-fuel mixture from the carburetor to more than one cylinder.

intake valve port: The portion of the engine that provides a path from the carburetor or intake manifold to the intake valve head.

integrally machined valve seat: A machined portion of the cylinder block that provides the sealing surface for a valve.

interference angle: The intentional deviation from a specification of two mating machine components to improve seating quality after a sufficient break-in period.

internal combustion engine: An engine that generates heat energy from the combustion of a fuel inside the engine.

internal energy: The sum of all energy in a substance, including potential and kinetic energy.

internally vented carburetor: A carburetor that has the fuel bowl vent located between the air filter and the venturi of the carburetor.

International Organization for Standardization (ISO): A nongovernmental international organization comprised of national standards institutions of over 90 countries (one per country).

interpersonal skills: Strategies and actions which allow a person to communicate effectively with other individuals in a variety of situations.

ionization gap: The distance between the ignition armature pole and the secondary pole in the spark tester.

jet: A fuel-limiting device that regulates fuel flow to the emulsion tube.

jug: An engine component in which the cylinder and cylinder head are cast as a single unit.

kinetic energy: The energy of motion.

kinetic friction: The friction exhibited by a moving mass.

labyrinth system: A set of baffles and/or a foam insert used to reduce sudden movement of fuel in the fuel tank resulting from motion of the application.

Lambda excess air factor (λ factor): A numerical value assigned to represent the stoichiometric ratio of atmospheric air to any hydrocarbon fuel.

lamination stack: An electrical component that consists of thin iron layers used to focus and control the lines of magnetic flux.

lead-acid battery: A battery that stores electrical energy using lead cell plates and sulfuric acid (H_2SO_4).

lever: A simple machine that consists of a rigid bar which pivots on a fulcrum (pivot point) with both resistance and effort applied.

L-head engine: An engine that has valves and related components located in the cylinder block.

limited angle torque (LAT) motor: A direct current (DC) motor used to control governor system components in an electronic governor system.

liquid-cooled engine: An engine that circulates coolant through cavities in the cylinder block and cylinder head to maintain the desired engine temperature.

load: A device that uses electricity, such as the starter motor, lights, or other application accessories.

low-inertia flywheel: A flywheel that uses the load to provide the minimum engine inertia required.

low-oil condition: A condition when the amount of oil in the engine is less than the amount required to maintain the proper oil film to lubricated surfaces.

magnet: A material that attracts iron and produces a magnetic field.

magnetic field: An area of magnetic force created and defined by lines of magnetic flux surrounding a material in three dimensions.

magnetic flux: The invisible lines of force in a magnetic field.

magnetism: An atomic level force derived from the atomic structure and motion of certain orbiting electrons in a substance.

Magnetron® ignition system: An ignition system that uses electronic components in place of breaker points and a condenser.

main bearing: A bearing that supports and provides a low friction bearing surface for the crankshaft.

main (high-speed) circuit: The path from the fuel bowl to the emulsion tube created by the fixed orifice jet in the carburetor.

malfunction: The failure of a system, equipment, or part to operate as designed.

manual friction speed control: A control mounted on the engine that uses a friction lever to maintain an engine at a specific top no-load speed.

manual switch: A switch that is operated by a person.

margin: The surface of a valve joining the valve face and the top surface of the valve head.

market octane number (MON): The octane number that affects engine knock at high speed and performance in severe operating conditions and under load.

Material Safety Data Sheet (MSDS): Printed material used to relay hazardous material information from the manufacturer, importer, or distributor to employer and employees.

maverick air: Undesirable, unaccounted air entering the engine through leaks caused by worn, loose, or failed engine components.

maximum BHP curve: A graphic representation that indicates engine power developed with the throttle in WOT position across the usable engine rpm range in a test laboratory.

mechanical compression release system: A compression release system that incorporates a weighted lever or arm attached to the camshaft or cam gear.

mechanical governor system: A governor system that uses a gear assembly that meshes with the camshaft or other engine components to sense and maintain the desired engine speed.

mechanical switch: A switch that is operated by the movement of an object.

mesh screen fuel filter: A single plastic screen that strains out particles in the fuel tank or suspended in the fuel.

methanol: An alcohol additive that is distilled from methane gas and used in gasoline.

methyl tertiary butyl ether (MTBE): A nonalcohol oxygenate additive derived from a chemical reaction of methanol and isobutylene and used in gasoline.

micrometer: A hand tool used to make very accurate inside, outside, or depth measurements.

micron (μ): A unit of area measurement equal to one thousandth of a millimeter (.001 mm).

muffler: An engine component fitted with baffles and plates that subdues noise produced from exhaust gases exiting the combustion chamber.

multiple-barrel carburetor: A carburetor that contains more than one venturi.

multiple-cylinder engine: An engine that contains more than one cylinder.

multi-viscosity oil: An oil that has the characteristics of two viscosity ratings for the required flow at low ambient temperatures and has adequate oil film protection at high operating and/or ambient temperatures.

National Fire Protection Association (NFPA): A national organization that provides guidance in assessing hazards of the products of combustion.

National Institute for Occupational Safety and Health (NIOSH): A national organization that acts in conjunction with OSHA to develop and periodically revise recommended exposure limits for hazardous substances or conditions in the workplace.

needle and seat: Components used together that provide a tapered seal to regulate the flow of fuel into the carburetor.

neutrons: The parts of the atom that are neutral, and have no electrical charge.

nitric oxide (NO): An oxide of nitrogen that is created in small amounts in nature and is somewhat toxic and colorless.

nitriding: A piston ring surface treatment process that uses a thermal process in which nitrogen and some carbon are absorbed into the piston ring surface.

nitrogen dioxide (NO$_2$): An oxide of nitrogen that is created in the combustion chamber during the instantaneous increase in an advancing flame front.

nonferrous metal: A metal that does not contain iron.

N-type material: A portion of the silicon crystal that has an excess of electrons and a deficiency of protons.

nucleus: The center of the atom, which consists of protons and neutrons.

Occupational Safety and Health Administration (OSHA): A federal agency that requires all employers to provide a safe environment for their employees.

octane: The ability of a fuel sample to resist engine knock.

octane rating: A numerical value assigned to gasoline that indicates the ability to eliminate knocking and pinging in an operating engine.

Ohm's law: A law that states the relationship between voltage, current, and resistance in any circuit.

oil failure: The deterioration of oil and its viscosity through oxidation, heat, and the accumulation of solids over time.

oil ring: The ring located in the ring groove closest to the crankcase that is used to wipe excess oil from the cylinder wall during piston movement.

oily waste can: A can designed for safe containment of rags and paper soiled with flammable materials.

180° opposed-twin engine: An engine that has two horizontal cylinders opposite each other.

one-piece valve: A valve that is constructed from a single piece of austenitic steel.

open installation: An engine installation that does not have any barriers which impede access of air to the engine.

operation: The engine operation mode in which the engine has completed the startup mode and operates normally.

Outdoor Power Equipment Institute (OPEI): The national trade association representing manufacturers of consumer and commercial outdoor power equipment and their major components.

overcurrent condition: A condition that occurs when the amount of current flowing in a circuit exceeds the design limit of the circuit.

overhead valve (OHV) engine: An engine that has valves and related components located in the cylinder head.

overheating: A cause of engine failure from an engine component material that has distorted beyond a specific yield point.

oversized engine: An engine that is typically capable of producing 20% – 30% more power than required by the application at full load.

overspeeding: A cause of engine failure from damage resulting from excessive engine rpm.

oxides of nitrogen (NOx): A term assigned to several different chemical compounds consisting of nitrogen (N$_2$) and oxygen (O$_2$).

parallel circuit: A circuit that has two or more paths (branches) for current flow.

parasitic load: Any load applied to an engine that is over and above the frictional load of an engine, such as a lawn mower blade.

partial enclosure installation: An engine installation that allows partial access to engine components.

partially-ducted engine: An air-cooled engine in which cooling air is provided by the blower housing and ambient air flow.

pathway: A conductor (commonly copper wire), which connects different parts of the circuit.

peripheral coating: A coating that is applied only to the wear surface of the piston ring.

permanent (hard) magnet: A magnet that retains its magnetism after a magnetizing force has been removed.

permissible exposure limit (PEL): The OSHA limit of employee exposure to chemicals.

personal protective equipment (PPE): Safety equipment worn by a small engine service technician for protection against safety hazards in the work area.

phase modulated cooling fan: A cooling fan that has blades spaced at different distances from each other.

phosphating: A piston ring surface treatment process that changes the outer surface of the piston ring to phosphate crystals.

pilot jet: A carburetor component that contains a fixed orifice jet that meters and controls fuel flow to the idle circuit of the carburetor.

pinion gear: A gear on the starter motor that follows the helix to engage and drive the flywheel ring gear.

piston: A cylindrical engine component that slides back and forth in the cylinder bore by forces produced during the combustion process.

piston head: The top surface (closest to the cylinder head) of the piston which is subjected to tremendous forces and heat during normal engine operation.

piston pin: A hollow shaft that connects the small end of the connecting rod to the piston.

piston pin bore: A through hole in the side of the piston perpendicular to piston travel that receives the piston pin.

piston ring: An expandable split ring used to provide a seal between the piston and the cylinder wall.

piston windows: A series of small holes machined into the oil ring groove surface of the piston.

plasma spraying (PSP): A thermal process in which hardfacing metal in powder form is molten with an electric arc and propelled to coat the desired surface.

pleated paper fuel filter: A paper filter element that consists of multiple folds or pleats to strain out particles suspended in the fuel.

pneumatic governor system: A system which uses force from moving air (pneumatic force) produced by rotating flywheel fins to sense engine speed.

point plunger: An ignition system component that holds the movable point contact surface away from the fixed contact surface to interrupt current flow to the primary winding.

polarity: The state of an object as negative or positive.

polarity-sensitive circuit: A circuit that does not operate properly when exposed to the wrong polarity.

polymer: A molecule consisting of repeating structural units that have been chemically formulated to perform in a specific manner.

portable power tool: A power tool that can be transported by the operator.

potential energy: Stored energy a body has due to its position, chemical state, or condition.

power: The rate at which work is done.

power curve: A graphic representation of horsepower and torque output of a specific engine in a test laboratory.

power event: An engine operation event in which the compressed charge is ignited and hot expanding gases force the piston head away from the cylinder head.

power take-off (PTO): An extension of the crankshaft that allows an engine to transmit power to an application.

power tool: A tool that is electrically, pneumatically, or hydraulically powered for greater efficiency when performing service and repair tasks.

preignition: An undesirable engine condition which occurs when a small portion of a combustion chamber component or a particle in the combustion chamber becomes excessively heated and ignites the charge as it enters the combustion chamber.

pressure: A force acting on a unit of area.

pressure filtration lubrication system: An engine lubrication system in which a pump is used to circulate oil in a limited area of the engine.

pressure lubrication system: An engine lubrication system in which a pump is used as the primary component to circulate oil throughout the entire engine.

primary winding: A coil that induces voltage in the secondary winding.

primer system: A rubber bulb that is depressed to force a metered amount of fuel into the venturi to help start a cold engine.

process: A sequence of operations that accomplishes desired results.

projection-tip welded valve: A valve that is constructed from austenitic steel that has approximately .09″ of hardened steel welded on the end of the valve stem.

prony brake dynamometer: A dynamometer for measuring engine torque using a brake that exerts pressure on a spring scale.

protons: The parts of the atom that have a positive electrical charge.

P-type material: A portion of the silicon crystal that has an excess of protons and a deficiency of electrons.

pulsating DC: DC voltage produced by rectifying (removing) one-half of an AC sine wave.

pulse: Half of a cycle.

pushrod: An engine component that transfers motion from the tappet to one side of the rocker arm in overhead valve engines.

quick-check method: A troubleshooting method that isolates the cause of a malfunction by focusing on common problems identified by the manufacturer, product history, and/or service technician experience.

race: The bearing surface in an antifriction bearing that supports rolling elements during rotation.

radial load: A load applied perpendicular to the shaft.

radiation: Heat transfer that occurs as radiant energy without a material carrier.

radiator: A multi-channeled container that allows air to pass around the channels to remove heat from the liquid within.

radiator cooling fan: A device that pulls or pushes cooling air through a radiator.

reactivity hazard: The degree of susceptibility of materials to release energy by themselves or by exposure to certain conditions or substances.

rebound compression test: A test for engine compression in which the flywheel is spun counterclockwise (backwards) and checked for the amount of rebound force.

recommended maximum operating BHP curve: A graphic representation that indicates engine power at 85% of maximum BHP.

rectifier: An electrical component that converts AC to DC by allowing the current to flow in only one direction.

reed valve valving system: A two-stroke cycle engine system that uses a valve made from thin spring steel that opens and closes with pressure changes in the crankcase to control air-fuel mixture flow.

reformulated (oxygenated) gasoline (RFG): Gasoline that contains chemical additives to increase the amount of oxygen present in the gasoline blend.

regulator/rectifier: An electrical component that contains one or more diodes and a zener diode.

Reid vapor pressure (RVP) test: A test used to determine the pressure produced from the vaporization process.

remote speed control: A control that allows the operator to vary engine speed from a remote location.

repowering: The process of replacing the original engine on a piece of equipment with another engine.

research octane number (RON): The octane number that affects engine knock at low to medium speed.

reserve capacity: The amount of time a battery can produce 25 A at 80°F.

resistance (R): The opposition to the flow of electrons.

resistive load: An applied load that reduces the possibility of the alternator system delivering full amperage through the circuit.

resonance: The state of the vibration wave frequency being equal to the natural vibration wave frequency of the component.

retainer groove: The part of a valve that is recessed for mounting the valve spring retainer.

rewind starting system: A mechanical starter that commonly consists of a rope, pulley, and return spring used to manually rotate the crankshaft to start an engine.

ring compressor: A tool that is used to compress piston rings for installation in the cylinder bore.

ring expander: A tool that expands the piston ring uniformly and causes no permanent distortion to the piston ring.

ring groove: A recessed area located around the perimeter of the piston that is used to retain a piston ring.

ring lands: The two parallel surfaces of the ring groove which function as the sealing surface for the piston ring.

rocker arm: An engine component that acts as a pivoting device for opening and closing overhead valves.

rod bearing: A bearing that provides a low-friction pivot point between the connecting rod and the crankshaft and the connecting rod and piston.

rod cap: The removable section of a two-piece connecting rod that provides a bearing surface for the crankpin journal.

rotary engine: An internal combustion engine that operates using the rotating motion of a rotor.

rotary valve valving system: A two-stroke cycle engine system that uses a rotating flat disk with a section removed to control air-fuel mixture flow.

rotating screen: An engine component attached to the outer side of the flywheel that prevents harmful foreign matter from entering the path of cooling air to the engine.

running surface: The portion of an engine component which interacts with a lubricated mating engine bearing surface during operation.

SAE viscosity rating: A number based on the volume of a base oil that flows through a specific orifice at a specified temperature, atmospheric pressure, and time period.

safety cabinet: A double-walled steel cabinet specifically designed for storage of flammable liquid containers.

safety can: A UL-approved container not exceeding 5 gal. that has a spring-loaded lid on the spout to prevent the escape of explosive vapors, but allows relief of internal pressure.

safety glasses: Glasses with impact-resistant lenses, reinforced frames, and side shields.

scavenging: The process of using the introduction of the fresh air-fuel mixture to help remove exhaust gases from the cylinder in a two-stroke cycle engine.

scoring: The result of scratching an engine component surface caused by a foreign object or undesirable transfer of metal from metal-to-metal contact between bearing surfaces.

secondary winding: A coil in which high voltage is induced for use at the spark plug.

seizure: The joining of engine components at the bearing or running surface caused by excessive heat, pressure, and/or friction.

self-inductance: A magnetic field that is created around a conductor whenever current moves through the conductor.

separator: An antifriction bearing component used to maintain the position and alignment of rolling elements.

sequential method: A troubleshooting method that isolates the cause of a malfunction by starting at one end of a system and progressing to the other end with sequential checks.

series circuit: A circuit that has two or more components connected so that there is only one path for current flow.

series/parallel circuit: A circuit that contains a combination of components connected in series and parallel.

service replacement engine: A replacement engine that has the operating characteristics and required features of the original engine but is not supplied by the OEM.

shaft orientation: The axis of the crankshaft as vertical or horizontal.

short circuit: An undesirable complete circuit path that bypasses the intended path and has very little resistance.

shutdown: The engine operation mode when the engine is shut down after operation.

sidedraft carburetor: A carburetor that has the air intake opening above the fuel bowl and parallel to a horizontal plane.

side loading: The application of a (typically) undesirable unilateral force to an engine component or components.

signature wear pattern: An area impacted by abrasive particles having specific appearance and dimensional characteristics.

silica: A compound of the elements silicon (Si) and oxygen (O_2).

silicon controlled rectifier (SCR): A semiconductor that is normally an open circuit until voltage is applied, which switches it to the conducting state in one direction.

sintered iron: A powdered iron compound that is heated and compressed to form the desired shape.

sintered iron valve guide: A separate machined valve guide insert manufactured from a powdered iron compound that is heated and compressed to form the desired shape.

60° V-twin engine: An engine that has two cylinders forming a V shape angled at 60° to a horizontal plane.

skirt: The portion of the piston closest to the crankshaft that helps align the piston as it moves in the cylinder bore.

slinger: A splash lubrication system component used on vertical crankshaft engines consisting of a spinning gear with multiple paddles cast into the plastic gear body.

sludge: The accumulation of a semi-solid, highly viscous oil material found in the crankcase of some internal combustion engines.

small engine: An internal combustion engine that converts heat energy from the combustion of a fuel into mechanical energy generally rated up to 25 horsepower (HP).

Society of Automotive Engineers (SAE): A network of engineers, business executives, educators, and students from more than 80 countries who share the interest of advancing engineering of mobile systems.

solenoid: A device that converts electrical energy into linear motion.

spark arrester: A component in the exhaust system that redirects the flow of exhaust gases through a screen to trap sparks discharged from the engine.

spark gap: The distance from the center electrode to the ground electrode on the spark plug.

spark ignition engine: An engine that ignites an air-fuel mixture with an electrical spark.

spark plug: A component that isolates the electricity induced in the secondary windings and directs a high voltage charge to the spark gap at the tip of the spark plug.

spark tester: A test tool used to test the condition of the ignition system on a small engine.

special hazard: The extraordinary properties and hazards associated with a particular material.

specific gravity: A comparison of the mass of a given sample volume compared to an equal volume of water.

speed-sensing device: A governor system component attached through linkage to the throttle plate of the carburetor to sense changes in engine speed.

splash lubrication system: An engine lubrication system in which oil is directed to moving parts by a splashing motion.

split-half method: A troubleshooting method that isolates the cause of a malfunction by splitting parts of a system in half until the cause is isolated.

spontaneous combustion: Self-ignition caused by chemical reaction and temperature buildup in waste material such as used oily rags.

spring rate: The force necessary to stretch a governor spring one unit of length in in., mm, or other units from its free length.

stainless steel: A ferrous alloy primarily consisting of chromium or nickel.

standard: An accepted reference or practice.

standards organizations: Organizations, often affiliated with governmental organizations, that coordinate the development of codes and standards among member organizations.

starter motor: An electric motor that drives the engine flywheel when starting.

starter solenoid: An electrical switch with internal contacts opened or closed using a magnetic field produced by a coil.

startup: The engine operation mode in which the engine cranks, the air-fuel mixture is drawn into the cylinder and compressed, and ignition occurs.

static friction: The force needed to accelerate or initiate the movement of a stationary mass.

static leak: An undesirable discharge of gasoline which occurs when the engine is not operating.

stationary power tool: A power tool that is commonly installed in a fixed position.

stator: An electrical component that has a continuous copper wire (stator winding) wound on separate stubs exposing the wire to a magnetic field.

stoichiometric ratio: The specific air-fuel ratio (by weight) of atmospheric air to fuel at which the most efficient and complete combustion occurs.

stroke: The linear distance a piston travels inside the cylinder from the cylinder head end (TDC) to the crankshaft end (BDC).

sump: A removable part of the engine crankcase that serves as an oil reservoir and provides access to internal parts.

surging: The undesirable slow changing of engine rpm in a cyclical pattern when set at a desired speed.

switch: Any component that is designed to start, stop, or redirect the flow of current in an electrical circuit.

symmetrical: A shape in which one-half is the mirror image of the other half.

systems approach method: A troubleshooting method that isolates the cause of a malfunction by dividing the engine into separate systems and subsystems.

taper-faced compression ring: A piston ring that has approximately a 1° taper angle on the running surface.

tappet: An engine component that rides on the camshaft and pushes the bottom of the valve stem to open the valve.

technical societies: Organizations composed of groups of engineers and technical personnel united by professional interest.

temperature: The intensity of heat.

temporary (soft) magnet: A magnet that can only become magnetic in the presence of an external magnetic field.

test tool: A measurement tool used to test the condition or operation of an engine component or system.

thermal conductivity: The ability of a material to conduct and transfer heat.

thermal distortion: An asymmetrical or nonlinear thermal expansion of a material.

thermal efficiency: A measurement that compares the amount of chemical energy available in fuel converted into heat energy used to produce useful work.

thermal expansion: The expansion of a material when it is subjected to heat.

thermal growth: The increase in size of a material when heated, with little or no change back to original dimensions.

thermostat: A valve placed between the radiator and the engine block on liquid-cooled engines that regulates the flow of coolant.

three-port valving system: A two-stroke cycle engine system that uses three ports to control the flow of air-fuel mixture and exhaust gases.

threshold limit value (TLV): An estimate of the average safe airborne concentration of a substance that represents conditions under which it is believed that nearly all workers may be exposed day after day without adverse effect.

throat: The main passage in the carburetor which directs air from the atmosphere and air-fuel mixture to the combustion chamber.

throttle plate: A disk that pivots on a movable shaft, regulating air and fuel flow in a carburetor.

throw: The measurement from the center of the crankshaft to the center of the crankpin journal, which is used to determine the stroke of an engine.

top dead center (TDC): The point at which the piston is closest to the cylinder head.

top no-load speed: The top speed setting an engine achieves without any parasitic load from equipment components.

torque: A force acting on a perpendicular radial distance from a point of rotation.

torque curve: A graphic representation that indicates maximum engine torque produced at a specific rpm.

total enclosure installation: An engine installation that does not allow a person to touch the engine by hand nor see the engine by normal line-of-sight vision.

trade associations: Organizations that represent producers and distributors of specific products.

transitional hole reservoir: A cavity that supplies fuel to the idle mixture screw and orifice and the transitional holes.

trigger: A magnetic pick-up located near the crankshaft pulley that senses and counts crankshaft rotation.

troubleshooting: The systematic elimination of the various parts of a system or process to locate a malfunctioning part.

troubleshooting chart: A logical listing of problems and recommended actions.

true idle: The carburetor setting when the throttle plate linkage is resting against the idle speed adjusting screw after idle air-fuel mixture adjustment.

two-piece-stem welded valve: A valve that is constructed from an austenitic steel valve head welded to a hardened valve stem. The two pieces are joined by friction welding.

two-stroke cycle engine: An internal combustion engine that utilizes two distinct piston strokes to complete one operating cycle of the engine.

undersized engine: An engine which consistently works close to maximum capacity with little or no reserve power.

Underwriters Laboratories Inc. (UL®): An independent organization that tests equipment and products to verify conformance to national codes and standards.

updraft carburetor: A carburetor that has the air intake opening below the fuel bowl.

useful life: The period of time after the break-in period when most small engines operate as designed.

valve: An engine component that opens or closes at precise times to allow the flow of air-fuel mixture into the cylinder and to allow the flow of exhaust gases from the cylinder.

valve face: The machined surface of a valve that mates with the valve seat to seal the combustion chamber.

valve guide: An engine component that aligns a valve stem in a linear path.

valve head: The large end of the valve that contains the margin and the valve face.

valve interface: The point of contact between the valve face and the valve seat.

valve neck: The part of a valve joining the valve head and valve stem.

valve overlap: The period during engine operation when both intake and exhaust valves are open at the same time.

valve rotator: A mechanical device on a valve used to rotate the valve each time it opens to provide even wear and distribution of heat.

valve seat: The machined stationary surface that mates with the valve face to seal the combustion chamber.

valve seat insert: A separate machined engine component pressed into the cylinder block that provides the sealing surface for a valve.

valve spring: A compression spring that closes the valve and holds it tightly against the valve seat.

valve spring retainer: An engine component that compresses and secures the valve spring on the valve stem.

valve stem: The long part of a valve that aligns the valve.

valve train: The part of an internal combustion engine that includes components required to control the flow of gases into and out of the combustion chamber.

valving system: The system on an internal combustion engine that controls the flow of gases into and out of the combustion chamber.

vaporization: The process in which a liquid is sufficiently heated to change states of matter from liquid to a vapor.

vapor lock: The stoppage of fuel flow caused by internal pressure of a fuel vapor bubble that equals or exceeds the ambient fuel pressure.

vapor pressure: The pressure exerted by vapor above the surface of a liquid in a closed container.

venturi: A narrowed portion of a tube.

viscosity: The internal resistance to flow of a fluid.

volatility: The propensity of a liquid to become a vapor.

volt (V): The unit of measure for electrical pressure difference between two points in a conductor or device.

voltage: The amount of electrical pressure in a circuit.

voltage regulation system: A system that controls the amount of voltage required to charge the battery with a regulator/rectifier.

voltage source: A battery or some other voltage-producing device.

volumetric efficiency: The ratio of volume available in the engine, and the actual volume filled during operation.

water dynamometer: A dynamometer used to measure engine torque using load produced by a water pump.

water jacket: A series of interconnected cavities cast into the engine block and cylinder head for the circulation of coolant.

water pump: An engine component that moves coolant through passages of a liquid cooling system.

wear-out period: The period of time after the useful life of the engine, when normal wear failures begin to occur.

wear ratio: A comparison of the rate of material loss from the piston ring and the cylinder bore over an extended period of time.

Welch plug: A hemispherically curved metal cover that expands and seals to the shape of a cavity when impacted.

wet bulb primer system: A primer system consisting of a rubber bulb filled with fuel connected to the fuel bowl by a drilled passage.

wiper ring: The piston ring with a tapered face located in the ring groove between the compression ring and oil ring.

work: Movement of an object by a constant force to a specific distance.

yield point: The limit at which the material can be exposed to thermal and/or mechanical stress and still return to its original size and chemical composition.

zener diode: A semiconductor that senses voltage to measure the state of battery charge at the battery terminals.

INDEX

Italicized numerals refer to illustrations.

abrasive ingestion, 92, 234-239
abrasive particle, 234
abrasive wear, 248
accident report, 39
AC sine wave, 149
AC voltage, 149-*150*
AC voltage test, *227*
adhesive wear, 248
adiabatic process, 75-76
adjustable fixed speed control, 255
adjustable orifice jet, 118
afterfire, 198
air bleed, 112
air-cooled engine, 4, 177, 230
 cooling systems, 177-181
air density, 116
air guide, 179
air pressure dynamics, 106
alcohol, 105
alternating current, 148
alternator, 155
 operation, *157*
altitude compensation, 116
aluminum, 55
 valve guide, 83
American National Standards Institute, 25
American Petroleum Institute, 26
American Society of Agricultural Engineers, 25-26
amperes, 148
anti-afterfire solenoid, 198, 218
antifreeze, 181
antifriction bearing, *54*
antiknock index, 102
applied load, 125
applied pressure, 88
area, 12
armature, 136
asperities, 184-*185*, 239
Associated Equipment Distributors, 27
ASTM International, 26
atom, 145
austenitic steel, 64, 78
autoignition, 102
automatic switch, 151
axial load, 54, 257

babbit, 55
back protection, 33
barrel-faced compression ring, 89
battery, 162-163
 loading device, 225
bearing journal, 51
bearings, *54*-55
Bernoulli's principle, 107-*108*
blower housing, 179
blown head gasket, 243
body, 11-*12*
bonding, 28
booster fan, 259
bore, 47

bottom dead center, 48
boundary lubrication, 241
bowl-style carburetor, *197*
bowl vent, 109
brake horsepower, 70
brake mean effective pressure, 73
brass valve guide, 83-84
breakage, 246-247
breakaway clutch, *171*
breaker point ignition system, *164*-167
breaker points, 164
break-in, 90
 period, 93, 214
bridge rectifier, 161
British thermal unit, 10
bronze, 55
brushes, 171

calorie, 10
cam gear, 64
cam lobe, 64
camshaft, 64
Canadian Standards Association, 25
capacitor, 159
carbon monoxide, 29-30, 100
carburetor, 107
 design, 117-120
 leakage, 222-225
 operation, 109
  principles, 107-116
 service procedures, 121
charge, 57
charging system, 155-163, 227
check ball, 118
check valve, 124
chemical hazard, 32
chemistry, 17
choke, 111
chromium plating, 92
circuit, 146
cleaning tank, 37-*38*
code, 25
coefficient of thermal expansion, 84, 175
cohesion, 76
coil, 156
cold cranking amps, 163
cold engine starting, 111-112
combination engine failure, 247
combustible liquid, 28
combustion, 58, 98
commutator, 171
compression, 57, 75
 event, 57
 ignition, 204-205
  engine, 1
  problems, 76
 ratio, 58
 release system, 95
 ring, 52, 89
 system, 75-96, 195-196
condenser, 166
conduction, 9-*10*, 177
conductor, 145
connecting rod, *53*
Consumer Product Safety Commission, 25
container labeling, 34

convection, *10*, 177
convolution, 94
cooling air discharge, 180
cooling air plenum, 180
cooling fan, 178
cooling fin, 48, 179
cooling system, 4, 202-203, 268-269
 service procedures, 192
counterweight, 51
crankcase, 48
 breather, *49*, 95
crankgear, 51
crankpin journal, 50
crankshaft, 50-*51*
current, *148*
cycle, 149
cylinder block, 47
cylinder bore, 47, 94-95
 design, 94-95
 distortion, 244
 finish, 95
cylinder design, 4
cylinder head, 50
cylinder leakdown test, 215
cylinder leakdown tester, 45-*46*

damping diaphragm, 124
DC amperage test, *228*
DC voltage, 150-*151*
decibel, 32, 256
Department of Defense, 25
Department of Transportation, 25
depletion region, 158
detonation, *77*
diaphragm, 119
diesel engine, 65-68
 components, 66-67
 service procedures, 206
Diesel, Rudolf, 6
digital multimeter, 44
dinitrogen monoxide, 101
diodes, *158*-159
dipper, 188
direct current, 148
displacement, 48, 194
distillation test, 103
dither effect, 137
documentation, 214-*215*
downdraft carburetor, 117
drawings and diagrams, *153*
dry bulb primer system, 112
DU™, 54
duty cycle, 202
dynamic leak, 224
dynamometer, 70

ear muffs, 32
earplugs, 32

316 SMALL ENGINES

ear protection, 32
easy-likely method, 208
eddy current, 156
  dynamometer, 72
efficiency, 72
elastic limit, 139
electrical prefixes, 153
electrical principles, 145-155
electrical symbols, 153
electrical system, 145-172, 199-202, 265-266
electric dynamometer, 70
electricity, 145
electric starting system, *170*
electrolyte, 162
electronic governor systems, 136-137
electrons, 145
elliptical shape, 86
emergency plan, 38-*39*
emulsion tube, 107, 113
  well, 112
energy, 7
  conversion principles, 7-17
engine, 1
  block, 47-50
  classification, 1-2
  components, 47-56
  ducting, 179-180
  emissions, 100
  envelope, 251
  failure, 233-247
  flywheel, 259-260
  footprint, 251
  heat, 173-174
  life, 260
  materials, 174-177
  noise, 256
  output, 69-73
  power, 251-254
  selection, 250-263
  speed, 254-256
environmental operating conditions, 261-263
Environmental Protection Agency, 24
Equipment and Engine Training Council, 26
ethanol, 105
evacuation system, 30
excessive-oil condition, 242
exhaust event, *60*
exhaust manifold, *203*
exhaust valve seat insert, *244*-245
external combustion engine, 1
external vent carburetor, 116
eye protection, 31

face shield, 31
failure analysis, 233-248
ferroxiding, 93
fire extinguishers, 27-*28*
  classes, 27-28
fire safety, 27-30
flame arrestor, 28
flame front, 59
flammability hazard, 34
flammable liquid, 28
flash point, 28-*29*
float, 114
flow chart, 213
flywheel, *56*
  ring gear, 171
foot protection, 33
force, 11-*12*
  differential, 132
foreign matter, 121
fossil fuel, 97
four-stroke cycle diesel engine operation, 68
four-stroke cycle engine, *3*, 56-61
four-stroke cycle engine valving systems, 63-65
free electron, 146
free length, 130
free piston ring gap, 88

freeze plug, 203
frequency, 149, 256
fretting, 257
friction, 184, 239
  bearing, 54
  horsepower, 70
  welding, 79
fuel, 97-106
  bowl level regulation, 114
  bowl vent design, 114
  cup, 119
  filter, 122-123, 264
  lines, 264-265
  pump, 123-124
  solenoid, 205
  system, 197-198, 262-265
  safety, 271
  tank, 263-264
full-wave rectification, 161-*162*
fully-ducted engine, 179-180
fuse, 153

galling, 240
gasoline, 97
  additives, 100
generator, 149
gerotor oil pump, *189*
glass thermometer, 10-*11*
glaze breaking, 92
glow plug, *67*
goggles, 31
governed idle, 141
  spring, 142
governed speed, 128
government agencies, 22
governor blade, 129
governor droop, 137, 253
  control, 137
  curve, 137, 253-*254*
governor return spring, 136
governor sensitivity, 139-143
governor spring, 128
governor system, 125-144, 198-199
  components, 128
gravity feed fuel system, 123
gunpowder engine, 4

half-wave rectifier, 158-*159*
hand protection, 32
hand tools, 40
  safety, 40
hardfacing, 79
harmonic hunting and surging, 221
hazardous material, 33
  disposal, 38
header pipe, 203
head gasket, *50*, 243
head pressure, 123
health hazard, 34
heat, 8-10
  induced expansion, 175
  transfer, 9
helix, 170
hemoglobin, 29
hertz, 149
high-inertia flywheel, 259
honing, 91
Hooke's Law, 139
horsepower, 16
hot soak back, 181
hot spot, 244
hunting, 138, 219
Huygens, Christian, 4
hydrocarbon emissions, 101

hydrocarbon molecule, 98
hydrometer, 162, 226
hysteresis, 142

idle circuit, 110-*111*
ignition, 1
  armature, 164
  event, 58-59
  systems, 164-169, 229-230
impression pattern, 236
indicated horsepower, 70
induction, *155*
  principle, 155
inductive field coil, 136
inertia, 56
inherent pressure, 88
injection pump, 66
injector, 66-*67*
in-line engine, 193
insufficient lubrication, 239
insulator, 149
intake event, 56-*57*
intake manifold, *197*
intake valve port, 236
integrally machined valve seat, 85
interference angle, 80-81
internal combustion engine, 1
internal energy, 173
internal vent carburetor, 114-115
International Organization for Standardization, 25
interpersonal skills, 211
investigation, 211-212
ionization gap, 229
isolation, 212-213

jet, 110
jug, 50

kinetic energy, 7-8
kinetic friction, 239

labyrinth system, 262
Lambda excess air factor, 99
lamination stack, 156-*157*
Langen, Eugen, 6
lead-acid battery, 162
Lebon, Eugene, 6
Lenoir, Etienne, 6
lever, *14*
L-head engine, 50
limited angle torque, 136
liquid-cooled engine, 4, 181, 231
  cooling systems, 181-184
load, 146
long governor lever arms, 136
low-inertia flywheel, 259
low-oil condition, 239-241
low oil level warning system, 191-192
lubrication, 184-192
  systems, 188-191, 203-204
  service procedures, 192

Index **317**

magnet, 154
magnetic field, 154
magnetic flux, 154
magnetism, *154*-155
magneto ignition system, 230
Magnetron® ignition system, 167-169
main bearing, *55*
main circuit, 110
malfunction, 207
manual friction speed control, 255
manual switch, 151
margin, 63
market octane number, 102
Master Service Technician program, 19
Material Safety Data Sheet, 34-*36*
maverick air, 76, 235
maximum BHP curve, 253
measuring horsepower, 70-72
mechanical compression release system, 96
mechanical governor adjustment, 143
mechanical governor systems, 132-136
mechanical switch, 151
mesh screen fuel filter, 122-123
methanol, 105
methyl tertiary butyl ether, 105-106
micrometer, 40
micron, 123
Model P engine, 7
muffler, 256
multiple-barrel carburetor, 117
multiple-cylinder diesel engines, 204-205
multiple-cylinder engine design, 193-195
multiple-cylinder engine displacement, 194-195
multiple-cylinder engines, 193-206
 service procedures, 205-206
multiple-cylinder engine systems, 195-204
multi-viscosity oil, 186

National Electrical Code®, 26
National FFA Organization, 27
National Fire Protection Association, 26
National Institute for Occupational Safety and Health, 24
needle and seat, 114
neutrons, 145
Newcomen steam engine, 5
Newcomen, Thomas, 5
NFPA Hazard Signal System, 34
nitric oxide, 101
nitriding, 93
nitrogen dioxide, 101
nonferrous metal, 55
N-type material, *158*
number of strokes, *3*
nucleus, 145

Occupational Safety and Health Administration, 22
octane, 102
Ohm's law, 153
oil characteristics, 185-188
oil failure, 243
oil pressure regulation, 191
oil ring, 53, 90
oil selection, 188
oil standards, 185, *187*
oily waste can, *29*
180° opposed-twin engine, 193-*194*
120 V starter motor operation, 172
one-piece valve, 78-79
open installation, *269*
operation, 212
Otto, Nikolaus, 6

Outdoor Power Equipment and Engine Service Association, 27
Outdoor Power Equipment Institute, 26
overcurrent condition, 152
overhead valve engine, 50
overheating, 243
oversized engine, 260
overspeeding, 221, 245-246
oxides of nitrogen, 101

parallel circuits, 151-*152*
parasitic load, 127, 202, 249
partial enclosure installation, *269*
partially-ducted engine, 179-180
parts cleaning, 37-38
parts removal, 121
pathway, 146
performance problems, 219-222
peripheral coating, 93
permanent magnet, 154
permissible exposure limit, 36
personal protective equipment, 30
phase modulated cooling fan, 178
phosphating, 93
pilot jet, 114
pinon gear, 170
piston, 51-*52*, 85-94
 design, 86-87
 dynamics, 88
 head, 51-*52*
 pin, 52-*53*
  bore, 52
 ring, 52, 88-90
  break-in period, 93
  dynamics, 90-91
  inertia, 90
  installation, 93
  materials, 92
  rotation, 91
  twist, *90*
 surface treatment, *88*
 window, 87
plasma spraying process, 79-80
pleated paper fuel filter, 123
pneumatic governor geometry, 131-132
pneumatic governor system, 128-129
point plunger, 164
polarity, 148
 -sensitive circuit, 168
polymer, 186
potential energy, *7-8*
power, 15-16
 curve, *253*
 event, *59*
 take-off, 51, 256-259
 tools, 40
  safety, 44
preignition, 77-*78*
premature wear, 248
press fit, 85
pressure, 12-13
 filtration lubrication system, 189-190, 232
 lubrication system, *190*, 232
primary winding, 164
primer system, 112
private organizations, 26
process, 207
projection-tip welded valve, 78-79
prony brake dynamometer, *72*
protective clothing, 30
protons, 145
P-type material, *158*
Pulsa-Jet carburetor, 119
pulsating DC, 159
pulse, 149
pushrod, 64

quick-check method, *210*

race, 54
radial load, 54, 257
radiation, *10*, 177
radiator, 10, *182*
 cooling fan, 182
reactivity hazard, 34
rebound compression test, *215*
recommended maximum operating BHP curve, 253
rectifier, 150
reed valving system, 65
reformulated gasoline, 105
regulator/rectifier, 159-*160*
Reid vapor pressure test, 104
remote speed control, 255
repowering, 249-250
research octane number, 102
reserve capacity, 163
resistance, 148-149
resistive load, 228
resonance, 142, 267-268
respiratory protection, 32
retainer groove, 63
rewind starting system, *170*
ring compressor, 93-*94*
ring expander, 93-*94*
ring groove, 52, 87
ring lands, 52, 87
rocker arm, 64
rod bearing, *55*
rod cap, 53
rotary engine, 69
rotary valve valving system, 65
rotating screen, 178
running surface, 79

SAE viscosity rating, 185
safety cabinet, 28
safety can, 28-*29*
safety color coding, 24
safety considerations, 270-271
safety glasses, 31
Savery, Thomas, 4
scavenging, 62
scoring, 240
secondary winding, 165
seizure, 240
self-inductance, 165
separator, 54
sequential method, 211
series circuit, *151*
series/parallel circuit, *152*
service procedures, 143-144
service replacement engine, 251
shaft orientation, 4
short circuit, 149
shutdown, 212
sidedraft carburetor, 117
side loading, 83
signature wear pattern, 234
silica, 234
silicon controlled rectifier, 168
sintered iron, 83
 valve guide, 84-85
60° V-twin engine, 193-*194*
SkillsUSA-VICA, 27
skirt, 52
slinger, 189
sludge, 243
small engine, 1
 development history, 4-7
 industry, 17-20
 operation safety, 21-22
Society of Automotive Engineers, 26
solenoid, 156
sound levels, 32
spark arrester, 271

spark gap, 165
spark ignition engine, 1
spark plug, *165*
spark tester, *45*, 229
specific gravity, 162, 226
specific hazard, 34
speed-sensing device, 128
splash lubrication system, 188-189
split-half method, 208-*209*
spontaneous combustion, 29
spring rate, 130
 change, 139
stainless steel, 78
standard, 25
standards organizations, 25
starter motor, *170*
starter solenoid, 171
starting problems, 217-219
starting system, 170-172, 225-227
startup, 212
static friction, 239
static leak, 222
stator, 157
student organizations, 27
stoichiometric ratio, 98
stroke, 14, 47
sump, 49
surging, 138, 219
switch, 151
systems approach method, *210*

taper-faced compression ring, 89
tappet, 64
technical societies, 25-26
temperature, 10
 scales, *11*
  Celsius, 11
  Fahrenheit, 11
temporary magnet, 154
test tools, 44
tetraethyl lead, 103
thermal conductivity, 51, 83, 175
thermal distortion, 176
thermal efficiency, 72
thermal expansion, 83, 175
thermal growth, 86, 176
thermostat, 183-*184*
three-port valving system, 65
threshold limit value, 36
throat, 109
throttle plate, 110
throw, 14, 50

tools, 39-46
top dead center, *48*
top no-load speed, 127
torque, *13*
 curve, 253
total enclosure installation, *269*
trade associations, 26
training organizations, 26-27
transitional hole reservoir, 114
Trevithick, Richard, 6
trigger, 199
troubleshooting, 207-232, 233
 chart, *213*
 compression systems, 215-216
 cooling systems, 230-231
 electrical systems, 225-230
 fuel and governor systems, 217-225
 lubrication systems, 231-232
 methods, 207-211
 steps, 211-215
true idle, 111, 219
12 V starter operation, 171-172
two-cylinder ignition system, 169
two-piece-stem welded valve, 79
two-stroke cycle engine, *3*, 61-62
 applications, 62
 valving systems, 65

undersized engine, 260
Underwriters Laboratories Inc., 26
updraft carburetor, 117
useful life, 214

Vacu-Jet carburetor, 117-119
valve, 63, 78-82
 dynamics, 81-82
 face, 63
 guide, 64, 82-85
  design, 83
  failure, 245
 hardfacing, 79-80
 head, 63
  design, 80-*81*
 interface, 81
 location, 64-65

neck, 63
overlap, *60*
resurfacing service procedures, 96
rotator, 63
seat, 63, *85*
 insert, 85
 spring, 63
 retainer, 62-*63*
 stem, 63
 train, 56
valving systems, 62-65
vaporization, 103
vapor lock, 104
vapor pressure, 28
venturi, 108-109
vibration, 266-268
viscosity, 185
volatility, 103
volt, 156
voltage, 146-147
 regulation, 159-161
 system, 159
 source, 146
volumetric efficiency, 72

water dynamometer, *70*
water jacket, 181
water pump, 182-*183*
Watt, James, 5
wear-out period, 214
wear ratio, 92
Welch plug, 114
wet bulb primer system, 112
wiper ring, 53, 89
work, 14-*15*

yield point, 96, 243

zener diode, 159

Using the Small Engines CD-ROM

Before removing the CD-ROM, please note that the book cannot be returned for refund or credit if the CD-ROM sleeve seal is broken.

System Requirements

The *Small Engines* CD-ROM is designed to work best on a computer meeting the following hardware requirements:

- 200 MHz Pentium processor or better
- Microsoft® Windows® 95, 98, 98 SE, Me, NT®, 2000, or XP operating system
- 64 MB of free available system RAM (128 MB recommended)
- 90 MB of available disk space
- 800 × 600 16-bit (thousands of colors) color display or better
- Sound output capability and speakers
- CD-ROM drive

Adobe® Acrobat® Reader™ software is required for opening many resources provided on the CD-ROM. If necessary, Adobe® Acrobat® Reader™ can be installed from the CD-ROM. Microsoft® Windows® 2000, NT®, or XP™ users who are connected to a server-based network may be required to log on with administrative privileges to allow installation of this application. See your Information Systems group for further information. Additional information is available from the Adobe web site at www.adobe.com. The Internet links require Microsoft® Internet Explorer™ 3.0 or Netscape® 3.0 or later browser software and an Internet connection.

Opening Files

Insert the CD-ROM into the computer CD-ROM drive. Within a few seconds, the start screen will be displayed. Click on START to open the home screen. Information about the usage of the CD-ROM can be accessed by clicking on USING THIS CD-ROM. The Chapter Quick Quizzes™, Illustrated Glossary, Media Clips, and Reference Material can be accessed by clicking on the appropriate button on the home screen. Clicking on the American Tech web site button (www.go2atp.com) or the American Tech logo accesses information on related educational products. Unauthorized reproduction of the material on this CD-ROM is strictly prohibited.

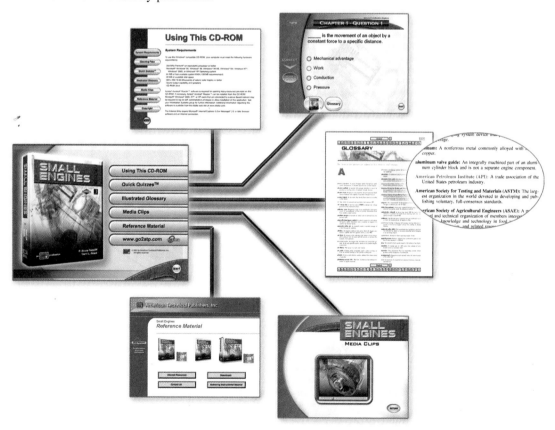

Microsoft, Windows, Windows NT, and Internet Explorer are either registered trademarks or trademarks of Microsoft Corporation in the United States and/or other countries. Adobe, Acrobat, and Reader are trademarks of Adobe Systems Incorporated. Netscape is a registered trademark of Netscape Communications Corporation. Quick Quizzes is a trademark of American Technical Publishers, Inc.